AF798225

Lesser-Known Languages of South Asia

Trends in Linguistics
Studies and Monographs 175

Editors

Walter Bisang
Hans Henrich Hock
Werner Winter
(main editor for this volume)

Mouton de Gruyter
Berlin · New York

Lesser-Known Languages of South Asia

Status and Policies, Case Studies and Applications of Information Technology

edited by
Anju Saxena
Lars Borin

Mouton de Gruyter
Berlin · New York

Mouton de Gruyter (formerly Mouton, The Hague)
is a Division of Walter de Gruyter GmbH & Co. KG, Berlin.

⊚ Printed on acid-free paper which falls within the guidelines
of the ANSI to ensure permanence and durability.

Library of Congress Cataloging-in-Publication Data

Lesser-known languages of South Asia : status and policies, case studies and applications of information technology / edited by Anju Saxena, Lars Borin.
p. cm. – (Trends in linguistics. Studies and monographs ; 175)
Includes bibliographical references and index.
ISBN-13: 978-3-11-018976-6 (hardcover : alk. paper)
ISBN-10: 3-11-018976-3 (hardcover : alk. paper)
1. Linguistic minorities – South Asia. 2. Sociolinguistics – South Asia. 3. South Asia – Languages – Variation. 4. Communication and technology – South Asia. I. Saxena, Anju, 1959– II. Borin, Lars.
P40.5.L562S645 2006
306.440954–dc22
2006021785

ISBN-13: 978-3-11-018976-6
ISBN-10: 3-11-018976-3
ISSN 1861-4302

Bibliographic information published by the Deutsche Nationalbibliothek

The Deutsche Nationalbibliothek lists this publication in the Deutsche Nationalbibliografie; detailed bibliographic data are available in the Internet at http://dnb.d-nb.de.

© Copyright 2006 by Walter de Gruyter GmbH & Co. KG, D-10785 Berlin
All rights reserved, including those of translation into foreign languages. No part of this book may be reproduced or transmitted in any form or by any means, electronic or mechanical, including photocopy, recording or any information storage and retrieval system, without permission in writing from the publisher.
Cover design: Christopher Schneider, Berlin.
Printed in Germany.

Preface

This volume grew out of our work on exploring the possibilities of using technology – specifically the modern *information and communication technologies* (ICT), and particularly *language technology* – in support of language documentation, language learning, and language maintenance, especially as these apply in the context of languages and cultures of South Asia, with particular emphasis on lesser-known South Asian languages.

Six of the papers presented here (those by Allwood, Borin, Grinevald, Nathan and Csató, Noonan, Singh) are extensively revised versions of papers read at a panel on “Globalization, technological advances and lesser-known languages in South Asia” organized by Anju Saxena in connection with the *18th European Conference on Modern South Asian Studies* at Lund University, Sweden, 2004, while the remaining 11 contributions have been solicited specifically for this volume.

The work on this volume has been funded in part by the Swedish Research Council (*Vetenskapsrådet*).

We would like to thank the series editor, Professor Werner Winter, for his encouragement and support during the preparation of this volume. We would also like to thank Birgit Sievert at Mouton de Gruyter for her advice at all stages of our long and sometimes crooked path from the first manuscript to the finished book, and John Wilkinson for his help in preparing the camera-ready copy.

Finally, a small point of orthography: When we write about the perhaps most salient aspects of modern ICT in this volume, we have decided to follow *Wired News* (among others) in not capitalizing the words *internet* and *web* (Tony Long: It’s Just the ‘internet’ Now, *Wired News* Aug. 16, 2004. <http://www.wired.com/news/culture/0,64596-0.html>, accessed June 13, 2006).

Anju Saxena
Lars Borin

Contents

Information and communication technologies and languages of South Asia

Introduction

Anju Saxena

> [I]t is alright to be Native, to speak the Native language, and to use Native tools and implements in play and work. After all, our technology was made by our ancestors to edify our Native worldviews. Please, what ever you do, do NOT give to the youngsters the idea that modern technology has an answer for everything. It does not. Use it merely as a tool, and use it minimally and judiciously. Remind the students, that technological tools are intensive in the use of natural resources and energy. To accept technology blindly is to negate the painful works to revitalize our Native languages and cultures. (Kawagley 2003: ix–x)

1. Going, going, gone: Vanishing languages and cultures[1]

The increasing globalization in the twentieth century, with a small group of nations dominating the scene, has had an adverse effect on the maintenance of social and cultural traditions of many communities. The pull factor (good employment opportunities, standard of living, etc.) and the push factor (larger and better trained and equipped armies, more modern weapons, etc.) have conspired to make some groups socio-economically dominant, and as a consequence promoted the cultures and languages of these groups over those of other, non-dominant groups (Crystal 2000; Nettle and Romaine 2000), to such an extent that the existence of a large number of smaller languages is threatened.

According to one estimate (Krauss 1996), 3000 of today's 6000 languages will disappear in this century, if no special measures are taken. Issues relating to language death, endangerment and threat to language diversity have come to the foreground of linguistic discussion (Krauss 1992, 1996; Hale 1992a) and efforts to revitalize endangered languages and to halt or prevent language death have been the themes of several conferences (including a UN conference; see Bradley and Bradley 2002a).

The term *language shift* refers to a situation where the use of a language is replaced by the use of another (usually a socio-economically or numerically dominant language). The end product of language shift is complete replacement, or language death, but it is normally a gradual process, where a shift in progress can affect a language in terms of the number of its speakers, the functional domains in which it is used and the degree of competence in the language (Dressler and Wodak-Leodolter 1977; Dorian 1989; Brenzinger 1992; Craig 1992; Grinevald 1997, 1998; Grenoble and Whaley 1998; Nettle and Romaine 2000; Bradley and Bradley 2002b). Linguists have noted the existence of language death[2] and language shift for quite some time (e.g., Swadesh 1948; Weinreich 1953: 106–110). However, since the 1960s increasing attention has been paid to language shift by linguists, who have been interested in studying the linguistic structure of the languages involved in language shift situations, where adjustments at all levels (phonological, lexical, grammatical) have been observed. In this connection, linguists have also been interested in examining if linguistic systems of dying languages ("obsolescent languages") show patterns which are just the opposite of creolization or first language acquisition (e.g., Dorian 1981; Dressler and Wodak-Leodolter 1977; Mithun 1989; Romaine 1989; Schmidt 1985; Trudgill 1978).

Factors such as migration, industrialization, urbanization, globalization, religion, government policies (e.g. the choice of the medium of instruction in schools, laws relating to language policies) and changing patterns of economy have been pointed out as potentially contributing to language shift and language death.

Social changes brought about by factors such as these may influence an individual or a speech community to revise his or its perceptions of his own self or of his language and/or their perception of the language of the other group or of the world. This may lead individuals or speech communities to change their pattern of language choice. Language shift is, in many cases, closely tied to ethnicity. In language shift situations, the language shift tends to take place when speakers want to leave behind a stigmatized ethnic identity and adopt a positive ethnic identity of some other group as a possible means for upward social mobility. A shift in the language choice patterns then becomes a means – a tool – for upward mobility (Dorian 1981).

Thus, one important factor in language shift – perhaps the most important factor – is arguably that of speaker (community) attitudes, which in turn are rooted in economical or political realities.[3] It is worth noting at this juncture that attitudes reveal themselves not so much in words as in actions, since the two often seemingly contradict each other. See the discussion of prior ideological clarification as the essential beginning for any program dealing with language and cultural preservation in Dauenhauer and Dauenhauer (1998: 62–66). Winter (1993) presents the relevant cases of (1) a Hualapai language revival activist and schoolteacher, who, while actively working for the use of Hualapai in school, nevertheless spoke only English to her children at home,[4] and (2) a Bantawa couple who worked actively to promote Bantawa in various ways, but communicated with each other and with their children in Nepali and English.[5] Winter comments (1993: 311):

> What is to be observed in both cases is a conflict between wanting to do something for the language and wanting to improve the chances of the children to succeed in the macrosociety of which they are, and always will be, part. The linguist observing this state of affairs may feel regret at what is happening here; but if it is a fact that maintaining a small language at the expense of a major or national one means severely reducing prospects of an economically satisfactory life for one's children, does one have a right to blame the parents?

In the terminology of Freilich (1991), this represents an attempt to use *smart* means to achieve *proper* goals. By these terms, Freilich aims to capture the oft-observed tension in all kinds of human communities between on the one hand that which culture, in the form of tradition, requires of us – this is what is "proper" – and on the other hand "smart" actions – which break the letter of proper rules – and which are brought about by the pragmatics of a continually changing social environment in which we have to survive. This is a generally useful distinction, although this is not as clear in the case of language as in the other manifestations of culture discussed by Freilich, mainly because the only effective way of achieving the "proper" goal of preserving the language seems to be by actually using it. As a general strategy it goes some way toward explaining how a situation such as that cited above might come about, however. The speaker simply may not be aware that language constitutes a special case. This distinction still seems useful, since, arguably, there are ways in which smart means can be used to achieve proper goals in this sense, the creative use of new technologies possibly being one such (see section 4 below).

Language death is not a new phenomenon. Languages have disappeared all through recorded history. Classic examples are Gothic, Sumerian and Hittite, to mention a few, and in the past five hundred years we have lost half of the known languages of the world (Sasse 1992). But what makes this issue especially grievous in modern times is the changing world scene. Factors such as internationalism and globalization, a modern supraregional economy and media of mass communication have intensified the situation where a small group of politically and economically dominant communities and their languages manifest too great a power on a large number of small communities. Hale (1992a: 1) elaborates the differences between the earlier language death phenomenon and the situation we are facing today:

> [L]anguage loss in the modern period is of a different character, in its extent and in its implications. It is part of a much larger process of LOSS OF CULTURAL AND INTELLECTUAL DIVERSITY in which politically dominant languages and cultures simply overwhelm indigenous local languages and cultures, placing them in a condition which can only be described as embattled. The process is not unrelated to the simultaneous loss of diversity in the zoological and botanical worlds. [emphasis in the original]

Language death arguably affects even the prerequisites for maintaining biodiversity (Skutnabb-Kangas 2000; Nettle and Romaine 2000). According to Skutnabb-Kangas, language diversity is disappearing at a faster rate than biodiversity. Her prognosis for year 2100 is (expressed as percentage of diversity lost): Biodiversity 2% but linguistic diversity 50% (optimistic forecast) and biodiversity 20% but linguistic diversity 90–95% (pessimistic forecast), highlighting the urgency of the matter.

Despite the depressing facts about the degree of language loss, there are some positive signs. The 1990s brought language endangerment to the forefront of the linguistic and political arenas, and some first steps were taken in order to turn the tide. This includes efforts by some communities involving local, national and international organizations and institutions.[6] The Hualapai Bilingual/Bicultural Education Program (Peach Springs, Arizona) is basically a local program which has been instrumental and effective in developing regional and national movements influencing Native American languages and their communities, e.g., the initiation of the American Indian Languages Development Institute and the Native American Languages Act (McCarty and Watahomigie 1999; see also presentations of various projects in the series of books published by Northern Arizona University: Cantoni 1996; Reyhner 1997; Reyhner et al. 1999, 2000, 2003; Burnaby and Reyhner 2002). Revitalization efforts are going on in smaller as well as larger communities (e.g. Mayan: England 1992, 1998; Rama: [Grinevald] Craig 1992; Grinevald 1998, 2005a; Hawai'ian: Warschauer and Donaghy 1997; Wilson 1999). There have been publications such as Dorian 1989, Grenoble and Whaley 1998, Crystal 2000, Nettle and Romaine 2000, Hinton and Hale 2001, Bradley and Bradley 2002b, *UNESCO's Red book on endangered languages*,[7] as well as conferences (e.g. the Endangered Languages Symposium organized by the Linguistic Society of America 1991), and the establishment of funding programs, such as the Hans Rausing Endangered Languages Project (HRELP) at London University's School of Oriental and African Studies.

This situation – a general loss of linguistic and cultural diversity and occasional efforts to counter the trend, i.e. using modern information and communication technologies – is prevailing everywhere in the world, including South

Asia. However, the linguistic situation in South Asia has been a bit out of focus in recent literature on language shift and language endangerment (Payne 1999; Ostler and Rudes 2000), with some notable exceptions (e.g. Abbi 1997; Saxena 2004). The aim of the present volume is to discuss the status of the lesser-known languages in South Asia and to discuss how modern technology can be a tool in documenting these languages and in spreading awareness about them. Issues that arise while applying technology developed using primarily Western literate languages to these for the most part oral languages will also be taken up here. The volume contains articles on the linguistic situation of South Asia, both general overviews portraying individual South Asian countries (Rahman, Singh, Turin), and case studies of particular South Asian language communities and/or sociolinguistic situations (Abbi, Kohistani and Schmidt, Zeisler). A number of articles raise issues of the impact of modern information and communication technology on lesser-known languages in general, and on South Asian language communities in particular (Annamalai, Bradley, Noonan, Renganathan and Schiffman), whereas others describe linguistic and cultural documentation work being carried out for South Asian languages (Hardie et al., Michailovsky), and some of the ethical issues raised in connection with linguistic fieldwork and language documentation (Grinevald). Finally, some of the contributions illustrate how cutting-edge information and communication technologies can be brought to bear on the problems of lesser-known language documentation and maintenance (Allwood, Borin, Nathan and Csató, Trosterud).

A note on terminology

Many terms are used in the literature to refer to the languages that are the focus of this volume. *Minority languages*, *indigenous languages*, and *endangered languages* are the terms most often met with in the linguistics literature, and in Indian literature the term *tribal languages* appears.[8] Elsewhere, e.g. in the language technology literature, one encounters terms such as *lesser used languages* (a term used officially in the European Union), *less prevalent languages*, *small(er) languages*, *low-density languages*, *vernacular languages*, *dialects*, *lesser-known languages*, and *less frequently taught languages*. See also Grinevald's article in this volume. In the more computer science-oriented presentations of work on language technology, one is sometimes confronted with the revealing "pseudo-term" *non-English languages*. This confusing multiplicity of terms is due at least to the different backgrounds of the scholars working in this area and also to the weights accorded the different criteria for classifying languages as distinct from, e.g., English.

In addition to this, many, perhaps all, of these terms are loaded in some way or another with pejorative connotations, or are ideologically charged in some way. Thus, it is no easy task to choose a general term for an introductory chapter such as this. However, we have decided to opt for *lesser-known languages*, as it is a relatively untainted term.

2. Language and linguistic diversity in South Asia

One sees some interesting patterns in different parts of the world concerning the direction of language shift. In the Americas and Australia the shift has mainly been to the languages of the colonial rulers (Spanish, Portugese, French and English) whereas in some other regions such as in Africa there is often a shift towards a non-colonial language (e.g. Amharic in Ethiopia, Bambara in Mali and Swahili in Zaire/Congo). In South Asia, some locally dominant languages (Hindi, Urdu, Nepali to mention a few, beside English, the colonial language) are gaining ground at the expense of the lesser-known languages.

The Indian subcontinent has a long history of linguistic diversity and multilingualism, spanning more than three millenia. Languages spoken in this region belong to at least four major language families: Indo-European (mostly Indo-Aryan), Dravidian, Tibeto-Burman and Austro-Asiatic. Societal multilingualism is an established tradition in South Asia, where not all of languages which are spoken in one community are employed in all spheres of activity (Pandit 1972). Despite this stable multilingualism, language death is not uncommon in the South Asian context.

As is typical of most of South Asia, speakers of lesser-known languages in India – the largest and most populous country in the region (population 1080 million in 2005) – are already or are increasingly becoming bilingual. Concerning the 114 languages mentioned in the Indian census, the rate of bilingualism recorded in the past four censuses indicates that bilingualism has doubled in 30 years from 9.7% in 1961 to 19.44% in 1991 (Bhattacharya 2002). While speakers of lesser-known languages learn the language(s) of the dominant group, the reverse is usually not the case. Whereas many adult Kinnauri[9] speakers, for example, speak Kinnauri as their mother tongue, and many elders living in the region are strongly monolingual, children and young adults are in very large numbers active bilinguals, with a preference for Hindi or the regional Indic variety. Many young people migrate outside this area for education and employment purposes, where the lingua franca is not their mother tongue. Such social situations have important linguistic consequences for these languages. Indigenous languages with no written tradition and with no or very little political and/or economic power at the local and national level fall by the wayside *en route* to modernity either completely, or are given up in particular contexts.

While some of the languages (such as Hindi, Tamil and Bangla) have a long written literary tradition and there has been much work done on these languages, very little is known about many, perhaps most, languages of this region. A case in point is a language such as Great Andamanese, with only a handful of speakers and almost no documentation (for details, see Anvita Abbi's article "Vanishing voices: A typological sketch of Great Andamanese" in this volume). Similarly, there is a great disparity in the number of speakers. While the then 18 scheduled languages[10] constitute 96.29% of the total population, the remaining 96 non-scheduled languages (including 0.07% who speak "other languages", defined as those languages which have less than 10 000 speakers) are spoken by 3.17% of the population according to the 1991 census. (Bhattacharya 2002: 58)

It is impossible to say anything concrete about the extent of language endangerment in India. Information on language at the national level has been collected since 1881 as part of the Indian census, held every 10 years, but census reports provide almost no concrete information about languages with less than 10 000 speakers (in other words, about endangered languages). Further, motivations for the distinction between language and dialect are not always clear. The census figures are based on self-reporting by language users, so that if a particular language is provided as the mother tongue in the census returns by an individual or by a group, this may at times be a reflection of loyalty more than an indication of actual language proficiency (Southworth 1978). Furthermore, how and which languages are taken into consideration is a complicated matter. The 10 400 mother tongue names returned in the 1991 census of India are reduced to 113 languages, plus one other mother tongues category for all languages with less than 10 000 speakers (cf. the 386 languages listed for India in the *Ethnologue*).[11] However, many names are discarded in the process, on grounds that are not always clear (Annamalai 2003). Udaya Narayana Singh's contribution to this volume, "Status of lesser-known languages in India", presents an overall picture of lesser-known languages of India and their current status, focusing on the constitutional provisions that exist in India with respect to minor and minority languages, language policy issues and problems with implementation of official decisions. Population growth, economic growth, urbanization, literacy and education are mentioned as factors which slow down the process of implementing laws and policies furthering the use of lesser-known languages in India today.

Considering the fact that even today around 80% of the population in India still live in rural areas, this may lead one to believe that this multitude of languages are well and thriving. This, however, is far from always the case, largely because of extralinguistic factors, such as the medium of instruction in schools, social mobility, administrative language, modern media such as television, etc.

However, there are also some success stories, such as that of Santhali in the newly formed 28th state of Jharkhand. Languages belonging to three language families are spoken in this region: Indo-Aryan (e.g. Sadari, Hindi, Bengali), Dravidian (Kurux, Malto) and Austro-Asiatic (e.g. Mundari, Ho, Santhali). Hindi and English are dominant languages used widely in public spheres, while the lesser-known languages are primarily used for in-group communication. They are not used in government offices, state legislation, business or legal matters. In Jharkhand, one sees two opposing trends: On the one hand, there are signs of promoting English (e.g. the Jharkhand government's proposal to introduce English as a subject from first grade onwards in schools) and on the other hand, one sees attempts to strengthen and to make more visible some lesser-known languages – in particular Santhali – of this region. Kurux, Mundari, Ho and Santhali have been introduced as the medium of instruction in primary schools, as well as provided as optional subjects in secondary schools. These languages are also offered at Ranchi University at graduate and post-graduate levels. There are also other similar efforts to make these languages more visible (e.g. there are 3 newspapers and 20 magazines published in Santhali by private organizations). There have also been efforts by various organizations to include Santhali written in the Olchiki script[12] as the official language of Jharkhand in the VIII Schedule of the Indian Constitution. (Mohan 2002: 230–240)

The linguistic situation in Nepal is also one of great language diversity – the *Ethnologue* lists 120 languages in Nepal, spoken by a population of 28 million (in 2005). As in so many other places, we are faced here with a situation where lip service is paid to language diversity, but where in reality the situation is one where there is one dominant language, viz. Nepali, an Indo-Aryan language. According to Winter (1993), there are signs of decreasing language diversity also in Nepal, earlier so that larger languages would encroach on smaller ones in the same region, but increasingly – following in the wake of national centralization and a growing supraregional economy – so that Nepali tends to take over as the language of all walks of life. However, beginning in the 1990s, there seems to be a growing awareness in Nepal about the situation of lesser-known languages, and more enlightened language policies are being formulated. Progress is slow, however.

"Minority language policies and politics in Nepal" is the theme of Mark Turin's article in this volume. Turin points out that at the policy level, there are some positive developments, where there is a shift from the "one nation, one language" policy in the 1950s to "an acknowledgement of the multi-ethnic and multi-lingual nature of the country post-1990. But these policies have not been put to practice to the extent one wishes for. Recognizing the important role script and literacy play (including the fact that the state provides more resources to written languages), Turin describes trends noticeable in Nepal today. While some organizations advocate either Devanagari or Tibetan scripts, there are communities who probably in their effort to establish their own identity are trying to devise their own new scripts.

The linguistic scene in Pakistan is similar to the Indian and Nepalese situation described above. Pakistan, like India and Nepal, is a multilingual state with 6 major and about 59 minor languages, spoken by a population of 162 million (in 2005). Tariq Rahman in his article "Language policy, multilingualism and language vitality in Pakistan" provides an overview of the linguistic scene in Pakistan with special reference to the unequal status of English and Urdu on the one hand, and lesser-known languages on the other. He further discusses some factors which have contributed to this unequal status. Government policies, according to him, is one significant factor. He attributes to English and Urdu the status of the languages of power in Pakistan. English is considered a symbol of power, sophistication and prestige, whereas small minority languages have a negative image associated with them. This trend is leading to language death in some cases and marginalization in others. Rahman advocates the promoting of additive multilingualism as a means to improve the status of these marginalized languages.

A concrete case study of the changing linguistic scene in Pakistan is presented in Razwal Kohistani and Ruth Laila Schmidt's article "Shina in contemporary Pakistan" in this volume. The focus of their article is on Shina, an Indo-Aryan language of the Dardic subgroup spoken in the Karakorams and the western Himalayas. They describe the increasing marginalization of the use of Shina, attributing it to factors such as modern education, advancements in the media, and communication. They point out that Urdu and English – which are dominant languages in the whole of Pakistan – as well as Pashto (the dominant language of the region) are gaining ground at the expense of Shina. In its urban center, Gilgit, Shina "has suffered a loss of prestige" and in the rural areas Shina is used (at least at present) both in public and private domains, but they fear that this relegation of the use of Shina to rural areas and private domains is preventing Shina from developing a standard language and literature. There are, however, some forces working against this development in the region (e.g. intellectuals who work in favor of Shina and Islamic missionaries who target the grassroots of the population).

3. Loss of linguistic diversity and the need for language documentation

There is a growing awareness about the negative consequences of language death and the concomitant loss of linguistic diversity. From the linguist's point of view, language diversity is essential for linguistic theory building and for a scientific study of mind and language (e.g. Hale 1992b, 1998), for which it is imperative that we have access to data from languages representing rich and diverse linguistic structures, underscoring the need for documentation and preservation of languages.

A language is a reflection of the community that speaks it. It embodies the philosophy and the world-view of its people. In communities which lack a writing system, this knowledge is handed down orally from one generation to the next. When a language dies, we lose not only the linguistic knowledge of that community, but also the knowledge about its culture:

> The most important relationship between language and culture that gets to the heart of what is lost when you lose a language is that most of the culture is in the language and is expressed in the language. Take it away from the culture, and you take away its greetings, its curses, its praises, its laws, its literature, its songs, its riddles, its proverbs, its cures, its wisdom, its prayers. The culture could not be expressed and handed on in any other way. What would be left? When you are talking about the language, most of what you are talking about is the culture. That is, you are losing all those things that essentially are the way of life, the way of thought, the way of valuing, and the human reality that you are talking about. (Fishman 1996: 81)

The loss of artistic and intellectual resources accompanying the loss of language has been addressed in the literature by a number of linguists. Mithun (1998), for example, presents some linguistic features of Central Pomo and Mohawk to illustrate how some specific ways that these languages conceptualize the world will be lost, if the languages are lost. On a similar note, Woodbury (1998) presents some cases to illustrate that the loss of a language implies the inability to express particular concepts. Cup'ik Eskimo has a series of affective suffixes with translations as 'poor dear N; poor dear (subject) does V'; 'darned N; darned (subject) does V'; 'funky N; funky (subject) does V'; and 'shabby old N; shabby old (subject) does V' (Woodbury 1998: 240). In English there are no affixes expressing these meaning(s), so one is forced to use lexical items, if anything. Woodbury conducted an experiment where a speaker first told the story in Cup'ik and then narrated the same story in English and finally provided a sentence-by-sentence translation. The results of this experiment showed that in the sentence-by-sentence translation there were no words expressing the meanings expressed by affective suffixes and in the free English narration there were only a few items expressing the meaning of the affective suffix, suggesting that the interpretation of these affective suffixes can, at the most, be captured only poorly in the English translation.

Similarly, the disappearance of a language may also imply loss of culture-specific information. Like many other smaller communities, the Mohawk people believe that they do not cease to be Native Americans if they do not speak their language. Jocks (1998) demonstrates convincingly how if a community does not have a rich knowledge of its cultural tradition manifested in its language, that community may become a caricature of itself, as it were. Traditional ceremonies, for example, may not only become formalized rituals: there is also a risk that translations of traditional ceremonies, for instance, may implicitly bring with them conceptions that outsiders have of these indigenous communities. He illustrates his point by pointing out differences in the conceptualization of knowledge in English and Mohawk: In English, knowledge is something which one can POSSESS, whereas in Mohawk, knowledge is an ACTIVITY (something one does and which must be maintained).[13]

One such unique linguistic/cultural configuration is described by Anvita Abbi in her paper "Vanishing voices: A typological sketch of Great Andamanese", in the form of a case study of Great Andamanese. Previous studies suggest that Great Andamanese could represent the remaining linguistic link to pre-Neolithic Southeast Asia. Great Andamanese has 13 speakers, highlighting the urgent need to document and describe this language. In this paper Abbi presents the results of her pilot study, outlining the phonological, morphological and syntactic features of Great Andamanese. Abbi's article is illustrative of many languages which are in danger of extinction because of changing socio-cultural patterns – languages which we know almost nothing or very little about.

Against this background, it is perhaps not surprising that much emphasis in recent times has been put on the need for documentation of lesser-known languages, especially endangered languages. Earlier, and to some extent still today, we often see linguists referring to their chief activity as description of languages. These are related, but not identical, activities. Conceptually, documentation precedes description:

> LANGUAGE DOCUMENTATION provides a record of the linguistic practices of a speech community, such as a collection of recorded and transcribed texts. LANGUAGE DESCRIPTION, on the other hand, provides a systematic account of the observed practices in terms of linguistic generalizations and abstractions, such as in a grammar or analytical lexicon. (Bird and Simons 2003: 557)

Logically, then, documentation in this sense can be the basis for description, but not vice versa. The products of documentation – including linguistic descriptions – it is increasingly realized, can be used to support languages that are still used, whereas mere description without documentation cannot be used to revitalize languages where there are no or almost no living linguistic practices left. There is at least the hope, however, that description and documentation together – "preservation for the record" in the words of Allwood (this volume) – could be used to accomplish this.

Unlike traditional linguistic descriptions, then, where the secondary products of the primary linguistic materials – grammars, dictionaries, presentation of theoretical analyses of various linguistic phenomena, etc. – were in focus and the linguistic data itself was not seen as primarily interesting,[14] in language documentation the focus is on primary linguistic material in a representative spectrum of genres, with an emphasis on naturally occurring discourse in different speech situations. Another objective is to include not only linguistic material but also material which provides some insights into the cultural aspects of these societies. This means that language documentation in fact has much in common with modern corpus linguistics (see Borin's contribution in this volume).

It should follow from Bird and Simons's characterization, quoted above, that literacy automatically implies documentation of a language, provided that writing and its products can be considered part of "the linguistic practices of a speech community". In this sense, then, language documentation has been going on for a very long time, at least in some cases, on clay tablets, on papyrus scrolls, on runestones, on bark, on wood, on leather, on paper, on bricks, on cloth, etc. In the same way, language description has a long history, but a peculiar one in the case of languages other than the classical languages or the new national languages of Europe (in the case of which it is perhaps better to speak of "language prescription"). The flip side of this particular coin is, again, that most of the lesser-known languages everywhere – South Asia being no exception in this regard – are non-written languages.

Woodbury (2003) attributes this recent interest in linguistic documentation to three elements, namely, an increasing awareness about diversity among and within languages not as a kind of aberration, but as an intrinsic definitional feature of language, of the threat of language endangerment, and of technological advances opening new possibilities for documenting linguistic data. He also points to a growing realization among linguists that primary linguistic data have never been properly theorized, but remain largely epiphenomenal to the generalizations expressed in grammatical discourse, in a way which in other sciences would be considered quite naive (Woodbury 2003: 40).

Austin (2003) emphasizes the need to talk about guidelines, e.g. ethics, relations with the language community, our responsibility to the community, to researchers and to the discipline (Grinevald 2005b). In her article "Worrying about ethics and wondering about 'informed consent': Fieldwork from an Americanist perspective", Colette Grinevald discusses a number of ethical (and at their core eminently practical) issues that arise in connection with linguistic fieldwork, driving home the point that linguistic fieldwork, in particular fieldwork on languages facing extinction, will have to deal with "a complexity of pressures which academia and financing foundations may have very little sense of as yet", because "fieldwork projects are not laboratory experiments", and forging a long-term working relationship with the language community "is one of the most challenging of the multiple responsibilities that fall on the fieldworkers, who are academics usually raised and trained far away from the realities of the field".

The need to discuss ethical guidelines has arisen from a realization that linguists and other fieldworkers are just as likely to be caught in the trap of ethnocentrism as anybody else, but that awareness-raising is one way of avoiding this. At the same time, indigenous language communities have realized and increasingly begun to question the way their cultures and languages are portrayed by outsiders:

> With the growing interest in things Indian in the United States and around the world, Native American culture has become a highly saleable commodity ... While this commercialization of Indian culture might seem to make good business sense to the Anglo-American majority, many native people experience it as an expropriation of their heritage by the dominant society. This taking is understood to involve the alienation, popularization and corruption of native traditions and imagery through their unauthorized reproduction and commercial exploitation by non-Indians. There is widespread consensus among native spokespeople that such 'cultural appropriation' is as potentially damaging to the survival of native ways of life as the expropriation of Indian lands in the nineteenth century, or the assimilationist strategies pursued by the Indian Schools. (Howes 1996: 138)

4. The role of technology in language preservation and loss

Modern technology – here I include both the somewhat older broadcast (analog) mass media technologies radio and televison, and the newer (digital) so-called information and communication technologies (ICT), i.e., computers, the internet, cell phones, interactive digital cable television, etc. – have been depicted as both foe and friend with respect to non-mainstream cultures and lesser-known languages. The former view is reflected in Krauss's (1992) characterization of television as "cultural nerve gas". Many researchers and other observers perceive that modern mass media pose a threat to diversity, forcing everything that comes in their way into the same cultural and linguistic straitjacket. E. Annamalai in his article on "The impact of technology on language diversity and multilingualism" describes how changes in the socio-cultural structure of a community (including the introduction of new technology) has a strong impact on its language, drawing on his work on the Andamanese language (Annamalai and Gnanasundaram 2001).

On the other hand, especially the most recent information and communication technologies (ICT) are often seen as holding great promise for the documentation, protection and promotion of language diversity, creating unprecedented opportunities for small language communities (e.g. Bredin 1996; Cazden 2003).

In order to discuss the role of technology *vis-à-vis* lesser-known languages, it will be appropriate to keep separate certain different aspects of modern information and communication technology, viz. its *form* (what *could* faithfully be conveyed by it), its *content* (what *is actually* conveyed by means of this technology), and its *uses* (more generally how technology can be potentially beneficial to small languages and cultures). But first of all, of course, one needs access to computers and the skills to use them, which is generally less likely to be the case in lesser-known language communities (McHenry 2002), illustrating another aspect of what has been called the "digital divide".

The form of ICT is relevant at least in two respects. Firstly, we are still very far from the hypothetical ideal state where texts in any (literary) language can be input, stored, processed, and presented on equal terms with all other languages in word processors, on the web, in email, in chat rooms, etc. This has to do with developments in the areas of input methods (e.g. for scripts with large character inventories), character coding and rendering, and software for natural language processing. It has to be emphasized at this point that the issue of development or non-development in these areas is not primarily a technical issue (although there is also a technical dimension to it), but has everything to do with policy and a will to have things be a particular way.

David Bradley's article "Lisu orthographies and email" in this volume reports the case of the use of Lisu on the internet and a revision this media has necessitated in the Lisu writing system. Lisu is a Tibeto-Burman language spoken in India, Burma, Thailand and China. It has a Latin-based orthography. The writing system uses upper case letters, upright and inverted. A revised version of this orthography has been devised for the internet and Bradley reports that its use is gradually spreading.

Renganathan and Schiffman's article "The impact of technological advances on Tamil language use and planning" highlights the complex and intertwined nature of the status of the concerned languages and its effect on the application of technological advances in South Asia. The focus of their article is on Tamil (though not a lesser-known language, one which faces competition from English in some public domains). Tamil, like other languages in South Asia, faces a challenging situation. These languages have to struggle for their survival and use in fields such as science and technology, where English has been (and still is) the dominant language. Language activists push for the use of these languages in all domains, including university education. This is in sharp contrast to the prevailing situation in higher education institutions which promote English, e.g. by using English as the medium of instruction, by (explicitly or implicitly) encouraging academic publications in English (rather than in Tamil, for instance). This obviously hampers or slows down the application of recent technological advances to Tamil. Despite this, some efforts have been made in the fields of Tamil computing and language technology. Some examples concern the creation and use of technical vocabularies in Tamil and the development of localized software.

The second important aspect of the form of ICT in this context is the circumstance that the technology is still predominantly geared toward the written language. Thus, only literary communities can make full use of it (whereas many lesser-known language communities are exclusively or primarily oral; see, e.g., Bernard 1996; Buszard-Welcher 2001).

It is frequently remarked in the literature that literacy is a prerequisite for the long-term survival of a language in the modern world. Bernard (1996) and Borin (this volume) make a useful distinction between two quite different usages of written language, noting that many languages of the world have been written, often by linguists but sometimes even by native speakers, without developing a literary tradition (Bernard 1996: n.p.). Only if a language is literary, rather than merely written, will it stand a chance in the long term, it is claimed. Michael Noonan in his article on "The rise of ethnic consciousness and the politicization of language in west-central Nepal" argues that standardization is a necessary component of a literary language, meaning that a standard orthography be devised, that a uniform spelling of words be introduced, that a canonical form be selected from among variants used by speakers, etc. Noonan observes that ethnic consciousness is a relatively recent phenomenon in west-central Nepal. Despite some official rhetoric on this matter, not much is done in reality to preserve and promote other languages than Nepali in education and other domains. He recounts the case of two cousins in Nepal. Both are fluent speakers of Chantyal, a Tibeto-Burman language, and who regularly exchange emails in Nepali – in which both are also fluent – written in Latin script transcription on a keyboard configured for English. Somehow the notion of instead writing the emails in Chantyal never occurred to the two correspondents, presumably because, in Noonan's words, "people ordinarily write the languages they were taught to write in school". Some lesser-known languages in Nepal, including Chantyal, have written forms of which their users are aware, but just like earlier, they are still hardly used in the school system (see also Bernard 1996). Noonan here calls attention to a chicken-and-egg quandary involving the relationship between the availability of primary education in a language and a standardization of that language, concluding that this is ultimately a political matter, but that the will to realize that this is so, let alone act on this realization, still seems to be lacking in Nepal.

Indeed, present-day language technology as we meet it in the form of spelling and grammar checking software relies on the existence of a standardized orthography. Ultimately, standardization means that some of the diversity in the language is eliminated. This issue is obviously not that straightforward. Bettina Zeisler in her article "Why Ladakhi must not be written – Being part of the Great Tradition: Another kind of global thinking" presents the illuminating case of a local mainly spoken language which faces competition not only from the officially dominant language, but also from within its own group. Ladakhi is spoken in the north of the Indian state of Jammu and Kashmir. It is not only under strong pressure from the official state language (Kashmiri), but also from the elitist attitudes of Ladakhi Buddhist scholars, who advocate literacy and literature only in Classical Tibetan – which many feel ought to be used for all writing, but which in practice only a few individuals master – and who work against promoting literacy and literature in Ladakhi. According to Zeisler, the classical orthography and grammar which represent some ninth century varieties – about as close to Ladakhi as Latin is to modern Spanish – are not suitable for writing Ladakhi, but at the same time, there are strong protests against using Ladakhi for literacy and literature by those who want to maintain the high status of Classical Tibetan.

Even though many linguists feel that it is self-evident that language standardization is all for the good, there are also dissenting opinions. Bernard (1996) feels that the same kind of market mechanisms that (over a period of several centuries) resulted in the regularization of the orthographies of languages like English should also be allowed to work for new literacies, whereas Östman (2001) questions the impartiality and universal validity of the principles commonly used to argue for language standardization: "If the Hualapai feel that there is nothing wrong with writing the Hualapai word for 'water' in a number of different ways, then that feeling and decision should be respected." (Östman 2001: 52; see also Foley 2003).

Turning now to the content of ICT, we find that it has two facets which are particularly pertinent in the context of lesser-known – and in particular endangered – languages. Firstly, there is the general circumstance that content here, as in media in general, is predominantly in "mainstream" languages, conveying the values, norms and attitudes of their cultures. About two thirds of the content of the web is in English, although less than half the online population are native speakers of English. Thus, ICT – together with the older mass media – is not culturally neutral as to its content, but instead provides immersion in majority languages and cultures to an unprecedented extent, and also at times provides inappropriate models for the use of the same technology for lesser-known languages (Cazden 2003). Secondly, although especially the internet is a democratic medium in the sense that lesser-known language communities may cut out the middleman and use this medium to spread information about themselves, to exchange information and to organize themselves in their own terms, this also comes with concomitantly greater risks of misappropriation: On the web, anyone can claim to represent a particular community (Warschauer 1998; McHenry 2002), and there is no reason to believe that this will happen less frequently in the new digital world than in the old non-digital one (cf. Howes 1996).

Keeping these potential stumbling blocks in mind, lesser-known language communities and researchers have endeavored to put modern technology to creative and culturally appropriate uses for their languages (Bredin 1996; Nettle and Romaine 2000, chapter 8; Buszard-Welcher 2001; McHenry 2002; Cazden 2003).

There have been top-down (i.e., by government agencies) as well as bottom-up efforts (i.e., the by speech communities themselves) in promoting lesser-known languages. ICT can play an important role in maintaining and promoting linguistic diversity, for instance, in documenting lesser-known languages and cultures and also in making information available to both speakers of these languages and outsiders. The web makes it easier to spread awareness about lesser-known languages and their communities. It also provides more flexible and easier means of communication within and outside the community (thus increased opportunities for the active use of languages). Certain communities in the Americas gathered for the first time by internet to organize themselves. An all-Hawai'ian language computer environment (with on-screen menus, messages, etc. only in Hawai'ian) and the Leokī chatroom has allowed a geographically dispersed community of Hawai'ian medium school classes to keep in touch electronically using the language (Warschauer and Donaghy 1997; Warschauer 1998). Finally, the availability of this modern, cool technology in a language confers prestige to that language, raising its status in the eyes of its users and others. In this vein, David Nathan and Éva Csató in their article in this volume, "Multimedia: A community-oriented Information and Communication Technology", emphasize the importance of turning field research results into products which immediately support communities speaking endangered languages in their efforts to maintain their linguistic and cultural heritage. They describe three different genres of ICT products for documentation of community language heritage and language learning designed by Nathan and Csató for and together with the endangered language communities and delivered to these communities.

Language documentation as described here has been made possible more than anything else by modern information and communication technology. This technology has brought about a digital revolution in the way that primary linguistic data can be recorded, stored, annotated, retrieved and correlated (including high-quality sound and video recordings; see Hinton 2001). Further, it provides the means to present older written material and analog recordings in more modern media as well, thereby making their information accessible in new ways. For instance, traditional paper dictionaries can be scanned and stored in lexical databases (Corris et al. 2002), enabling access in the reverse direction (target language to source language), or the production of a reverse direction word list on paper (Miyashita and Moll 1999).

We are, at present, witnessing some positive efforts in documenting lesser-known languages in South Asia, using information and communication technology. In their article "Corpus-building for South Asian languages", Hardie et al. describe their work in the EMILLE project on building a South Asian language corpus. The goal of the project – which was largely achieved – was to create a combination of corpora (monolingual written, monolingual spoken and multilingual parallel written, with English as the source language) of a number of South Asian languages representing the Indo-Aryan and Dravidian language families. The completed EMILLE corpora consist of about 92.5 million words of written corpora in 13 languages, 2.6 million words of spoken corpora in 5 languages, and 1.2 million words of parallel corpora in 6 languages, making a respectable total of 96.3 million words. The article illustrates some of the difficulties which tend to beset work on languages which deviate from the Western European "norm" in various respects:

- poor availability of electronic texts, both in amount and variety
- a plethora of text and character encodings
- different linguistic tradition with regard to normativity vs. "pure description" (especially relevant for the spoken language corpora)

– lack of language technology resources and tools for corpus analysis and annotation

Boyd Michailovsky in his article "Digitized resources for languages of Nepal" presents an overview of available IT resources for languages of Nepal. This includes tools for the coding and rendering of Nepalese languages and scripts, spoken and written corpora with special focus on annotated speech recordings and dictionaries and wordlists. (LACITO in France has initiated an archive which contains texts of lesser-known languages (including some languages of Nepal, e.g. Hayu and Limbu). This archive contains transcriptions with time-aligned sound recordings, linked to glosses and translations, all available on the web. For portability, the archive is designed using standard formats and is accessible via standard web browsers.

The Chintang and Puma Documentation Project, carried out by Universität Leipzig, Germany, together with Tribhuvan University, Nepal, aims to provide a rich linguistic and ethnographic documentation of two highly endangered but almost totally undocumented languages in eastern Nepal, Chintang and Puma. Documentation includes language practices in context, together with transcripts with rich linguistic and ethnographic annotations. The project also includes a detailed study of language acquisition (for Chintang) over a period of approximately two years, the purpose of which is to gain insights on the micro-process of language endangerement, the role of bilingualism and trilingualism in this process, and the social and psychological mechanisms that lead to language death.

A particular kind of ICT which ought to be particularly relevant in this connection is *language technology*. Lars Borin in his article "Supporting lesser-known languages: The promise of language technology" in this volume presents a short introduction to language technology. Recently, there has been a good deal of concern about the creation of language technology resources for other languages than English and a few others, and especially for lesser-known languages. Proposed methods for the automatic acquisition of linguistic knowledge by computer potentially allow for the rapid creation of such resources with minimal human work, which if realized would be very useful. However, current such methods – like language technology in general – have arguably been shaped by the typological and other traits of the most explored language, namely English, which is in many respects an atypical language from a linguistic point of view. There is a need to test and refine these methods on a number of structurally diverse languages, making South Asia a good testing ground, in order for us to get a better understanding of the generality or language-specificness of these methods. Such experiments could be coordinated with general documentation efforts going on in South Asia, resulting in the embryo of language technology resources for some lesser-known South Asian languages, as well as general methods for turning language documentation into linguistic description in the most economical way. This point is also emphasized by Jens Allwood in his article "Language survival kits", where he reiterates some cogent arguments in favor of efforts to preserve linguistic diversity. He further points to some ways in which modern technology and especially language technology can be brought to bear on this problem, namely primarily by supplying the basic tools making up the "language survival kits" outlined in the article.

A potentially useful way of looking at this issue is proposed by Trond Trosterud in his article on "Grammatically based language technology for lesser-known languages" in this volume, where he points out that the development of (at least certain kinds of) language technology applications can be seen as equivalent to doing basic linguistic descriptive work. In this way, the results of this work will be both a detailed formal linguistic description of some aspect of the language – morphology and some syntax in Trosterud's examples – and the beginning of basic language technology tools for the language.

5. Towards a pooling of knowledge

One important aim of this volume is to make available in one place articles belonging to areas of research that so far do not interact to any significant extent, namely those dealing with traditional South Asian descriptive linguistics and sociolinguistics, with documentary linguistics, intellectual and cultural property and fieldwork ethics, and with language technology. Researchers working in the areas of documentary linguistics and language technology have slowly become aware of each other in the last few years, and of how work in the other area could be potentially useful in furthering their own aims (see Borin's and Trosterud's articles in this volume). Similarly, the insights of documentary linguistics are slowly making their way into traditional descriptive linguistics and sociolinguistics, largely because of documentation funding initiatives such as those described above. However, the potential for synergy among these areas of research is almost limitless. In juxtaposing this assortment of seemingly quite disparate articles here, we wish to provide the reader, not so much with a do-it-yourself recipe for applying modern technology to the problem of language shift in South Asia today, but rather with some basic knowledge about the problems involved and some directions from which solutions could be forthcoming, a toolbox rather than a blueprint, if you like. Hopefully these articles will give you both a glimpse of the shape of things to come, and enough information so that you can contribute to the shaping of that future.

Notes

1. I would like to thank Colette Grinevald for her input. The preparation of this volume was partly funded by a Swedish Research Council/SIDA–Swedish Research Links project and a conference grant by the Swedish Research Council.
2. Various other terms (e.g. *language murder*, *language suicide* and *language extinction*) have been used in this context. Nettle and Romaine (2000) eschew the use of the term "language suicide" as in most cases there are external factors forcing speakers to shift as their only means of survival. There is no consensus as to what is meant by language death (when a language should be considered dead). A commonly held view is that if a language does not have any active speakers, the language is considered dead/extinct. McLendon (1980: 147–148) provides a strikingly apt simile for the process of language shift:

   > ... like a social gathering where some people leave early without affecting the interactions of the rest of the participants much or even being noticed. But after a certain time, more and more people leave ... At some point the few remaining participants realize that the majority of the participants at this event are gone and it must be defined as over even though some participants are left. Just as suddenly the few surviving speakers of a language discover they no longer have sufficient occasions which permit the use of the language because so few other individuals speak it and for a variety of reasons, such as lack of contact because of distance, or lack of compatibility or downright dislike, they rarely talk with the few individuals who are still able to speak. They do not turn mute, however. Rather they turn to the contacting language in an ever-expanding number of speech situations, and the 'dying' language ceases to be spoken not from lack of speakers but from lack of use.
3. But at the same time, it is important to keep in mind that attitudes of a speech community are not completely determined by these external factors; there are numerous observations showing that even under comparable external conditions, two speech communities may react diametrally differently (Dorian 1998).
4. Hualapai (Hwalbáy, Hwalbá:y, Walapai; see Östman 2000: 48–49) is an indigenous North American language, a Yuman language spoken in Arizona.
5. Bantawa is a Tibeto-Burman language spoken in Nepal. Nepali is the official state language of Nepal.
6. Officially, national and local authorities usually support the rights of so-called minority groups (including the use of one's own language), but in practice, such official views – sometimes even taking the form of laws or other regulatory documents – are not often implemented, partly because of limited resources and partly because of lack of genuine interest, or simply because at heart decision makers subscribe to an assimilationist ideology (Skutnabb-Kangas 1990; Dorian 1998;

Kawagley 2003), what some researchers (e.g. Spolsky 2004) have termed "ideological monolingualism".

7. This book lists endangered languages according to region. Some information is available at: <http://www.helsinki.fi/~tasalmin/nasia_report.html#Smansi>.
8. The term *tribal* is primarily used in the Indian context to refer to those languages which are listed as "tribal languages" in the Constitution of India (Article 342). The use of the term tribal in this sense is purely an administrative term – devoid of any linguistic motivation or basis. A community has been labelled as tribal in the Constitution of India because of a number of factors, factors such as historical, socio-economic and cultural (and language may be included as a subdomain of culture), but no linguistic motivation has been provided for treating or not treating a language as a tribal language.
9. Kinnauri is a Tibeto-Burman language spoken in the eastern part of the Indian state of Himachal Pradesh, and also in the neighboring region in China.
10. They are Assamese, Bengali, Gujarati, Hindi, Kannada, Kashmiri, Konkani, Malayalam, Manipuri, Marathi, Nepali, Oriya, Punjabi, Sanskrit, Sindhi, Tamil, Telugu, Urdu. Today the scheduled languages number 22. See Udaya Narayana Singh's article in this volume for further details.
11. *Ethnologue* figures are cited from the web version <http://www.ethnologue.com> which reflects the 14th edition (published in 2000) of the printed *Ethnologue* at the time of writing of this introduction.
12. There are, at present, five different scripts to write Santhali. In Bihar, it is written in Devanagari, in West Bengal it is written in Bengali, in Orissa it is written in Oriya, Christians write in Roman and it is also written in Olchiki, a native Santhali script.
13. In the words of Östen Dahl (p.c.), when it comes to arguing for language preservation, most linguists seem to turn Whorfian. And indeed it seems that linguistic relativity *à la* Sapir and Whorf – "Facts are unlike to speakers whose language background provides for unlike formulation of them" (Whorf 1956: 235) – must be invoked in order for the kinds of arguments just cited here to hold water.
14. Truth be told, this is still often the case; it may even be generally considered detrimental to an academic career in linguistics to indulge in too much primary data collection, i.e., linguistic documentation (Grenoble and Whaley 2002; Grinevald 2001).

References

Abbi, Anvita (ed.)
1997 *Languages of Tribal and Indigenous Peoples of India: The Ethnic Space*. Delhi: Motilal Banarsidass.

Annamalai, E.
2003 The opportunity and challenge of language documentation in India. In *Language Documentation and Description*, vol. 1, Peter K. Austin (ed.), 159–167. London: Hans Rausing Endangered Languages Project.

Annamalai, E., and V. Gnanasundaram
2001 Andamanese: Biological challenge for language reversal. In *Can Threat-*

ened Languages be Saved?, Joshua A. Fishman (ed.), 309–322. Clevedon: Multilingual Matters.

Austin, Peter K.
2003 Introduction. In *Language Documentation and Description*, vol. 1, Peter K. Austin (ed.), 6–14. London: Hans Rausing Endangered Languages Project.

Bernard, H. Russell
1996 Language preservation and publishing. In *Indigenous Literacies in the Americas: Language Planning from the Bottom up*, Nancy H. Hornberger (ed.), 139–156. Berlin: Mouton de Gruyter. (References here are to the electronic version: <http://nersp.nerdc.ufl.edu/~ufruss/commodit.html>).

Bhattacharya, S. S.
2002 Languages in India: Their status and function. In *Linguistic Landscaping in India with Particular Reference to the New States*, N. H. Itagi, and Shailendra Kumar Singh (eds.), 54–97. Mysore: CIIL.

Bird, Steven, and Gary Simons
2003 Seven dimensions of portability for language documentation and description. *Language* 79 (3): 557–582.

Bradley, David, and Maya Bradley
2002a Conclusion: Resources for language maintenance. In *Language Endangerment and Language Maintenance*, David Bradley, and Maya Bradley (eds.), 348–353. London: RoutledgeCurzon.

Bradley, David, and Maya Bradley (eds.)
2002b *Language Endangerment and Language Maintenance*. London: RoutledgeCurzon.

Bredin, Marian
1996 Transforming images: Communication technologies and cultural identity in Nishnawbe-Aski. In *Cross-Cultural Consumption: Global Markets, Local Realities*, David Howes (ed.), 161–177, London/New York: Routledge.

Brenzinger, Matthias (ed.)
1992 *Language Death: Factual and Theoretical Explorations with Special Reference to East Africa*. Berlin: Mouton de Gruyter.

Burnaby, Barbara, and Jon Reyhner (eds.)
2002 *Indigenous Languages across the Community*. Flagstaff: Northern Arizona University. Online edition: <http://jan.ucc.nau.edu/~jar/books.html>.

Buszard-Welcher, L.
2001 Can the web help save my language? In *The Green Book of Language Revitalization in Practice*, Leanne Hinton, and Ken Hale (eds.), 331–345. San Diego: Academic Press.

Cantoni, Gina (ed.)
1996 *Stabilizing Indigenous Languages*. Flagstaff: Northern Arizona University. Online edition: <http://jan.ucc.nau.edu/~jar/books.html>.

Cazden, Courtney B.
2003 Sustaining indigenous languages in cyberspace. In *Nurturing Native Languages*, Jon Reyhner, Octavia V. Trujillo, Roberto Luis Carrasco, and Louise Lockaid (eds.), 53–57. Flagstaff: Northern Arizona University.

Corris, Miriam, Christopher Manning, Susan Poetsch, and Jane Simpson
2002 Dictionaries and endangered languages. In *Language Endangerment and Language Maintenance*, David Bradley, and Maya Bradley (eds.), 329–347. London: RoutledgeCurzon.
Craig [= Grinevald], Colette
1992 A constitutional response to language endangerment: The case of Nicaragua. *Language* 68 (1): 17–23.
Crystal, David
2000 *Language Death*. Cambridge: Cambridge University Press.
Dauenhauer, Nora Marks, and Richard Dauenhauer
1998 Technical, emotional, and ideological issues in reversing language shift: Examples from Southeast Alaska. In *Endangered Languages: Language Loss and Community Response*, Lenore A, Grenoble, and Lindsay J. Whaley (eds.), 57–98. Cambridge: Cambridge University Press.
Dorian, Nancy C.
1981 *Language Death: The Life Cycle of a Scottish Gaelic Dialect*. Philadelphia: University of Pennsylvania Press.
1998 Western language ideologies and small-language prospects. In *Endangered Languages: Language Loss and Community Response*, Lenore A. Grenoble, and Lindsay J. Whaley (eds.), 3–21. Cambridge: Cambridge University Press.
Dorian, Nancy C. (ed.)
1989 *Investigating Obsolescence: Studies in Language Contraction and Death*. Cambridge: Cambridge University Press.
Dressler, Wolfgang, and Ruth Wodak-Leodolter (eds.)
1977 *Language Death* (= IJSL 12). The Hague: Mouton.
England, Nora C.
1992 Doing Mayan linguistics in Guatemala. *Language* 68 (1): 29–35.
1998 Mayan efforts toward language preservation. In *Endangered Languages: Language Loss and Community Response*, Lenore A. Grenoble, and Lindsay J. Whaley (eds.), 99–116. Cambridge: Cambridge University Press.
Fishman, Joshua A.
1996 What do you lose when you lose your language?. In *Stabilizing Indigenous languages*, Gina Cantoni (ed.), 80–91. Flagstaff: Northern Arizona University.
Foley, William A.
2003 Genre, register and language documentation in literate and preliterate communities. In *Language Documentation and Description*, vol. 1, Peter K. Austin (ed.), 85–98. London: Hans Rausing Endangered Languages Project.
Freilich, Morris
1991 Smart rules and proper rules: A journey through deviance. In *Deviance: Anthropological Perspectives*, Morris Freilich, Douglas Raybeck, and Joel Savishinsky (eds.), 27–50. New York: Bergin & Garvey.
Grenoble, Lenore A., and Lindsay J. Whaley
2002 What does Yaghan have to do with digital technology? *Linguistic Discovery* 1 (2). Online journal:
<http://linguistic-discovery.dartmouth.edu/>.

Grenoble, Lenore A., and Lindsay J. Whaley (eds.)
1998 *Endangered Languages: Language Loss and Community Response*. Cambridge: Cambridge University Press.

Grinevald, Colette
1997 Language contact and language degeneration. *Handbook of Sociolinguistics*, Florian Coulmas (ed.), 257–270. Oxford: Blackwell.
1998 Language endangerment in South America: A programmatic approach. In *Endangered Languages: Language Loss and Community Response*, Lenore A. Grenoble, and Lindsey J. Whaley (eds.), 124–160. Cambridge: Cambridge University Press.
2001 Encounters at the brink: Linguistic fieldwork among speakers of endangered languages. In *Lectures on Endangered Languages*, vol. 2, O. Sakiyama (ed.), 285–313. Kyoto, Japon, ELPR Publication series C002.
2005a Why Rama and not Rama Cay Creole? In *Language Documentation and Description*, vol. 3, Peter K. Austin (ed), 196–224. London: Hans Rausing Endangered Languages Project.
2005b Globalization and language endangerment: poison and antidote. The HRELP Annual Public Lecture (Hans Rausing Endangered Languages Project at SOAS), London, 11 February 2005.

Hale, Ken
1992a On endangered languages and the safeguarding of diversity. *Language* 68 (1): 1–3.
1992b Language endangerment and the human value of linguistic diversity. *Language* 68 (1): 35–42.
1998 On endangered languages and the importance of linguistic diversity. In *Endangered Languages: Language Loss and Community Response*, Lenore A. Grenoble, and Lindsay J. Whaley (eds.), 192–216. Cambridge: Cambridge University Press.

Hinton, Leanne
2001 Audio-video documentation. In *The Green Book of Language Revitalization in Practice*, Leanne A. Hinton, and Ken Hale (eds.), 316–329. San Diego: Academic Press.

Hinton, Leanne, and Ken Hale (eds.)
2001 *The Green Book of Language Revitalization in Practice*. San Diego: Academic Press.

Howes, David
1996 Cultural appropriation and resistance in the American Southwest: Decommodifying 'Indianness'. In *Cross-Cultural Consumption: Global Markets, Local Realities*, David Howes (ed.), 138–160, London/New York: Routledge.

Jocks, Christofer
1998 Living words and cartoon translations: Longhouse 'texts' and the limitations of English. In *Endangered Languages: Language Loss and Community Response*, Lenore A. Grenoble, and Lindsay J. Whaley (eds.), 217–233. Cambridge: Cambridge University Press.

Kawagley, Angayuqaq Oscar
2003 Nurturing native languages. In *Nurturing native languages*, Jon Reyhner, Octavia V. Trujillo, Roberto Luis Carrasco, and Louise Lockard (eds.), vii–x. Flagstaff: Northern Arizona University.

Krauss, Michael
1992 The world's languages in crisis. *Language* 68 (1): 4–10.
1996 Status of native American language endangerment. In *Stabilizing Indigenous Languages*, Gina Cantoni (ed.), 16–21. Flagstaff: Northern Arizona University.
McCarty, Teresa L., and Lucille J. Watahomigie
1999 Reclaiming indigenous languages. *Preservation on the Reservation (and Beyond)* [= *Common Ground*, Fall 1999]. Online edition: <http://www.cr.nps.gov/aad/cg/fa_1999/Language.htm>.
McHenry, Tracey
2002 Words as big as the screen: Native American languages and the internet. *Language Learning & Technology* 6 (2) [Nicholas Ostler, and Jon Reyhner (eds.), *Special Issue on Technology and Indigenous Languages*]: 102–115. Online journal: <http://llt.msu.edu/>.
McLendon, Sally
1980 How languages die: A social history of unstable bilingualism among the Eastern Pomo. In *American Indian and Indoeuropean Studies: Papers in Honor of Madison S. Beeler*, Kathryn Klar, Margaret Langdon, and Shirley Silver (eds.), 137–150. The Hague: Mouton.
Mithun, Marianne
1989 The incipient obsolescence of polysynthesis: Cayuga in Ontario and Oklahoma. In *Investigating Obsolescence: Studies in Language Contraction and Death*, Nancy C. Dorian (ed.), 243–257. Cambridge: Cambridge University Press.
1998 The significance of diversity in language endangerment and preservation. In *Endangered Languages: Language Loss and Community Response*, Lenore A. Grenoble, and Lindsay J. Whaley (eds.), 163–191. Cambridge: Cambridge University Press.
Miyashita, Mizuki, and Laura A. Moll
1999 Enhancing language material availability using computers. In *Revitalizing Indigenous Languages*, Jon Reyhner, Joseph Martin, Louise Lockard, and W. Sakiestewa Gilbert (eds.), 113–116. Flagstaff: Northern Arizona University.
Mohan, Shailendra
2002 Linguistic landscape and social identity: A case of Jharkhand. In *Linguistic Landscaping in India with Particular Reference to the New States*, N. H. Itagi, and Shailendra Kumar Singh (eds.), 230–240. Mysore: CIIL and Mahatama Gandhi International Hindi University.
Nettle, Daniel, and Suzanne Romaine
2000 *Vanishing Voices: The Extinction of the World's Languages.* Oxford: Oxford University Press.
Ostler, Nicholas, and Blair Rudes (eds.)
2000 *Endangered Languages and Literacy*. Bath, UK: The Foundation for Endangered languages.
Östman, Jan-Ola
2000 Ethics and appropriation – with special reference to Hwalbáy. In *Issues of Minority Peoples*, Frances Karttunen, and Jan-Ola Östman (eds.), 37–60. Publications No. 31, Department of General Linguistics, University of Helsinki.

Pandit, P. B.
1972 *India as a Sociolinguistic Area*. Poona: University of Poona.
Payne, Doris
1999 Review of Grenoble and Whaley 1998. *Journal of Linguistics* 35 (3): 618–624.
Reyhner, Jon (ed.)
1997 *Teaching Indigenous Languages*. Flagstaff: Northern Arizona University.
Reyhner, Jon, Gina Cantoni, Robert N. St. Clair, and Evangeline Parsons Yazzie (eds.)
2000 *Revitalizing Indigenous Languages*. Flagstaff: Northern Arizona University. Online edition: <http://jan.ucc.nau.edu/~jar/books.html>.
Reyhner, Jon, Joseph Martin, Louise Lockard, and W. Sakiestewa Gilbert (eds.)
1999 *Learn in Beauty: Indigenous Education for a New Century*. Flagstaff: Northern Arizona University. Online edition: <http://jan.ucc.nau.edu/~jar/books.html>.
Reyhner, Jon, Octavia V. Trujillo, Roberto Luis Carrasco, and Louise Lockard (eds.)
2003 *Nurturing Native Languages*. Flagstaff: Northern Arizona University. Online edition: <http://jan.ucc.nau.edu/~jar/books.html>.
Romaine, Suzanne
1989 Pidgins, creoles, immigrant, and dying languages. In *Investigating Obsolescence: Studies in Language Contraction and Death*, Nancy C. Dorian (ed.), 369–383. Cambridge: Cambridge University Press.
Sasse, Hans-Jurgen
1992 Theory of language death. In *Language Death: Factual and Theoretical Explorations with Special Reference to East Africa*, Matthias Brenzinger (ed.), 7–30. Berlin: Mouton de Gryuter.
Saxena, Anju (ed.).
2004 *Himalayan Languages. Past and Present*. Berlin: Mouton de Gruyter.
Schmidt, Annette
1985 *Young People's Dyirbal: An Example of Language Death from Australia*. Cambridge: Cambridge University Press.
Skutnabb-Kangas, Tove
1990 Legitimating or delegitimating new forms of racism – the role of researchers. *Journal of Multilingual and Multicultural Development* 11 (1–2): 77–100.
2000 *Linguistic Genocide in Education – or Worldwide Diversity and Human Rights?* Mahwah, NJ & London, UK: Lawrence Erlbaum Associates.
Southworth, Franklin C.
1978 On the need for qualitative data to supplement language statistics: Some proposals based on the Indian census. *Indian Linguistics* 39: 136–154.
Spolsky, Bernard
2004 *Language Policy*. Cambridge: Cambridge University Press.
Swadesh, Morris
1948 Sociologic notes on obsolescent languages. *International Journal of American Linguistics* 14 (4): 226–235.
Trudgill, Peter
1978 Creolization in reverse: Reduction and simplification in the Albanian dialects of Greece. *Transactions of the Philological Society 1976–7*, 32–50. Oxford: Basil Blackwell.

Warschauer, Mark
1998 Technology and indigenous language revitalization: Analyzing the experience of Hawai'i. *Canadian Modern Language Review* 55(1). (References here are to the electronic version: <http://www.utpjournals.com/jour.ihtml?lp= product/cmlr/551/551-Warschauer.html> or <http://www.gse.uci.edu/markw/revitalization.html>).

Warschauer, Mark, and Keola Donaghy
1997 Leokī: A powerful voice of Hawaiian language revitalization. *Computer Assisted Language Learning* 10 (4): 349–361.

Weinreich, Uriel
1953 *Languages in Contact*. [Publications of the Linguistic Circle of New York, No. 1]. Page references to reprinted edition 1963. The Hague: Mouton.

Whorf, Benjamin Lee
1956 Languages and logic. In *Language, Thought, and Reality: Selected Writings of Benjamin Lee Whorf*, John B. Carroll (ed.), 233–245. Cambridge, Massachusetts: MIT Press.

Wilson, William H.
1999 Return of a language: Hawaiian makes a remarkable comeback. *Preservation on the Reservation (and Beyond)* [= *Common Ground*, Fall 1999]. Online edition: <http://www.cr.nps.gov/aad/cg/fa_1999/Language.htm>.

Winter, Werner
1993 Some conditions for the survival of small languages. In *Language Conflict and Language Planning*, Ernst Håkon Jahr (ed.), 299–314. Berlin: Mouton de Gruyter.

Woodbury, Anthony
1998 Documenting rhetorical, aesthetic, and expressive loss in language shift. In *Endangered Languages: Language Loss and Community Response*, Lenore A. Grenoble, and Lindsay J. Whaley (eds), 234–258. Cambridge: Cambridge University Press.
2003 Defining documentary linguistics. In *Language Documentation and Description*, vol. 1, Peter K. Austin (ed.), 35–51. London: Hans Rausing Endangered Languages Project.

Language situation and language policies in South Asia

Status of lesser-known languages in India

Udaya Narayana Singh

1. Introduction

India accounts for 2.4% of the world's land surface with a total land-area of 2 973 190 square kilometres,[1] but it is obviously a densely populated area with 16% of the world's population living here (Heitzman and Worden 1996). Consequently, it has always been a home for a large number of languages. For instance, Census 1961 reports a total of 1652 "mother tongues", out of which 184 had more than 10 000 speakers (R. A. Singh 1969). The figures have changed in later census reports.[2] The encyclopaedic *People of India* series of the Anthropological Survey of India (K. S. Singh 1992), identified 75 "major languages" out of a total of 325 languages used in Indian households. *Ethnologue* (Gordon 2005), too, reports India as home for 398 languages, including 387 living and 11 extinct languages. Since as early as in the 1990s, India was reported to have at least 32 languages with a large population base of one million plus speakers. In fact, all seven countries of South Asia put together are considered as the third most linguistically populous area (Nettle 1999), after Papua New Guinea in Asia and the African region of Ivory Coast to Tanzania; South Asia is comparable only with Mexico in the new world (Grimes 1993).[3]

It is estimated that there are about 700–1000 languages spoken in the South Asian region, belonging to at least four major language families – Indo-European (most of which belong to one sub-branch, Indo-Aryan), Tibeto-Burman, Dravidian, and Austro-Asiatic. Multilingualism is not a new phenomenon in the Indian context. Even Sir George Grierson's (1903–1923) twelve-volume *Linguistic Survey of India* – material for which was collected in the last decade of the 19th century, had identified 179 languages and 544 dialects. One of the early Census reports also showed 188 languages and 49 dialects (Census 1921). But, despite this, recent social changes such as technological advances, urbanization and globalization are rapidly changing the linguistic tapestry of India – upsetting, in some ways, the linguistic equilibrium.

Since Independence the Indian government has made pronouncements in favour of linguistic diversity and promotion of less privileged groups (including languages) by means of introducing language policies and laws, but partly because of social factors, such as large population growth, low literacy level, disparity between rich and poor, between urban and rural areas, effects of these government policies have not been as visible as one would have liked to.

The aim of this paper is to present an overview of the linguistic situation in India today, beginning with some background information on the linguistic demography of India. The focus in the next section will be on government efforts to promote and maintain linguistic diversity and document as well as support less privileged groups (including languages), including a discussion on the constitutional provisions for smaller language communities. This will be followed in section three by a discussion of some factors complicating the implementation of these language policies. Section four will focus on some noticeable trends visible today relating to lesser-known languages. These observations are based on a comparative study of census reports of the last several decades.

Although the focus here is on the linguistic scene in India, some pointers concerning South Asia will be made because the socio-linguistic scene in India is similar to other South Asian countries on several fronts.

2. Languages in India

Languages spoken in the South Asian region belong to at least four major language families: Indo-European (most of which, 74.24%, belong to its sub-branch Indo-Aryan), Dravidian (with 23.86% speakers), Austro-Asiatic (1.16%), and Sino-Tibetan (0.62%) as pointed out by Baldridge (1996, 2002; see also Gordon 2005).

The biggest chunk of languages and mother tongues belong to the Indo-Aryan sub-family of Indo-European languages. The immediate predecessor of Indo-Aryan is Indo-Iranian, the oldest specimens of which are available in the Zend-Avesta. Among the modern Indo-Aryan languages, Hindi and Bangla are the most well-known languages. Western Hindi is a Midland Indo-Aryan language, spoken in the Gangetic plain and in the region immediately to its north and south. Around it, on three sides, are Panjabi, Gujarati, Rajasthani. Eastern Hindi is spoken in Oudh and to its south. In the outer layer, we get languages such as Kashmiri, Lahnda, Sindhi, Gujarati, Marathi, in the northern and the western region, and Oriya, Maithili, Bengali and Assamese in the east.

The word *Dravidian* was first used by Robert A. Caldwell (1856), who introduced the Sanskrit word *Dravida* to designate the speech community. Among Dravidian languages, besides the four internationally known languages (Tamil, Telugu, Kannada and Malayalam), there are 26 languages by the current count, of which 25 are spoken in India and one (Brahui) is spoken in Baluchistan on the Pakistan-Afghanistan border. Spoken by more than 300 million people in South Asia, the antiquity of Dravidian languages is largely due to the rich grammatical and linguistico-literary tradition of Classical Tamil (Annamalai 2003). Even other major Dravidian languages possess independent scripts and literary histories dating from the pre-Christian era. The smaller Dravidian languages include Kolami-Naiki, Parji-Gadaba, Gondi, Konda, Manda-Kui, Kodagu, Toda-Kota, and Tulu (Krishnamurti 2003). The Northern Group of Dravidian languages is the smallest: Brahui, Malto, and Kudukh. The Central Group of Dravidian languages seem to be most widespread: Gondi, Konda, Kui, Manda, Parji, Gadaba, Kolami, Pengo, Naiki, Kuvi, and Telugu. The Southern Group includes Tulu, Kannada, Kodagu, Toda, Kota, Malayalam, and Tamil.

The Austric family of languages is divided into two branches, Austroasiatic and Austronesian, the latter formerly called Malayo-Polynesian. They are spoken in India, Southeast Asia, and the Pacific Islands. The Austroasiatic branch has three sub-branches: Munda, Mon-Khmer, and Vietnamese-Muong. The Munda languages in India are spoken in the eastern and southern parts of the country. Among the more well-known Munda languages are Santali, Mundri, Bhumij, Birhar, Ho, Tri, Korku, Khari, Juang, and Savara. The Munda speakers are found mostly in the hills and jungles, while the plains and valleys have some pockets inhabited by people speaking these languages.

The Tibeto-Burman family is a part of Sino-Tibetan languages, spread over a large area – from Tibet in the north to Burma in the south, and from the Ladkh wathrat of Kashmir in the west to the Chinese provinces of Szechuen and Yunnan in the east. Lepcha, Sikkimese, Garo, Bodo, Manipuri, and Naga are some of the better-known Tibeto-Burman languages. Only 4 071 701 people, or 0.62% of the Indian population speak these mother-tongues.

Several smaller languages such as Burushaski in the North-West are language isolates. Then there are separate families (Mallikarjun 2002) like Andamanese which includes quite a few diverse languages in the Andamans, and one could possibly also add six odd languages spoken in 22 odd Nicobar Islands under this group.

Thus, as becomes evident from this discussion, language families in India roughly coincide with a broad geographic division of the sub-continent. Indo-Aryan speakers are spread over northern and central regions, whereas the twenty-odd Dravidian groups are mostly located in the Southern peninsula. The Austro-Asiatic languages are spoken mainly in the East and Central India, whereas the Tibeto-Burman communities live in the northern Himalayan region (like Himachal Pradesh) as well as in the seven North-Eastern States.

3. Linguistic and social facts: A correlation

In this section, we would first present the states and union territories of India in terms of a few broad categories, from A through E. The division is purely based on the multilingual profile of these states, which is indicated in the last column (Column 7) of Table 1. The names of linguistic majority groups are listed first in Column 7 against each state (names of which appear in Column 2, whereas in Column 3 their percentage is shown. Columns 4 and 5 show the percentage of population of the first two linguistic minority groups state-wise, and their names are given within parentheses in Column 7. When a state has many smaller linguistic groups other than the first three groups, their percentages are mentioned in Column 6.

The states in set A – Kerala, Punjab, Gujarat, Haryana, Uttar Pradesh, Rajasthan, Himachal Pradesh, Tamil Nadu, West Bengal, and Andhra Pradesh – have a negligible percentage of minor speech groups in terms of population, with the majority language spoken by more than 85% inhabitants of the state. Under set B, one gets states where the majority language group accounts for over 70% of the population but one still finds a sizable linguistic minority. Set C has those states – Goa, Meghalaya, Tripura and Karnataka – that have been the hotbed of language tensions and riots. In many cases, this is due to the fact that they have had a dominating linguistic minority group, such as Bengali speakers in Tripura or the Marathi community in Goa. The tension in Karnataka came from an unexpected quarter – particularly from the bordering speech community of Marathi speakers, and this as well as many other tensions later had to do with control over scarce resources – like water (the Cauvery water sharing dispute with the neighbouring Tamil Nadu and the Tamil-Kannada tension, for example) or land, etc. Meghalaya had witnessed a similar tension due to large scale in-migration. After the creation of Meghalaya in 1972, the first violent demonstrations against the outsiders (which in this case meant the Bengalis, Marwaris, Biharis, and Nepalis) resulted in a number of deaths and arson in 1979, 1987, 1989, 1990 and again in 1992 (Maitra and Maitra 1995).[4] The linguistic tensions have been quite volatile in the set D states (Assam, Sikkim and Manipur), too, which seems to be due to their linguistic composition as well as inter-group attitudes. Assam, unlike most other areas of the Northeast, was better integrated with mainstream India prior to independence; but it has been segmented a number of times, and it has also witnessed large-scale in-migration for a long time – so much that the speakers of Assamese were almost going to become a linguistic minority in their own home state. Manipur has remained volatile and unstable because of a long border with Myanmar and also due to ethnic-linguistic tensions and feeling against the "outsiders". Set E is the most variegated geo-space in India with numerous tongues.

Table 1. Extent of multilingualism in the Indian states

Set	STATES (P. = Pradesh)	Major lang.	Minor 1 (%)	Minor 2 (%)	Others	LANGUAGES Major lg (+ Two Minor)
1	2	3	4	5	6	7
A.	Kerala	96.6	2.1	0.3	1.0	Malayalam (Tamil, Kannada)
	Punjab	92.2	7.3	0.1	0.4	Punjabi (Hindi, Urdu)
	Gujarat	91.5	2.9	1.7	3.9	Gujarati (Hindi, Sindhi)
	Haryana	91.0	7.1	1.6	0.3	Hindi (Punjabi, Urdu)
	Uttar P.	90.1	9.0	0.5	0.4	Hindi (Urdu, Punjabi)
	Rajasthan	89.6	5.0	2.2	3.2	Hindi (Bhili, Urdu)
	Himachal P.	88.9	6.3	1.2	3.6	Hindi (Punjabi, Kinnauri)
	Tamil Nadu	86.7	7.1	2.2	4.0	Tamil (Telugu, Kannada)
	West Bengal	86.0	6.6	2.1	5.7	Bengali (Hindi, Urdu)
	Andhra P.	84.8	8.4	2.8	4.0	Telugu (Urdu, Hindi)
B.	Madhya P.	85.6	3.3	2.2	8.9	Hindi (Bhili, Gondi)
	Bihar	80.9	9.9	2.9	6.3	Hindi (Urdu, Santali)
	Orissa	82.8	2.4	1.6	13.2	Oriya (Hindi, Telugu)
	Mizoram	75.1	8.6	3.3	13.0	Lushai (Bengali, Lakher)
	Maharashtra	73.3	7.8	7.4	11.5	Marathi (Hindi,Urdu)
C.	Goa	51.5	33.4	4.6	10.5	Konkani (Marathi, Kannada)
	Meghalaya	49.5	30.9	8.1	11.5	Khasi (Garo, Bengali)
	Tripura	68.9	23.5	1.7	5.9	Bengali (Tripuri, Hindi)
	Karnataka	66.2	10.0	7.4	16.4	Kannada (Urdu, Telugu)
D.	Sikkim	63.1	8.0	7.3	21.6	Nepali (Bhotia, Lepcha)
	Manipur	60.4	5.6	5.4	29.6	Manipuri (Thadou, Tangkhul)
	Assam	57.8	11.3	5.3	25.6	Assamese (Bengali, Boro)
E.	Arunachal	19.9	9.4	8.2	62.5	Nissi (Nepali,Bengali)
	Nagaland	14.0	12.6	11.4	52.0	Ao (Sema, Konyak)

We could now look at these states and fill in more details. In the seven states listed in category A, one finds not only very small segments of minor speech groups, it is mostly the case that even those few that appear as minor speech groups in columns 4 and 5 under a given state under A, appear as a major language elsewhere. A few examples will make the point clearer. Tamil (2.1%) and Kannada (0.3%) are minor languages in Kerala where the dominant language is Malayalam (96%). However, note that Tamil is spoken by 86.7% in Tamil Nadu and Kannada by 66.2% residents of Karnataka – where they function as dominant languages. Once again, in Goa, Kannada is a minority tongue spoken by only 4.6% of the population. In six states under A, Urdu appears as a minority language. It is not surprising that there would be a bond developed across the states among such minor groups (like Urdu) that are dominated in a number of states.

Set B has Hindi (Madhya Pradesh and Bihar), Oriya (Orissa), Lushai (Mizoram) and Marathi (Maharashtra) as major languages spoken by a very large segment of their inhabitants, but these states also have a large number of tribal communitites with their own languages. It is not surprising, therefore, that the first two (Madhya Pradesh and Bihar) have now been reorganized and have given rise to two new states, mostly dominated by several tribal groups. Thus, Chattisgarh was carved out of Madhya Pradesh and Jharkhand out of Bihar. Creation of these states were primarily meant to mark their separate ethno-linguistic identity. However, since there is not a single tribal language spoken in these two new states that has been developed in all respects, Hindi remains their state language.

Set C states have their fair share of both pronounced and hidden language tensions. The Konkani in Goa had to fight a long battle to claim their own position, as their language was always classified as a dialect of Marathi. But their predicament is that they speak a language that is written in four scripts – in Roman and Devanagari in Goa, in Kannada in Karnataka and in Malayalam in Kerala. Karnataka has had a long border row (with Maharastra), a battle over scarce resources such as water (with Tamil Nadu) and linguistic clashes with Marathi.

The linguistic tensions have been quite volatile in the set D states too. Set E is the most variegated geo-space in India with numerous tongues.

Consider the total picture now. Statistically, we have already seen in the preceding section that if we take the entire Indian sub-continent, the smallest group is the speakers of Austro-Asiatic languages, who make up approximately 1.16% of the population, most of whom live in a region extending from West Bengal through Bihar and Orissa into Madhya Pradesh. These groups earlier had no states to call their own, a status that has changed with the formation of three new states, Jharkhand, Chattisgarh and Uttaranchal, in 1999. The situation now is such that some states have more than one official language, with each language serving a specifically designated purpose or use in a certain region. For instance, the state of Bihar tried to quell the linguistic aspirations of its different speech communities by declaring Urdu as an additional official language in 1980, the main official tongue being Hindi.

4. Some relevant facts about India

4.1. Population growth

In the last official Census count, India's population was 1 027 015 247 as of March 31, 2001 (Census 2001). Table 2 here gives us a glimpse of the decadal variation of population in India during the last one hundred years.

Table 2. Decennial growth of population in India

Census year	Population (millions)	Decennial growth rate	Geometric growth rate
1901	238.40	–	–
1911	252.09	5.75	0.56
1921	251.32	–0.31	–0.03
1931	278.98	11	1.06
1941	318.66	14.22	1.34
1951	361.09	13.31	1.26
1961	439.23	21.51	1.98
1971	548.16	24.8	2.24
1981	683.33	24.7	2.22
1991	846.30	23.85	2.14

** Exclusive of Jammu & Kashmir
Source: Registrar General of India and Census of India (1921, 1951, 1961, 1971, 1981, 1991 and 2001)

A few more indicators on the demographic profile of India: The United Nations Population Division statistics[5] show that the population of India has taken only 34 years to increase from 500 million to 1 billion as against China's 33 years and the world average time taken to double this figure would be 454 years. There is no doubt that the demographic profile of India has changed very fast. Over one-third of Indians are younger than 15 years of age, by 2000 estimates.[6] Further, more than 70% of the population live in more than 550 000 odd villages.[7] Although 37.7% of alla Indians are now in the 0–14 age-group (1996), this high ratio will drop down to 27.7% only in 2016. In comparison, most Indians (55.6%) belong to the 15–59 age group[8] – the population of which will dramatically go up by 2016 – pushing the people in older age groups.[9]

4.2. Religion

Multilingualism being the order of the day in South Asia, it is not surprising that cultural habits, rituals, and belief-systems show an equal extent of plurality. Religion, caste, and language issues usually dominate the politics in South Asia. Although 83% of the population are Hindu, India also houses more than 120 million Muslims forming 14% of Indians – making it one of the world's largest Muslim populations. The population also includes the following smaller religious minorities: Christian 2.4%, Sikh 2%, Buddhist 0.7%, Jains 0.5%, other 0.4%.[10] Census 2001 has given details of the decadal growth rate of different religious groups in India which is worth reproducing here (see Table 3). The rate of decline is alarming among the Sikhs, whereas the figures for Christianity seem to be on the rise.

Table 3. Decennial growth of population by religious communities in India

	Population and decadal growth by religious communities in India; 1981–2001 (Excluding assam and Jammu&Kashmir)			
Religious[11] communities	Absolute increase in population[12] 1981–1991	1991–2001	Decadal growth rate (%) 1981–1991	1991–2001
All[13]	156 889 206	175 641 434	23.8	21.5
Hindus	124 805 159	134 677 636	22.8	20
Muslims	23 494 790	27 931 536	32.9	29.3
Christians	2 729 900	4 177 211	17	22.1
Sikhs	3 298 781	2 742 805	25.5	18.9
Buddhists	1 673 298	1 486 899	36	23.2
Jains	141 065	866 517	4	26
Others	364 884	348 540	13.2	111.3

4.3. Caste

The caste system in India reflects occupational and religiously defined hierarchies in this region. Traditionally, there are four broad categories of castes (*varnas*), including a category of outcastes, earlier called "untouchables" but now commonly referred to as "dalits", and special constitutional provisions have been made for these castes, generally known as the "Scheduled Castes". Similarly, there is also a separate list of "Scheduled Tribes". The Scheduled Caste and Scheduled Tribe population in India according to the 1991 census are 138 223 277 and 67 758 380 respectively, constituting 16.33% and 8.01% of India's total population respectively.[14] It may be noted that the proportion of Scheduled Caste and Scheduled Tribe population has increased considerably from 15.8% and 7.8% respectively in 1981.

5. Language and space

5.1. Creation of states based on linguistic principles

After India gained independence in 1947, it was suggested that the newly independent nation should have a federal system, composed of a limited number of states. The basis of their formation was to be linguistic – a region with one major language would comprise one state. Thus, Prime Minister Nehru appointed the States Reorganization Commission (SRC) in August 1953, with Justice Fazi Ali, K. M. Panikkar and Hridaynath Kunzru as members, to examine "objectively and dispassionately" the entire question of the reorganization of the states of the union. The States Reorganization Act was passed by parliament in November 1956, and it provided for 14 states and 6 centrally administered territories.[15] Some states were created from parts of others to unite members of a language group, as the whole approach was based on the linguistic principle. In 1956, thus, the government reduced the number of states from a total of 27 to 14. Even before the act was passed, there was already strong agitation for the partition of the Bombay state into two large states – one each for the Gujarati and Marathi speech communities. Since the SRC did not agree to that, language riots followed in both Bombay and Ahmedabad. Finally in 1960, Bombay was divided into two new states, Gujarat and Maharashtra. Once again, in November 1966, two states were formed out of one earlier state, Punjab. One remained the state of Punjab, where the majority spoke Punjabi with Sikhs as the dominating religious group, and the other entity with a predominantly Hindu population became the state of Haryana, where the majority spoke a variety of Hindi (often known as "Haryanavi").

5.2. Administrative divisions

At present, as per Census 2001 statistics, India is administratively organized into 35 States and Union Territories. Each of these units has under it divisions or units at several sub-levels. At the first level, there are Districts (593 in number) – further sub-divided into Sub-districts (5564).[16] Sub-districts are also called *Tehsils* or *Talukas*, *Mandals* (in Andhra Pradesh), Circles, C.D. Block (in Bihar, Tripura, Meghalaya, West Bengal, and Jharkhand), R.D. Block (in Mizoram), Commune *Panchayats* (in Pondicherry), Sub-divisions (in Arunachal Pradesh and Lakshadweep), and even Police Stations (in Orissa).

It may be noted that nearly 26.1% of the total population of the country live in the urban areas which have shown a phenomenal population explosion

– from 28.85 million in 1901 to 159.46 million in 1981 and 217 million in 1991.[17] Currently, there are 51 Cities, 384 Urban Agglomerates and 5161 Towns (2843 in 1951) in India. Out of the total urban population, about 138 million people, or 16 percent, lived in only 299 urban agglomerations (Census 1991). Only 24 metropolitan cities accounted for 51 percent of India's total population, with Bombay and Calcutta leading at 12.6 million and 10.9 million, respectively. This administrative organization of India provides some idea of the enormity of this region. Although states in India are organized according to languages (where each state has its own state official language), each state has also language communities who have other mother tongues than the official state language.

6. Language policies promoting linguistic diversity

6.1. India's linguistic diversity and the Indian Constitution

When the Constituent Assembly adopted the Constitution of India on November 26, 1949, there were 14 languages listed in the Eighth Schedule of the Indian Constitution. They were (in the order of number of speakers): Hindi, Telugu, Bengali, Marathi, Tamil, Urdu, Gujarati, Kannada, Malayalam, Oriya, Punjabi, Kashmiri, Assamese and Sanskrit. There have been three amendments to the Eighth Schedule during the last 55 years, the results of which have been as follows. Sindhi was included through the Constitution Amendment Bill No 21 in 1967, Konkani, Manipuri and Nepali (or Gorkhali) through Amendment Bill No. 71 in 1992, and Maithili, Santali, Bodo, and Dogri through Amendment Bill No. 100 in 2003. Thus, currently, there are 22 languages in the Eighth Schedule. Table 4 lists these languages together with the number of speakers of each of these languages (as given in Census 1991).

Shri Jaipal Singh had proposed that out of the 176 *Adivasi* (or tribal) languages (as in 1949), Mundari (with 400 000 speakers), Gondi (with 320 000 speakers) and Oraon (with 110 000 speakers) should be included in the 8th Schedule of the Constitution, because they were important and spoken by a larger number of people than some of the languages already included. He selected only these three out of many tribal languages so as not to overburden the Schedule, and he felt that they would "enrich the *Rashtrabhasha* [national language] of the country" (CAD 2003: 1439; see also Patra 1998). Rajasthani and Hindustani were two of the 14 languages proposed to be included in the list by Naziruddin Ahmad (CAD 2003: 1482). But this was not accepted. Syama Prasad Mookerjee requested for the inclusion of Sanskrit (CAD 2003: 1391). The amendment seeking the insertion of Sanskrit was later adopted (CAD 2003: 1486). Finally, however, only 14 languages were included.

Table 4. Scheduled languages in the Indian Constitution and their speakers

Sr. no.	Languages	Speakers	Percentage
1.	Assamese	13 079 696	1.55%
2.	Bengali	69 595 738	8.22%
3.	Bodo	1 221 881	15%
4.	Dogri	89 681	0.01%
5.	Gujarati	40 673 814	4.81%
6.	Hindi	337 272 114	39.85%
7.	Kannada	32 753 676	3.87%
8.	Kashmiri	56 693 (outside J&K)	N.A.(1991)
		3 174 684 (1981 fig)	0.48% (1981)
9.	Konkani	1 760 607	0.21%
10.	Malayalam	30 377 176	3.59%
11.	Manipuri	1 270 216	0.15%
12.	Marathi	62 481 681	7.38%
13.	Maithili	7 766 597	0.93%
14.	Nepali	2 076 645	0.25%
15.	Oriya	28 061 313	3.32%
16.	Punjabi	32 753 676	2.76%
17.	Sanskrit	49 736	0.01%
18.	Santali	5 216 325	0.62%
19.	Sindhi	2 122 848	0.25%
20.	Tamil	53 006 368	6.26%
21.	Telugu	66 017 615	7.80%
22.	Urdu	43 406 932	5.13%

Prior to the appearance of these 14 languages in the Schedule VIII, these languages had acquired different denominations at different stages of preparation and formulation of the Indian Constitution. The Congress party described them as "national languages". The original aim of naming some languages in the Constitution seemed to be to prepare a list of languages to be used in administration, expression of science and technology in independent India. M. Satyanarayana, a member of the Drafting Committee on the Language Resolution (with the permission of Jawaharlal Nehru) prepared a list of 12 languages – Hindi, Gujarati, Marathi, Kannada, Malayalam, Tamil, Telugu, Oriya, Bengali, Assamese, Punjabi and Kashmiri. Nehru added Urdu as the thirteenth language to the list.

The draft provisions on language prepared by K. M. Munshi and N. Gopalaswamy Ayyangar for discussion by the Indian National Congress (outside the Constituent Assembly) had made a provision for a "Commission" with a Chairman and representative members of different languages of Schedule VII-A for the progressive use of Hindi, and to restrict the use of English in various domains. At that juncture, this Schedule had Hindi, Urdu, Punjabi,

Kashmiri, Bengali, Assamese, Oriya, Telugu, Tamil, Malayalam, Canarese, Marathi, Gujarati, and English. It may be noted that English was a part of the Congress Party draft, and not a part of the draft of the Constituent Assembly. There were, however, no discussions in the Constituent Assembly on the criteria adopted for inclusion of languages in the Eighth Schedule.

In the Article 343 on "Official language of the Union", a provision was left to extend the use of English language "for all the official purposes of the Union" even after "a period of fifteen years", with a proviso that "the President may, during the said period, by order authorize the use of the Hindi language in addition to the English language and of the Devanagari form of numerals for any of the official purposes of the Union".

The first Prime Minister of India, Pt. Jawaharlal Nehru stated about the recognition of languages: "The makers of our Constitution were wise in laying down that all the 13 or 14 languages were to be national languages. There is no question of any one language being more a national language than the others ... Bengali or Tamil or any other regional language is as much a national language as Hindi" (Lok Sabha Debates: 11640, as quoted by Kumaramangalam 1965: 54). While addressing Parliament in 1963, Nehru described the languages of the Eighth Schedule as national languages. The Congress Working Committee meeting of April 5, 1954, recommended that examinations for the All-India Services to recruit government administrators should be progressively held in Hindi, English and the principal regional languages, and candidates may be given option to use any of the languages for the purpose of examinations (Kumaramangalam 1965: 44). The Congress Working Committee meeting of June 2, 1965, stated that "The Union Public Service Commission [UPSC, a national core of administrators] examinations will be conducted in English, Hindi and other national languages mentioned in the Eighth Schedule of the Constitution" (Kumaramangalam 1965: 100–101).

6.2. Non-major languages and the Indian Constitution

By the early 1950s, in Census 1951 and several other official documents, we find a mention of 14 languages recognized in the Indian Constitution as well as 23 major tribal languages and 24 other minority languages, speakers of each having crossed the 100 000 figure. It must be mentioned here that the terms "major" or "minor" language being used here are not used accordingly in the Census documents. The Official Language Resolution (3) of 1968 considered languages listed in the Schedule as major languages of the country. The Government of India (1986, 1992) documents considered them as Modern Indian languages. These languages are also identified as Scheduled Languages in (several) official contexts.[18] It is important to realize that the Indian Constitution does not define or use the words "minor" or "minority" languages, although there is a mention of "linguistic minorities". What we get from the Census 1961 document is an interesting array of different categories of languages (see Table 5).

Table 5. Attestation of mother-tongues as per Census 1961

Serial no.	Description	No of mother tongues	Total no. of speakers
1	No. of mother tongues returns of the country.	1652	438 936 918
2	No.of mother tongues attested in LSI classification[19]	572	436 224 545
3	No.of mother tongues not traced in LSI but tentatively classified.	400	426 076
4	No. of mother tongues attested in LSI but tentatively reclassified.	50	1 908 399
5	No. of mother tongues considered unclassified	527	62 432
6	Foreign mother tongues	103	315 466

6.3. Constitution provisions promoting linguistic diversity

The Constitution of India (see Kagzi 2001) promotes linguistic diversity, some examples of which are as follows. The first one concerns the language choice of debates in the Indian Parliament, as stated in Article 120:

> Art 120. Language to be used in Parliament. – (1) Notwithstanding anything in Part XVII, but subject to the provisions of article 348, business in Parliament shall be transacted in Hindi or in English;
>
> Provided that the Chairman of the Council or States or Speaker of the House of the People, or person acting as such, as the case may be, may permit any member who cannot adequately express himself in Hindi or in English to address the House in his mother tongue.

Theoretically, this could include all mother-tongues as listed by the Census of India reports. The same kind of provision is also made at the state level (under Art 210 under "The States"). Art 345 of the Indian Constitution states very clearly that "[s]ubject to the provisions of article 346 and 347, the Legislature of a State may be allowed to adopt *any one or more of the languages in use in the State* or Hindi as the language or languages to be used for all or any of the official purposes of that State." In this context, Article 347 is more explicit:

> Art 347: On a demand made in that behalf, the President may, if he is satisfied that a substantial proportion of the population of the State desire the use of any language spoken by them to be recognized by that State, direct that such language shall also be officially recognized throughout that State or any part thereof for such purposes as he may specify.

Further, because of both government and private initiatives, there are institutions (such as schools and colleges) providing facilities for the displaced language communities, making it possible for different language groups to live together and yet retain their own linguistic traditions. Such institutions have special status in legal terms. For instance, the following articles of the Indian Constitution under "Cultural and Educational Rights", and its seventh Amendment in 1956 are illustrative:

> Art 29. *Protection of interests of minorities.* – (1) Any section of the citizens residing in the territory of India or any part thereof having a distinct language, script or culture of its own shall have the right to conserve the same. (2) No citizen shall be denied admission into any educational institution maintained by the State or receiving aid out of State funds on grounds only of religion, race, caste, language or any of them.

> Art 30. *Right of minorities to establish and administer educational institutions.* – (1) All minorities, whether based on religion or language, shall have the right to establish and administer educational institutions of their choice.

There is yet another provision in the Constitution which allows these minority communities to express their grievances in their own languages. Art 350 is a case in point:

> Art 350. *Language to be used in representations for redress of grievances.* – Every person shall be entitled to submit a representation for the redress of any grievance to any officer or authority of the Union or a State in any of the languages used in the Union or in the State, as the case may be.

A special provision has also been made under Article 350A to provide smaller communities educational opportunities in their mother tongue:

> 350 A. *Facilities for instruction in mother-tongue at primary stage.* – It shall be the endeavor of every State and of every State and of every local authority within the State to provide adequate facilities for instruction in the mother-tongue at the primary stage of education to children belonging to linguistic minority groups; and the President may issue such directions to any State as he considers necessary or proper for securing the provision of such facilities.

Further, Constitutional provisions provide for equal opportunity for all groups and sections of the community. It acts as a guarantor and a levelling force. It is true that Act 19, the Official Languages Act, makes Hindi the official national language and recognizes the use of Hindi in all official domains. According to the official order from the Ministry of Home Affairs, "it is the duty of the Union to promote the spread of the Hindi Language and to develop it so that it may serve as a medium of expression for all the elements of the composite culture of India".[20] Note that it is not at the cost of linguistic pluralism that Hindi is sought to be promoted here. There are also provisions made (as seen above) in the Indian Constitution for safeguarding and promoting lesser-known languages and communities. A similar case in point is the establishment and creation of the "Commissioner for Linguistic Minorities" under Article 350B. The Commissioner for Linguistic Minorities has advisory or recommendatory powers. The report of the Deputy Commissioner of Minorities is popularly known as the "Minorities Commission Report". Apart from this, both at the Central and State governance levels, there are separate ministries and/or departments of tribal welfare – the aim of which is to look after the special needs of non-major communities.

> 350 B. *Special Officer for linguistic minorities.* – (1) There shall be a Special Officer for linguistic minorities to be appointed by the President. (2) It shall be the duty of the Special Officer to investigate all matters relating to the safeguards provided for linguistic minorities under this Constitution and report to the President upon those matters at such intervals as the President may direct, and the President shall cause all such reports to be laid before each House of Parliament, and sent to the Governments of the States concerned.

In the same vein, even in art and fiction diversity is promoted. As an evidence supporting this observation, one could consider the regular occurrence of movements and the way social work groups organize themselves to promote their cause. Such causes are also promoted by the fourth estate – both print and visual media.

7. Practical hurdles in implementing language policies

Despite these good intentions of the state, these language policies have not been implemented to the extent they were intended for originally. They did not have the kind of influence as one had wished for. However, there are many factors which have contributed to this state of affairs, including unmanageable population growth, slow as well as low economic growth, literacy problems, unplanned urbanization, and problems with education planning.

7.1. Population and economic growth

The sheer size of problems such as demographic pressures and low economic growth in South Asia poses a challenge for language planners, education managers and economists. One often wonders whether these countries in South Asia would be able to afford a plan that tries to bring some kind of parity among all languages – large and small – howsoever justified the plan may be about giving equal access to education among all linguistic groups. The economic cost of such a step would have to depend on the general economic condition of the entire South Asian region. It goes without saying that the total picture of economic performance does not seem to be very hopeful as yet. And this is true in spite of mammoth efforts put in by national governments as well as international agencies to alleviate poverty. The fact remains that their human development index of all South Asian nations is very low. A look at these few economic indicators as given in Table 6 will show the enormity of the task of bringing parity among different linguistic groups.

Table 6. Human development index (HDI) of South Asian nations

Indicators	India	Pakistan	Bangladesh	Sri Lanka	Nepal	Bhutan	Maldives
HDI	0.595	0.497	0.509	0.740	0.504	0.536	0.752
Total population (millions), 2002	1049.5	149.9	143.8	18.9	24.6	2.2	0.3
Pop. growth rate (%), 1975–2002	1.9	2.8	2.4	1.3	2.3	2.3	3.0
Urban pop. % of total, 2002 (1975)	28.1 (21.3)	33.7 (26.4)	23.9 (9.9)	21.1 (22.0)	14.6 (5.0)	8.2 (3.5)	28.4 (18.1)
Life expectancy index, 2002	0.64	0.60	0. 60	0.79	0.58	0.63	0.70

A comparison of all South Asian nations could be helpful to our understanding the challenges faced by India. As such, the density of population is greatest in Asia with more than 108 people per square kilometer, as compared to 23 in Latin America, 24 in Africa and 14 in North America (Kumar 1999).[21] Note that except for Sri Lanka and the Maldives, life expectancy figures are very poor, and these are the only two among South Asian nations that have a much higher human development index value. Around the year 1900, death rates in South Asia were higher because of regular appearance of endemic disease, epidemics and natural calamities like famines. During the period 1911–1921, the birth and death rates in undivided India were virtually equal – 48 per 1000 people. The advancement in curative and preventive medicine contributed to a steady decline in the death rate by the mid-1990s, although it is still the case that numerous calamities and disasters stalk this area. While the population growth rate now looks more respectable with 1.9%, the sheer number in absolute terms makes all efforts in education and language planning go haywire. According to Census 2001: "India adds almost the total population of Australia or Sri Lanka every year. A 1992 study of India's population notes that India has more people than all of Africa and also more than North America and South America together." With low economic growth and high population growth, the sheer cost-benefit analysis of any plan will make the task of the governments more difficult.

Because of the slow economic growth and large annual population growth, more people are getting pushed below the national poverty lines here. The human poverty index figures confirm this kind of consequence. Further, when such tendencies are seen, it is generally the case that the worst sufferers would be the illiterates, the rural folk, the women and the segment belonging to the subaltern including speakers of smaller language groups.

7.2. Education and literacy

With staggering adult illiteracy figures in most of these nations, ranging from 38.7% to 58.9%, the task before education planners seems daunting. The school enrolment has been nowhere near the goal of universal education – with half the school children dropping out even after getting into a primary school. Table 7 table shows trends in the area of education and literacy in South Asia.

Table 7. Education and literacy figures of South Asian nations

Indicators (2001–02 fig)	India	Pakistan	Bangladesh	Sri Lanka	Nepal	Bhutan	Maldives
Education index	0.59	0.40	0.45	0.83	0.50	0.48	0.91
Adult literacy rate (15+)	61.3[22]	41.5[23]	41.1	92.1	44.0	47.0[24]	97.2
Adult illiteracy (% age 15+)	38.7[25]	58.5[26]	58.9	7.9	56.0	NA	2.8
Combined school enrolm: all 3 levels (%),	55[27]	37[28]	54	65[29]	61	49[30]	78
% of public expenditure on education (99–01)	4.1	7.8[31]	2.3	1.3	3.4	5.2	4.0
Net primar. enrolment %	83[32]	35	87	105[33]	70[34]	NA	96
Children in grade 5 (%)	59[35]	NA	65	94[36]	78	91	NA

One of the most positive and dramatic improvements in literacy has been seen in what were previously considered highly backward states in India. For instance, between 1991 and 2001, literacy increased in Rajasthan by 22.5%, to an absolute literacy figure of 61%, in Chattisgarh by 22.3% to 65.2% literacy, and in Madhya Pradesh by 19.4% to 65.4% literacy (Census 2001). These figures compare very well to the national average of 13.75% growth rate in literacy in the decade 1991–2001 and 65.4% overall literacy in 2001.[37] Compare this with the 1947 figure for India which was abysmally low.[38] Still when we compare the literacy figures for India with those of Vietnam (92%), Sri Lanka (90%), Malaysia and Indonesia (84%), and Myanmar (74%), they are nowhere near. This is paradoxical in view of the fact that India is otherwise well developed in both general and science education.[39]

Compared to other South Asian nations[40] gender inequity in literacy remains a serious problem in India with female literacy at 54% trailing way behind male literacy in India which was 76% (Census 2001). The lowest female literacy recorded was in Bihar, but the widest gender gap in literacy was in Rajasthan. One of the troubling aspects of the literacy data was how industrially advanced states showed very poor literacy growth as compared to the national average of 13.75%, e.g. Gujarat (8.7% growth), and even Tamil Nadu, Karnataka and Punjab (at around 10–11%). On the other hand, industrially non-advanced states like Orissa and Uttar Pradesh reported slightly higher than average improvements. It seems that once a state achieves industrialization, economic solvency and near-national average literacy growth, it tends to slip off on further eradication (as shown by the 1% literacy growth rate in Kerala), suggesting that eradication of literacy among the final one third of the population will pose a huge challenge in India. This is clearly achievable if due attention is paid to lesser known speech communities of India.

It is perhaps important to also add that only 50 out of India's 94 major languages (out of a list of 118 languages with 10 000 plus speakers) were found to be used in the written domains by Padmanabha et al. (1989). This means that oralcy is the practice in many, many speech communities even in the 1990s.

8. A comparative study of selected census reports in respect of Indian languages

A comparative study of all VIII Schedule languages reveals that only some languages are showing a drastic increase in number of speakers over the last few decades. Let us first consider the decadal increase in terms of ratio of numbers of speakers of these languages to the total population of India to get an overall picture (see Table 8).

Table 8. Languages in the 8th Schedule: Comparative table of census data (% to total population)

Language names[41]	1961	1971	1981	1991
Hindi	34.90	38.04	38.71	39.85
Bengali	8.86	8.17	7.51	8.22
Telugu	9.85	8.16	7.41	7.80
Marathi	8.70	7.62	7.24	7.38
Tamil	7.99	6.88	–[42]	6.26
Urdu	6.10	5.22	5.11	5.13
Gujarati	5.31	4.72	4.84	4.81
Kannada	4.56	3.96	3.76	3.87
Malayalam	4.45	4.00	3.76	3.59
Oriya	4.11	3.62	3.37	3.32
Punjabi	2.86	2.57	2.87	2.76
Assamese	1.77	1.63	–[42]	1.55
Sindhi[43]	[0.18]	0.31	0.30	0.25
Nepali[43]	[0.15]	0.26	0.20	0.25
Konkani[43]	[0.17]	0.28	0.23	0.21
Manipuri[43]	[0.16]	0.14	0.13	0.15
Kashmiri	0.51	0.46	0.46	–[44]
Sanskrit	N	N	N	0.01

Here we see a tremendous upsurge in the Hindi mother tongue population between 1961 (34.9%) and 1991 (39.85%). This can be explained partly by a faster rate of population increase in the Hindi-speaking states, but it must also be due to the fact that Census 1961 listed many varieties (or so-called "dialects") of Hindi separately, which was modified while tabulating the 1991 figures. Yet another interesting feature is that of all four late entrants to the schedule (especially Sindhi, Nepali and Konkani). It is also plausible that many native speakers of these languages felt diffident earlier to write their mother-tongue and perhaps opted for the regional language instead. They felt confident about returning their schedules with a mention of their new-found identity. Thirdly, for some languages, the variation was not very much, across decades, e.g. for Bengali (except 1971 when Assam figures were unavailable due to language tensions), Punjabi, and Assamese. Fourthly, the drop down (from 0.51% to 0.46%) in the case of Kashmiri was due to the exodus of Hindu Kashmiri population, many of whom migrated outside India over a period of time, or hid their identity. But the variation is surprising in respect of Telugu, Marathi, Tamil, Kannada, Malayalam and Oriya.

In some accounts, we get a picture of 172 distinct languages, but as we have mentioned in the beginning, different accounts of India place the total number of languages somewhere between 118 – a number we get from Census 1991 – and 401 languages – the last figure coming from Grimes 1993. In fact, 118 are such languages as are spoken by over 10 000 people each. If we are to go by the number of "mother tongues", as per 1991 figures, we get an astronomically high figure because of what are called "rationalized" mother tongues and "other mother tongues" in the Indian Census terms. As it has already been mentioned, the 1961 Census Report which is still regarded as a more reliable reflection of Indian linguistic scene, had listed 1652 mother tongues, and a few hundred languages.

In many other respects, the Indian Census does not give details so that one could see whether the languages are lost more among male than among female speakers. The effect of migration on language attrition has been the subject of studies but the Census is generally silent about this aspect, too. As far as language attrition is considered, there is no possibility of correlating the figures with age or the urban/rural divide, either. However, the general trends reported in the studies by Pandit (1973, 1977), Rangila (1986), Bayer (1986) and others, are that there is an overwhelming tendency in India to retain languages across generations as well as in transplanted situations.

9. Discussion and concluding remarks

It may be profitable to compare the fate of lesser-known smaller languages in South Asia with their counterparts in the developed world. The decline in their number is a universal phenomenon (Campbell 1994). If one wishes to see the decline in the percentage of speakers of indigenous languages among the indigenous populations of specific countries such as Canada, one would notice that according to the statistical account presented by Burnaby and Beaujot (1986: 36), the percentage has gone down from 87.4% in 1951 to 26% in 1996. Although over 60 languages were originally spoken in Canada, according to Kinkade (1991: 158), at least 13% or 8 were extinct by 1990, and there are 23 languages in Canada that are "endangered" (38%) now, because they have few speakers under 50 years of age and almost no children are learning them. As for the number of Indigenous languages originally spoken in North America, Bright (1994) and Mithun (1999: 1) put it at around 300, but Chafe (1962) had counted 211 languages as still living in the USA in 1960. But even then, out of these only 89 (42%) had speakers of all ages, making it obvious that most of the other 58% languages are "endangered" or "near-extinct" (Campbell and Mithun 1979). Similar linguistic genocides had taken place in Australia, too (McConvell and Thieberger 2001).

Campbell (1997: 16) had predicted that 80% of the North American languages spoken at the turn of this century "will die in this generation". Similar thoughts were expressed by him even in Campbell (1994). Zepeda and Hill (1991: 136) estimate that 51 (approximatively 24%) of the 211 languages supposed to have been alive in 1960 have disappeared thirty years later. The Australian situation is equally disturbing. Out of about 300 languages as in 1800, there has been a decrease of 90% in the number of such speakers of all age groups who can speak fluently. The decline rate in indigenous people speaking their own languages has been from 100% in 1800 to 13% in 1996. Of the 20 languages categorized in 1990 as "strong", three should already be regarded as "endangered". All these cases paint a very grim picture of the smaller linguistic groups in the whole world. Language endangerment in other parts of the multilingual and developing world such as Latin America does not provide an exciting picture anyway (Grinevald 1998).

In civil societies, there should be no disagreement now that the linguistic minority communities also have a right to develop their own languages and writing systems. Their language rights include their right to have schools that would cater to their needs, and such other facilities. There is, therefore, a connection between linguistic (including scriptal) development and language rights – especially with respect to countries which may or may not be economically reasonably developed but which have communities within the country that are traditionally denied the fruits of development (cf. Phillipson 1992).[45] It is a known fact that in the field of human rights, of which language rights is only a part, there are two hierarchies – never explicitly proposed, but nevertheless taken for granted. The first of these hierarchies is that language rights is part of the basket of socio-cultural rights, which, when compared to an individual's basic civil and political rights, has always been placed at the lower end of a line of vertical relationship obtaining between different kinds of human rights. Secondly, even among the socio-cultural rights, language rights, relating to protection and promotion of one's language and script occupied the lowest priority. Further, most people would agree that although a newly emerging literate community can theoretically "choose" their processes of standardization, or patterns of development as a vehicle of modern communication, and their writing systems (maybe from among existing options, employed by the neighbouring or contiguous languages), they cannot evolve a fully developed and functional system of their own. Smaller and newly developing languages may have to consider taking the route to secondary standardization – based on a model that may suit them. This may naturally give rise to a kind of hierarchy of languages in terms of their date and methods of development. Such hierarchies can be the source for the greatest handicap for those involved in language and scriptal management in a given smaller language community. It is true that the state is expected to be a promoter and protector of minority linguistic interests in a multilingual setting (which should at least be the case in the democratic countries), but these hierarchies are often propelled by socio-cultural forces at play – rather than decided by the state.

However, even though there are models of governance available where such protections are ensured – constitutionally and legally, these provisions are often moulded or bent (or even sacrificed) by the representatives of the majority community who occupy positions of power by virtue of sheer arithmetic of number. The consolation is however that the UNESCO has given a lead in protecting, preserving and documenting the smaller language communities and their cultures.[46] It is also the case that some governments have begun adding constitutional provisions to safeguard the interest of the minority communities, as Nepal has done. Article 4.1 of the Constitution of Nepal (1990; see also Thuladar 1966) adopted on November 9, 1990, states very clearly that "Nepal is a multiethnic, *multilingual*, democratic, independent, indivisible, sovereign, Hindu and Constitutional Monarchical Kingdom." ("multilingual" highlighted for obvious reasons). In Article 6, "Language of the Nation", we read further that "(1) The Nepali language in the Devanagari script is the language of the nation of Nepal. The Nepali language shall be the official language. (2) All the languages spoken as the mother tongue in the various parts of Nepal are the national languages of Nepal." Here the second part is very crucial for the smaller language groups, although two crucial articles on "Fundamental Rights" and "Rights to Freedom" where "language" should have found place do not have it mentioned.[47] But this is made up by Article 18, under "Cultural and Educational Right", where it is stated:

> (1) Each community residing within the Kingdom of Nepal shall have the right to preserve and promote its language, script and culture.
>
> (2) Each community shall have the right to operate schools up to the primary level in its own mother tongue for imparting education to its children.

India responded to this move by not only mentioning numerous languages under the 8th Schedule of the Constitution and in various articles as described above, but also by setting up a Commissioner for Linguistic Minorities, the reports of which (42 bulky Annual Reports published so far) are an eye-opener for many other countries. The latest announcement in January 2005 by the present Government of the official decision to set up yet another entity called the Commission for Endangered Languages is another right move. As a preparation for these, the Planning Commission had long ago identified 75 primitive languages and tribes of India which need special protection. Besides this, the Commission has also been setting aside 10% of the budget of the entire Human Resources Development and other ministries to focus on programmes for the North-East where most of the smaller language groups live.

One would hope that this combined effort will have some positive impact on the life of minor and minority languages. That there is some effect in certain sectors already can be seen from the fact that it is not merely the major languages that participate in the nation's publication and information dissemination programme, as in 1971, there used to be 3954 newspapers in 35 languages, and the figure has since doubled in 2003, with Hindi (2507), Urdu (534), English (407), Marathi and Tamil (395 each) alone (4238 in total) surpassing the earlier figure.

That multilingualism and pluriculturalism have been highly respected in India in all ages (U. N. Singh 1987, 1990) is clear from many documents and evidences. In fact, even while talking about ancient Indian literature – which many of us confuse with only history of Sanskrit literature, we find scholars like Winternitz (1933) commenting that "[t]he history of Indian literature … not only stretches across great periods of time and an enormous area, but is also one which is composed in many languages". But there is no denying the fact that the country has also been a field of linguistic tension. Such tensions involving smaller languages can be seen even now. For example, even though 80 percent of all Indians – nearly 750 million (1995 estimates) – speak one or another among a few Indian languages, and even though Hindi is understood by close to 60%, there are still many other languages with a long literary history, grammatical and lexicographical tradition and rich literary heritage, and they are still in use in all modern means of communication. As a result, although the official language of India is Hindi, there is always a hidden tussle as well as open confrontation between supporters of Hindi as an official language who mostly oppose the use of English, and supporters of the regional languages who look to English as an alternative link between the Indian states.

Smaller languages often do not enter into this game that bigger linguistic groups play, because they are engaged in a battle of survival in the first place.

Notes

1. In fact, it is 3 287 590 square kilometers, including the territorial seas.
2. Census 1981 reported 112 mother tongues with more than 10 000 speakers.
3. Also see <http://www.hku.hk/linguist/program/world10.html>.

4. Groups such as the Federation of Khasi, Jaintia, and Garo People (FKJGP) and the Khasi Students' Union (KSU) came to the fore since 1990 mainly to uphold the rights of the "hill people" in the region. As a consequence, such linguistic and ethnic tensions recurred again and again.
5. See <http://www.indianngos.com/issue/population/statistics/statistics3.html>.
6. The details of which are: 0–14 years: 34% (male 175 228 164; female 165 190 951); 15–64 years: 62% (male 324 699 562; female 301 821 383); and 65 years plus: 4% (male 23 925 371; female 23 138 386).
7. See <http://www.indianngos.com/issue/population/statistics/>.
8. <http://www.indianngos.com/issue/population/statistics/statistics5.html>.
9. Calculated from the Report of the Technical Group on Population Projections.
10. See <http://www.goski.com/1india.htm>.
11. 1991 figures in case of all religious communities include an estimated population of 16 052 people in 33 villages in the Dhule district in Maharashtra state; Further details are not available.
12. Figures for 2001 exclude Mao, Maram, Paomata and Purul Sub-division of the Senapati district of Manipur.
13. The statement does not include the cases of "Religion not stated" category by 345 277 (1981–1991) and by 3 485 405 during 1991–2001.
14. <http://gbpihed.nic.in/envis/HTML/monograph2/SC-STPopulation.html>.
15. <http://www.indiansaga.info/history/postindependence/reorganization.html>.
16. <http://www.censusindia.net/results/admin.html>.
17. <http://www.cpcb-ecocities.com/e1622/e1628/e2110/index_eng.html>.
18. According to the well-known anthropologist, D. N. Majumdar (1961) as quoted in a Harvard University site <http://www.hsph.harvard.edu/Organizations/healthnet/SAsia/forums/Tribals/Tribals/M001.HTM>, a tribe is "[a] social group with territorial affiliation, endogamous with no specialization of functions, ruled by the tribal officers, hereditary or otherwise, united in language or dialect, recognizing the social distance from tribe or castes but without any stigma attached in the case of caste-structure following tribal traditions, beliefs, customs, illiberalization of natural ideas from alien sources, above all, consciousness of homogeneity of ethnical and territorial integration". Similar viewpoints can be seen in his later work such as Mujumdar and Madan (1973), or even earlier (Majumdar 1944).
19. LSI refers to *Linguistic Survey of India* (Grierson 1903–1923).
20. Office Memorandum No.F.5/8/65-O.L.-; Ministry of Home Affairs, Government of India.
21. Also see <http://www.ncte-in.org/pub/popedu/chp2.htm>.
22. Census data.
23. Data refer to a year other than that specified and are based on Census figures.
24. Data refer to a year or period other than that specified, differ from the standard definition or refer to only part of a country. See <http://hdrc.undp.org.in/APRI/hds/hdfct/bhutan.htm>.
25. Census data.
26. Data refer to a year between 1995 and 1999, and are based on Census figures.
27. Excluding the state of Tripura.
28. Data refer to a year other than that specified.
29. Preliminary UNESCO Institute for Statistics estimate, subject to further revision.

30. Because the combined gross enrolment ratio was unavailable, the Human Development Report Office estimate of 49% was used.
31. Data refer to UNESCO Institute for Statistics estimate when national estimate is not available.
32. Data refer to the 2000/01 school year.
33. Preliminary UNESCO Institute for Statistics estimate, subject to further revision.
34. Data refer to the 2000/01 school year. It was 85 during 1990–91.
35. Data refer to the 1999/2000 school year.
36. 1990–91 figure.
37. <http://members.tripod.com/~INDIA_RESOURCE/census.html>.
38. British India (11% literacy) and Princely Indian States (16%) in 1947; see for details <http://members.tripod.com/~INDIA_RESOURCE/census.html>.
39. It is important to relate it with Government spending on education (as a percentage of public expenditure). The figures for Malaysia (15.4%), Indonesia (9%) and Philippines (15.7%) are all much higher than education spending in India. (From Parul Malhotra, *Financial Express*, Feb 22, 2001.)
40. Bhutan (54% – 1996 fig), Nepal (50% – 2000 fig), Pakistan (45.4% – 1998 fig), Bangladesh (38.5% – 1996 fig), and Afghanistan (22% – 1990 fig), as per <http://india_resource.tripod.com/census.html>.
41. Name of the languages are in descending order of strength – 1991 Census.
42. Full figures for Tamil and Assamese for 1981 are not available as the census records for Tamil Nadu were lost due to floods and the 1981 Census could not be conducted in Assam due to the disturbed conditions then prevailing there.
43. For Sindhi, Nepali, Konkani and Manipuri, the figures for 1961 are mentioned within brackets as they were still not 8th Schedule languages.
44. Full figures for Kashmiri for 1991 are not available as the 1991 Census was not conducted in Jammu & Kashmir due to disturbed conditions.
45. Cf. Asia-Pacific Human Rights Network Report: <http://www.hri.ca/partners/sahrdc/hrfeatures/HRF13.htm>; also, Skutnabb-Kangas and Philipson (1995).
46. For further details on the Universal Declaration of Linguistic Rights, given in ten languages: <http://www.linguistic-declaration.org/>. UNESCO's MOST Programme gives access to a vast number of documents on language rights, language legislation, and linguistic minorities: <http://www.unesco.org/most/ln2pol.htm>.
47. In Art 11, Pt 3 (Fundamental Rights) the Right to Equality is detailed like this: "no discrimination ... on grounds of religion (dharma), race (varya), sex (linga), caste (jât), tribe (jâti) or ideological conviction (vaicârik) or any of these". But note that "language (bhâsâ)" is missing here again. Article 12, "Right to Freedom", has in the list "freedom of opinion and expression", and Article 13 is about "Press and Publication Right", but only oblique reference is made to language(s).

References

Annamalai, E.
2003 Review of Bh. Krishnamurti, *The Dravidian Languages. Frontline*, Volume 20, Issue 22, October 25 – November 07.

Baldridge, Jason
1996 Reconciling linguistic diversity: The history and the future of language policy in India. Undergraduate Honors Thesis, University of Toledo.
2002 Reconciling linguistic diversity: The history and the future of language policy in India. *Language in India*, Vol. 2, 3 May. <http://www.languageinindia.com/may2002/baldridgelanguagepolicy.html>.
Bayer, Jennifer
1986 *Dynamics of Language Maintenance among Linguistic Minorities: A Sociolinguistic Study of the Tamil Communities in Bangalore*. Mysore: Central Institute of Indian Languages.
Bright, William
1994 Native American Indian languages. In *Native North American Almanac*, Duane Champagne (ed.), 427–447. Detroit: Gale Research. Repr. in *Native America: Portrait of the Peoples*, Duane Champagne (ed.), 397–439. Detroit: Visible Ink, 1994.
Burnaby, Barbara, and Roderic Beaujot
1986 *The Use of Aboriginal Languages in Canada: An Analysis of 1981 Census Data*. Ottawa: Social Trends Analysis Directorate and Native Citizens Directorate, Department of the Secretary of State.
CAD
2003 *Constituent Assembly Debates (Legislative)*: 17.11.1947 to 24.12.1949. Microfilms, 4th ed. Delhi: Parliament Library and Reference, Research, Documentation and Information Service, Government of India.
Caldwell, Robert A.
1856 *Comparative Grammar of Dravidian or South Indian Family of Languages*. Madras: University of Madras. (2nd ed. 1875 as *A Comparative Grammar of the Dravidian Languages*. London.)
Campbell, Lyle.
1994 Language death. In *The Encyclopedia of Language and Linguistics*, R. E. Asher (ed.), vol. 4, 1960–1968. Oxford: Pergamon Press.
1997 *American Indian Languages: The Historical Linguistics of Native America*. Oxford: Oxford University Press.
Campbell, Lyle, and Marianne Mithun (eds.)
1979 *The Languages of Native America: Historical and Comparative Assessment*. Austin: The University of Texas Press.
Census
1921 *Census of India 1921*. New Delhi: Office of the Registrar General of India.
1961 *Census of India 1961*. New Delhi: Office of the Registrar General of India.
1971 *Census of India 1971*. New Delhi: Office of the Registrar General of India.
1981 *Census of India 1981*. New Delhi: Office of the Registrar General of India.
1991 *Census of India 1991*. New Delhi: Office of the Registrar General of India.
2001 *Census of India 2001. Provisional Population Totals: India*. Paper 1 of 2001. New Delhi: Office of the Registrar General of India.

Chafe, Wallace
1962 Estimates regarding the present speakers of North American Indian languages. *International Journal of American Linguistics* 28: 162–171.
Constitution of India
Constitution of India. Delhi: Government of India. <http://indiacode.nic.in/coiweb/welcome.html>.
Constitution of Nepal
1990 *Constitution of the Kingdom of Nepal.* (Into force on Friday the twenty-third day of the month of Kartik of the year 2047 Bikram Sambat; November 9, 1990.) Official translation in the *Himalayan Research Bulletin,* Vol. XI, Nos. 1–3, 1991. As given in University of Wuerzburg site at <http://www.uni-wuerzburg.de/law/np>.
Gordon, Raymond J., Jr. (ed.)
2005 *Ethnologue: Languages of the World.* 15th ed. Dallas, Texas: SIL International. Online edition: <http://www.ethnologue.com/>.
Government of India
1986 *National Policy on Education.* Delhi: Publication Division.
1992 *The 'Programme of Action'*. Delhi: Publication Division.
Grierson, George A.
1903–1923 *Linguistic Survey of India,* Vols I–XVI. Reprinted Delhi: Motilal Banarsidass.
Grimes, Barbara
1993 *Ethnologue: The World's Languages.* 12th edition. Dallas: Summer Institute of Linguistics.
Grinevald, Colette
1998 Language endangerment in South America: A programmatic approach. In *Endangered Languages. Language Loss and Community Response,* Lenore A. Grenoble, and Lindsay J. Whaley (eds.), 124–160. Cambridge: Cambridge University Press.
Heitzman, James, and Robert L. Worden (eds.)
1996 *India: A Country Study.* Washington D.C: Library of Congress. Online edition: <http://lcweb2.loc.gov/frd/cs/intoc.html>.
Kagzi, Mangal Chandra Jain
2001 Kagzi's *the Constitution of India as Amended Upto 83rd Amendment* with Complete Text and Statement of: Very Exhaustive Commentary. Delhi: India Law House.
Kinkade, Dale
1991 The decline of native languages in Canada. In *Endangered Languages,* R. H. Robins, and E. M. Uhlenbeck (eds), 157–176. Oxford: Berg.
Krishnamurti, Bhadriraju
2003 *The Dravidian Languages.* Cambridge: Cambridge University Press.
Kumar, Gyanendra
1999 Population situation. In *Population Education: Content,* Riaz Shakir Khan (ed.), Chapter II. Jamia Nagar, New Delhi: Institute of Advanced Studies in Education, Jamia Millia Islamia.
Kumaramangalam, S. Mohan
1965 *India's Language Crisis: An Introductory Study.* New Delhi: New Century Book House.

Lok Sabha Debates
Lok Sabha Debates: 14.5.1954 to 20.12.2002. Microfilm. Delhi: Parliament Library and Reference, Research, Documentation and Information Service, Government of India.
Maitra, Ramtanu, and Susan Maitra
1995 Northeast India: Target of British apartheid. <http://www.larouchepub.com>.
Majumdar, D. N.
1944 *The Fortunes of Primitive Tribes*. Lucknow: Universal Publishers.
1961 *Races and Cultures of India*. Bombay: Asia Publishing House.
Mallikarjun, B.
2002 Mother tongues of India according to the 1961 census. *Languages of India*, Vol 5, August.
McConvell, P., and N. Thieberger
2001 *Australia State of the Environment*. Technical Paper Series (Natural and Cultural Heritage), Series 2. Government of Australia: Department of the Environment and Heritage.
Mithun, Marianne
1999 *The Languages of Native North America*. Cambridge: Cambridge University Press.
Mujumdar D. N., and T. N. Madan
1973 *An Introduction to Social Anthropology*. Bombay: Asia Publishing House.
Nettle, Daniel
1999 *Linguistic Diversity*. Oxford: Oxford University Press.
Official Languages Act
The Official Languages Act. (1963; as amended 1967); Act no. 19. *Language in India*, 2.2. April, 2002.
Padmanabha, P., B. P. Mahapatra, V. S. Verma, and G. D. McConnell
1989 *The Written Languages of The World: A Survey of the Degree and Modes of Use* (2. INDIA, Book 1, Constitutional Languages, Book 2, Non-Constitutional Languages). Quebec: International Centre for Research on Bilingualism, Laval University Press, and New Delhi: Office of the Registrar General of India.
Pandit, P. B.
1973 *India as a Sociolinguistic Area*. Poona: Deccan College.
1977 *Language in a Plural Society*. Delhi: Deva Raj Chanana Memorial Lectures Committee & Manohar Book Depot.
Patra, K.
1998 *History and Debates of Constituent Assembly of India*. Delhi: Sangam Books Limited.
Phillipson, R.
1992 *Linguistic Imperialism*. Oxford: Oxford University Press.
Rangila, R. S.
1986 *Maintenance of Punjabi Language in Delhi*. Mysore: Central Institute of Indian Languages.
Singh, K. S. (ed.)
1992 *People of India*. 72 volumes; New Delhi: Anthropological Survey of India.

Singh, R. A.
1969 *Inquiries into the Spoken Language of India (from Early Time to Census of India 1901).* (Monograph No. 1.) New Delhi: Office of the Registrar General of India.
Singh, Udaya Narayana
1987 On some issues in Indian multilingualism. In *Perspectives in Language Planning*, Udaya Narayana Singh, and R. N. Srivastava (eds.), 153–65. Calcutta: Mithila Darshan.
1990 On language development: the Indian perspective. In *Proceedings of the 14th International Congress of Linguists*, Joachim Schildt Bahner, and Dieter Viehweger (eds.), 1460–1471. Berlin: Akademie Verlag.
Skutnabb-Kangas, Tove, and Robert Philipson (eds.)
1995 *Linguistic Human Rights: Overcoming Linguistic Discrimination.* Berlin: Mouton de Gruyter.
Tuladhar, Tirtha Raj
1966 Constitution *of* Nepal. Kathmandu: S.N.
Winternitz, M.
1933 *A History of Indian literature.* Calutta: University of Calcutta.
Zepeda, Ofelia, and Jane H. Hill
1991 The condition of Native American languages in the United States. In *Endangered Languages*, R. H. Robins, and E. M. Uhlenbeck (eds.), 135–155. Oxford: Berg.

Minority language policies and politics in Nepal

Mark Turin

1. Introduction

This article offers some structured reflections on language policies in Nepal and the associated politicization of linguistic identity.[1] I begin by addressing legislation dealing with linguistic diversity, and then discuss challenges to, and limitations of, the existing policies. Drawing on specific examples, I discuss the complexities of standardizing Nepal's spoken languages and the importance – for both the State and minority language communities – of developing orthographies and written traditions for Nepal's many tongues. Through this paper, I hope that policy-makers may develop a more nuanced understanding of the complexity of the ethnolinguistic fabric of modern Nepal, and that scholars will reflect for a moment on the formation and implementation of effective legislation for languages and their speakers.

2. Legal and constitutional context

Nepal's official linguistic policy has changed considerably over time. At present, it is in a state of flux due to the pressures exerted upon the state by ethnic advocacy movements and linguistic pressure groups on the one hand, and by the demands of the Maoist insurgents on the other.[2] The Maoist leadership have demanded that all languages and dialects spoken in Nepal be set on an equal footing and that in areas where ethnic communities are in a majority, these communities should be permitted to form their own autonomous governments. The Maoists also defend the right of every citizen of Nepal to receive a secondary level education in their stated mother tongue, even though most observers view these claims as unrealistic and unworkable. In order to better understand the background to such claims, it is necessary to look at the history of language policy in Nepal.

During Panchayat rule, which ended with the restoration of democracy in 1990, the State promoted a doctrine of "one nation, one culture, one language" and the national education policy of the time was largely intolerant of indigenous and minority languages. As illustrated by the following citation from a National Education Planning Commission report, the Panchayat era policy overwhelmingly favoured Nepali:

> … and it should be emphasised that if Nepali is to become the true national language, then we must insist that its use be enforced in the primary school… Local dialects and tongues, other than standard Nepali, should be vanished [*recte* banished] from the school and playground as early as possible in the life of the child. (College of Education 1956: 97, as cited in Gurung 2003)

In these years, the focus was on unity rather than on diversity, and the State's preference was that Nepal be a monolingual nation speaking only Nepali. Minority languages and linguistic rights were thus consciously disregarded. Since the Panchayat era, however, the Nepali government has made significant progress in recognizing the multi-ethnic and multi-lingual nature of the nation, as indicated by the content of the Constitution of Nepal:

(a) The Nepali language in the Devanagari script is the language of the nation. The Nepali language shall be the official language.
(b) All the languages spoken as the mother language in the various parts of Nepal are the national languages of Nepal. (Article 6, Part 1)

The ambiguity of the Constitution here is notable: while Nepali is the "language of the nation" and the "official language", mother tongues spoken by indigenous peoples are "the national languages of Nepal". Some commentators see the distinction as highly nuanced, while others are critical of what they perceive to be an intentional semantic confusion based on insincere rhetoric, and they reject the claim that the Constitution of Nepal is a forward-looking and robust document (Lawoti 2003). Continuing on in the Constitution, Article 18 of Part 3, in the section on Fundamental Rights, states that:

(c) Each community residing within the Kingdom of Nepal shall have the right to preserve and promote its language, script and culture.
(d) Each community shall have the right to operate schools up to the primary level in its own mother tongue for imparting education to its children.

While the combination of Articles 6 and 18 provides a solid constitutional bedrock for linguistic minorities to have access to mother tongue language instruction, it remains unclear from Article 18 (2) whether the "right to operate schools" is one which will be underwritten by government financial aid.

The constitutional guarantee of Article 18 was not entirely new for Nepal, even though its precise formulation in the post-democracy constitution of 1990 was a significant departure. Article 7 of the 1971 Education Act of Nepal already stated that:

(e) The medium of instruction in schools shall be the Nepali language.
(f) Provided that education up to the primary level may be imparted in the mother tongue.

On October 16, 2001, five Members of Parliament (MP) of the House of Representatives presented a Non-Governmental Bill Relating to the Management of Languages. While this bill followed up the provisions enshrined within the Constitution relating to issues of visibility, documentation, and cultural preservation, an important new recommendation was for a "three-language policy" including the mother tongue, a second language (Nepali), and an international language (most likely English). This recommendation was presented as being very much in line with emerging research and international best practices in education which demonstrate that trilingual education, when implemented with due care and attention in multilingual nations, may help to make children comfortable in a range of languages applicable in different social contexts.

3. Policy failure and challenges to the state

The constitutional ambiguity described above set the stage for a number of linguistic tensions in Nepal. There are no shortage of national and international provisions for what may be termed "linguistic rights", and many indigenous peoples' groups and activist organizations in Kathmandu are fully aware of these rights as enshrined in the Constitution, the Education Act and its Amendments, and the recommendations of the various governmental reports which address these issues. The real concern relates to the ability of such groups – and particularly the indigenous people and linguistic minorities of rural Nepal whom they claim to represent – to gain access to, and then effectively use, the legal system to defend their basic linguistic and social rights. Aside from one prominent case discussed immediately below, language activists do not commonly evoke legal provisions to defend their rights; and debates about language, ethnicity, and culture are generally not acted out in courts.

The case in question relates to a decision made by three local administrative bodies between August and November 1997 – the Kathmandu Metropolitan City, Dhanusha District Development Committee and Rajbiraj Municipality – to use local languages (Newar and Maithili respectively) as official languages in addition to Nepali. This right had been enshrined in the Local Self-Governance Act of 1999, which deputed to local bodies the right to use, preserve, and promote local languages. The decision by these three local bodies to use regional languages was legally challenged and cases were filed in the Supreme Court of Nepal, after which an interim order was issued on March 17, 1998, prohibiting the use of local and regional languages in government administration. This order led to much discontent and resentment among minority communities, and a number of action committees were promptly formed to address the ruling. On June 1, 1999, the Supreme Court nevertheless announced its final verdict and issued a *certiorari* declaring the decision of the local administrative bodies to use regional languages to be unconstitutional and illegal. The court's verdict raised serious questions about the sincerity of the government's commitment to the use of minority languages in administration, and further increased resentment among minority language communities. Public demonstrations and mass meetings were called, and the Nepal Federation of Nationalities (NEFEN) organized a national conference on linguistic rights on March 16–17, 2000 with support from the International Work Group on Indigenous Affairs (IWGIA). The proceedings of this conference were published in April 2000. Four resolutions were adopted during the conference, one of which demanded that:

> ...legal provisions be made to allow the use of all mother-tongues and the verdict of the court be declared void since it runs against the values of the present Constitution of Nepal which recognises all mother-tongues as "national languages" and the Local Autonomy Act [LSGA] of 2055 which contains provisions for the use, preservation and promotion of mother-tongues by local bodies. (Nepal Federation of Nationalities 2000: 8)

As illustrated by the above example, ethnolinguistic issues in Nepal are highly politicized and many activists feel powerless to guarantee their rights in the face of government opposition and hypocrisy. Disagreements also exist between different indigenous peoples' movements on the correct path to achieve equality. At opposing ends of the continuum are those advocates who propose working to change the system from within, and militant organizations who have allied themselves with the Maoist movement, believing that parliamentary debate will not deliver practical results at the grassroots level. The middle ground, however, is occupied by a plethora of organizations who support minority rights, but who are losing faith in the government's ability to bring about any meaningful change.

There is widespread concern among language activists and villagers from indigenous communities that despite the countless legal provisions respecting their fundamental linguistic rights, an institutional inertia exists regarding the emotive issues of mother tongue education and the access granted to minority communities to positions in government and the administration. Indigenous people and minority language communities have highly restricted access to the existing legal provisions to defend their rights, particularly in rural areas poorly serviced by infrastructure, and are intimidated by the very institutions which are meant to represent and protect them. As Sonia Eagle has written, "in Nepal, language issues may be seen as representative of the broader issues of powerlessness, prejudice, and inequality felt by minority groups throughout the country" (1999: 322).

While the situation is naturally complex, there are three principal reasons why linguistic minorities rarely resort to legal means to defend their rights. First, the machinery of government is still primarily controlled by high caste groups who have held power for the last 250 years, and have little incentive to change or relinquish control. Second, educated indigenous peoples in both urban and rural Nepal are reluctant to use official channels – legal or administrative – to redress inequalities since they believe the system itself to be weighted against their interests and their chances of success limited. This is a realistic concern, as illustrated by the rulings against Newar and Maithili illustrated above, particularly since fluency in spoken Nepali and a high degree of literacy are prerequisites for legal exchange, skills which many linguistic minorities still do not have. In a recently published paper, the British scholar Bryan Maddox illustrates how the most acute forms of linguistic inequality are experienced by the least educated and literate groups in society, and by a minority of monolingual communities who are not able to access the languages of power. Third, many indigenous peoples and linguistic minorities in rural areas are simply not aware of their rights, or if they are, they have no practical knowledge of how and where to best assert them. The above factors, combined with widespread discrimination against minority populations, have effectively inhibited the development and inclusion of ethnic and linguistic minorities within the Nepali nation.

Given the disjuncture between the legal and constitutional provisions for linguistic equality on the one hand, and the reality of the overwhelming dominance of Nepali on the other, it is easy to understand the frustrations and despair of activist groups representing minority communities. The crisis lies not in the formulation of policy, but in the ability and desire of the governing classes to actively change the status quo.[3]

4. The importance of orthography and written tradition in the formation of linguistic policy in Nepal

While all but eight of the many languages spoken in Nepal as mother tongues have no literate tradition, in its report to the government on April 14, 1994,

Nepal's National Language Policy Recommendations Commission presented a four-fold stratification of languages spoken in Nepal ranked on the basis of having a written form. At the top, in first position, were those languages with elaborate and well-attested written traditions, such as Nepali, Newar, Maithili, Limbu, Bhojpuri, and Awadhi. In second position came languages "in the process of developing a written tradition" such as Tamang, Gurung and various others (Sonntag 2001: 169). In third position came those languages without a written tradition, while the "dying" languages, such as Raute, were listed last. In this hierarchical caste-system of languages, script and literacy are the highest units of value, and "written languages" are accorded a higher status than spoken ones. The educational and linguistic agendas of the Nepalese state thus converge around the issues of script and orthography: languages with a written tradition and a history of literature are promoted and supported above endangered spoken forms.

Noting the Commission's ranking of languages according to their possession, development, or evolution of a written form, it comes as no surprise to learn that ethnoactivists and promoters of indigenous languages have adjusted their programmes accordingly. Language development activities, many of which seek national recognition and funding, now commonly include some of the following components: "graphization" or the establishment of an orthography and spelling conventions; "standardization" the process of making one speech variety a "super-dialectal" norm, and "modernization", the extension of the lexicon to cope with the experiences of the modern socio-linguistic world (Webster 1999: 556). Since the mid-1990s, the lexicalization of a language and the development, or resurrection, of a suitable script or set of orthographical conventions have become prerequisites for introducing a language into education as the medium of instruction, the latter being a primary aim of both many language activists and a major component of contemporary linguistic policy in Nepal. International donors are at present engaged in lengthy negotiations with His Majesty's Government of Nepal (HMG/N) to assure that the forthcoming five-year plan for education, dubbed *Education for All 2004–2009*, will address the needs of Nepal's ethnic and linguistic minorities. While the Core Document of EFA 2004–2009 prepared by the Ministry of Education and Sports (MOES) points out that "programmes that provide education in mother tongues will be encouraged in order to increase access of children from diverse linguistic groups" (2003: 18) and that the Curriculum Development Centre has "succeeded in developing curriculum and textbook materials in eleven minority languages" (2003: 25), donors and linguistic activists remain sceptical of the government's commitment to effective implementation of such pilot projects.

A few general issues relating to language documentation and lexicalization are worth noting. First, the process of standardization required for a pedagogical grammar, textbook, or dictionary necessarily results in a degree of language simplification. Just as divergent spellings of words and regional variations of speech were constrained by the standardization of English grammar and spelling by Samuel Johnson, so too the development of writing systems for Nepal's indigenous languages are resulting in the standardization of the spoken language and the concurrent elevation of one speech variety to a normative position above others. There are various dialects or speech varieties of Thakali and Tamang, for example, and in the process of developing a suitable writing system and corpus of pedagogical materials in the language, one variety (or a synthetic mixture of both) will necessarily be promoted as standard and representative. Given the highly diverse and heterogeneous ethnolinguistic tapestry of Nepal in particular, and the Himalayan region in general, the process of linguistic standardization can be expected to be complicated. Studies of identity politics have shown that minority groups the world over may sooner learn a national or international language than adjust their own speech forms to resemble that of their immediate neighbours.

Second, when oral languages are standardized and written forms are created, a speech community must either choose to use an existing script or to invent an entirely new one. Various scripts exist within Nepal, the two dominant ones being the Nepali, or Devanagari script, and the Tibetan script. Other languages with pre-existing and unique scripts include Newar, Limbu and Lepcha. Indigenous peoples speaking languages without a literate tradition generally choose between three options when developing a writing system: using the Devanagari script, using the Tibetan script, or devising a new script.

The strength of the Nepali/Devanagari script is that it is widely recognized and understood by citizens from different ethnic backgrounds, largely on account of the growth of primary education and the boom in print media since 1990. The disadvantage is that the phonetic basis of the Devanagari script imposes orthographical constraints on the sounds it is able to represent.[4] In addition, many of the indigenous communities in Nepal who speak Tibeto-Burman languages are reluctant to use a script derived from an Indo-Aryan language to which their language is genetically unrelated. The "Nepalification" through script or lexicon of indigenous Tibeto-Burman languages is strongly resisted by many more militant members of the ethnic movement in Nepal.

The advantage of the Tibetan script, on the other hand, is that it derives from a language in the same language family as many of Nepal's indigenous and unwritten Tibeto-Burman languages. Some phonological features of Nepal's extant Tibeto-Burman languages may therefore be more easily represented using the Tibetan script. At a symbolic and political level, 'Tibetanness' makes reference to a cultural heritage alternative to the dominant traditions embodied by Hindu Nepal. The disadvantages of choosing the Tibetan script, however, are overwhelming. Most of Nepal's Tibeto-Burman languages are far removed from modern spoken and written Tibetan, both in terms of grammar and phonology. Membership in the same language family in no way guarantees linguistic similarity or the applicability of one script for all languages in the group. The complex spelling rules of modern Tibetan are also entirely inapplicable to unwritten languages which have no classical literary form, as the Sherpa and Tamang communities of Nepal have learned at their peril.

Finally, some indigenous peoples of Nepal are developing new scripts for their mother tongues. While these attempts are laudable, they are also often unrealistic given the generally poor level of educational attainment of those involved in the process and the practical challenges in disseminating new scripts (publishing outlets, computer fonts, special schools). There are few professionally-trained lexicographers or linguists among those indigenous activists working on the development of scripts or compiling language corpora for Nepal's endangered languages. The desire for a script is an understandable aspiration for minority language communities given the psychological link often made between script = literate tradition = classical language = recorded history = cultural authenticity and power. Some linguistic activists in Nepal see the development of a script for their language as primarily important for the status that this will accord their community on the national stage, as in gaining a higher ranking in the Language Commission's table, rather than for any resulting mother tongue or bilingual education programme that may ensue.

The challenge of finding the "right" script can be illustrated through examples. Thangmi is a Tibeto-Burman language spoken by little more than 30 000 people, most of whom are resident in the Dolakha and Sindhupalcok districts to the east of Kathmandu. While most Thangmi speakers are reconciled to using a slightly modified form of the Devanagari script to write their mother tongue, and also believe that they never had their own unique writing system, some of the more active members of the community are eager to unearth any indication of a uniquely Thangmi script. I have often heard it said that the Thangmi language once had its own script but has since lost it, a kind of fall from linguistic grace.[5] Such a belief reflects the widespread, if mistaken, assumption that all "real" languages were once written as well as spoken and that only through recovering a lost script will the Thangmi language activists be able to validate their claims to linguistic antiquity and autochthony in the areas which they presently inhabit.

Tamang, on the other hand, is spoken by over 1 million people or 5.19% of the total population, making it one of Nepal's most widespread ethnic languages. The Nepal Tamang Ghedung, an ethnic organization representing Tamang concerns at a national level, writes its name in three scripts: Nepali (Devanagari) for the benefit of most ethnic Tamangs who are functionally literate and have passed through the Nepali education system; a modified Tibetan script (dispensing with the complicated spelling conventions) on account of the language's place in the Tibeto-Burman language family and also because a growing number of Tamang Buddhists are versed in the Tibetan script; and English for the international or western educated audience. Such a tri-scriptural approach, while catering to all parties, is clearly pragmatically unworkable as a long term solution.

5. Conclusion

Over the last half century, Nepal's approach to legislating language policy and accounting for linguistic rights has seen a marked improvement. Moving from a "one nation, one language" model promoted through the 1950s, there was a noticeable move towards encouraging and supporting Nepal's indigenous languages and the communities who speak them by the time that democracy was restored to Nepal in 1990. While the constitution of Nepal enshrines a number of linguistic rights for minorities, and while the government is signatory to various international agreements, few if any of the promises and constitutional rights have been actively pursued or implemented, and the government's commitment to linguistic rights continues to be theoretical rather than practical.

Regrettably, the disjuncture between rights and reality has only served to further politicize, and radicalize, the already embittered linguistic minorities, many of whom no longer believe government pledges on mother tongue education and bilingual classrooms. Furthermore, the extreme focus on writing systems and the associated push to develop suitable orthographies for spoken languages has done little to offer practical support for Nepal's home-grown diversity of spoken tongues. It is an unfortunate paradox that while previously unwritten languages are being standardized and are developing written forms, the number of mother tongue speakers of many of these languages continues to fall. It appears that some graphization programmes are missing the wood for the trees by emphasizing standardization and centralization rather than linguistic fluidity and dynamism which spoken languages need to survive.

Recognizing that many minority language communities have accepted the idea that a "proper" language must be written, I have addressed some of the motivations which inform decisions for or against the use of certain scripts in the representation of these languages. While it is likely that many of Nepal's minority languages will be reduced from communicative vernaculars to symbolic, albeit written, markers of identity within a generation, this loss should not overshadow language revival activities such as those described by Noonan in this volume and in Turin (in press). The cultural values and political valences attached to languages are dynamic and changing, rather like linguistic forms themselves. Scholars and policy makers would do well to recognise this and to develop analytical tools and legislative amendments which are robust and yet flexible enough to make sense of Nepal's shifting ethnolinguistic reality.

Notes

1. I am grateful to Professor Dr. George van Driem, Dr. Daniel Barker, Dr. Anju Saxena, and Sara Shneiderman for their valuable comments on earlier versions of this paper. Sections of this paper were presented at the *Agenda of Transformation: Inclusion in Democracy* conference in Nepal in April 2003, then under the title "The many tongues of the nation: ethnolinguistic politics in post-1990 Nepal".
2. In August 2004, the official number of dead passed the 10 000 mark, making Nepal's Maoist-State conflict the deadliest civil war in Asia at present (Newar 2004: 1).
3. Maddox concludes that a language policy simply based on the promotion of the mother tongue would not be subtle enough to respond to Nepal's linguistic diversity.
4. A recent paper by Michael Noonan, available as a downloadable PDF from his website <http://www.uwm.edu/~noonan/>, addresses recent adaptations of the Devanagari script for the Tibeto-Burman languages of Nepal.
5. Thangmi ritual practitioners or shamans, known as *guru*, narrate an origin tale in which Thangmi ancestors were once so close to starvation that they ate their religious texts out of desperation, thereby losing the original and unique Thangmi script and retaining only the spoken form of the language.

References

Eagle, Sonia
1999 The language situation in Nepal. *Journal of Multilingual and Multicultural Development* 20 (4–5): 272–327.

Gurung, Yogendra Bahadur
2003 *Indigenous Peoples Development Plan for Rural Water Supply and Sanitation (RWSS-II)*. Kathmandu: Institute for Social and Gender Equality.

His Majesty's Government of Nepal
2003 *Education for All 2004–2009: Core Document*. The Ministry of Education and Sports, Kathmandu, Nepal. 17 November 2003.
Lawoti, Mahendra
2003 Inclusive democratic institutions in Nepal. Paper presented at The Agenda of Transformation: Inclusion in Nepali Democracy conference. Social Science Baha, Kathmandu, 24–26 April, 2003.
Maddox, Bryan
2004 Language policy, modernist ambivalence and social exclusion: A case study of Rupendehi district in Nepal's Tarai. *Studies of Nepali History and Society* 8 (2): 205–224.
Nepal Federation of Nationalities
2000 *Proceedings of the National Conference on Linguistic Rights in Nepal.* Kathmandu: Triyuga Offset Press.
Newar, Naresh
2004 10,000+. *Nepali Times*, No. 209, 13–19 August 2004: page 1.
Sonntag, Selma K.
2001 The politics of determining criteria for the languages of education in Nepal. In *Droit et langue(s) d'enseignment: Law and Language(s) of Education*, Thomas Fleiner, Peter H. Nelde, and Joseph-G. Turi (eds.), 161–174. Bâle: Helbing and Lichtenhahn.
Turin, Mark
in press Rethinking Tibeto-Burman: Linguistic identities and classifications in the Himalayan periphery. In *Tibetan Borderlands: Proceedings of the Tenth Seminar of The International Association of Tibetan Studies*, Christiaan Klieger (ed.). Leiden: Brill.
Webster, Jeff
1999 The language development-language promotion tension: a case study from Limbu. In *Topics in Nepalese Linguistics*, Yogendra P. Yadava, and Warren W. Glover (eds.), 556–565. Kathmandu: Royal Nepal Academy.

Language policy, multilingualism and language vitality in Pakistan

Tariq Rahman

1. Introduction

Pakistan is a multilingual country. Its national language, Urdu, is the mother tongue of only 7.57 % of the people, although it is very widely used in the urban areas of the country. Its official status is the same as it was when the British ruled the country as part of British India. Apart from Urdu and English, the country has five major languages: Punjabi, Pashto, Sindhi, Siraiki and Balochi.

There are fifty-five other languages, some of them on the verge of extinction (see Appendix 1). The aim of this paper is to examine the language policy of Pakistan and to attempt to identify how it privileges certain languages, and to explore what political, social, educational, and economic consequences this policy entails.

Table 1. Major languages in Pakistan (Source: Census 2001: 107)

Languages	Percentage of speakers
Punjabi	44.15
Pashto	15.42
Sindhi	14.10
Siraiki	10.53
Urdu	7.57
Balochi	3.57
Others	4.66

2. Pakistan's language policies

There have been statements concerning language policy in various documents in Pakistan, including the different versions of the constitution, statements by governmental authorities in the legislative assembly debates, and, above all, the various documents relating to education policy which have been issued by almost every government. Language policies as seen in the 1973 Constitution of Pakistan are as follows:

(a) The National language of Pakistan is Urdu, and arrangements shall be made for its being used for official and other purposes within fifteen years from the commencing day.
(b) Subject to clause (a) the English language may be used for official purposes until arrangements are made for its replacement by Urdu.
(c) Without prejudice to the status of the National language, a Provincial Assembly may by law prescribe measures for the teaching, promotion and use of a provincial language in addition to the national language (Article 251).

The national language is Urdu (national languages were Urdu and Bengali from 1955 until 1971, when East Pakistan became Bangladesh) though this language is, and has always been, the mother-tongue of a minority of the population of Pakistan. This minority came from India, mostly after the creation of Pakistan in 1947, and is termed Mohajir (refugee or immigrant). The rationale for this privileged status of Urdu, as given by the government of Pakistan, is that Urdu is so widely spread that it almost holds the status of being the first language of all Pakistanis. Above all, it is a symbol of unity, helping to create a unified "Pakistani" identity. In this symbolic role, it serves the political purpose of resisting any ethnicity which could otherwise break the federation. As for the provision that other Pakistani languages may be used, it is explained that the state, being democratic and sensitive to the rights of the federal units, allows for the use of provincial languages, if desired.

As for the medium of instruction, the rationale is that Urdu, the most widespread urban language, is the language used in education. As English is useful in official and international language instances, it, too, is taught at the higher levels, especially to those who study science and technology.

2.1. Political consequences of Urdu's privileged status

One major consequence of Urdu's privileged status has been the ethnic resistance to this status. As mentioned earlier, Urdu is not the mother tongue of most Pakistanis. However, Urdu is indeed the most widely understood language and is perhaps the major medium of interaction in the urban areas of the country. Even ethnic activists agree that it could be a useful link between the various ethnic groups. However, it has faced resistance because it has been patronized, often in insensitive ways, by the ruling elite in the centre.

The story of this patronization is described in detail in several books (see Rahman 1996) but always fell short of what the more ardent supporters of Urdu demanded (for their position, see Abdullah 1976). In the beginning, since a very powerful section of the bureaucracy (being Mohajirs) spoke Urdu as its mother-tongue, there was an element of cultural hegemony concerning the special status of Urdu. The Mohajir's elite position, stated or implied, was that they were more cultured than speakers of other indigenous languages of Pakistan. Hence it was only natural that Urdu should be used instead of other less-privileged languages.

This created much resentment against Urdu and, indeed, may be said to have infused the element of personal reaction to or antagonism against the speakers of Urdu in the first twenty years of Pakistan's existence.[1] The main reason for the opposition to Urdu was, however, not linguistic or cultural. The main reason for the opposition Urdu faced in the provinces was because it was taken as the symbol of the central rule of the Punjabi ruling elite. The use of Urdu as an ethnic symbol is given in detail in Rahman (1996) but a brief recapitulation of major language movements may be useful.

The most significant consequence of the policy stating that Urdu would be the national language of Pakistan was its opposition by the Bengali intelligentsia, or what the Pakistani sociologist Hamza Alavi calls the "salariat" – people who draw salaries from the state (or other employers) and who aspire to jobs (Alavi 1988). One explanation for this opposition is that the Bengali salariat would have been at a great disadvantage if Urdu, rather than Bengali, would have been used in the lower domains of power, such as the media, administration, judiciary, education, and military. As English was the language of the higher domains of power and Bengali was a "provincial" language, the real issue was not linguistic. It was that the Bengali salariat was deprived of its just share in power at the centre and even in East Bengal, where the most powerful and lucrative jobs were controlled by the West Pakistani bureaucracy and the military. Furthermore, the Bengalis were conscious that money from the Eastern region, from the export of jute and other products, was predominantly financing the development of West Pakistan or the army which, in turn, was West Pakistani- (or, rather, Punjabi) dominated (Government of Bangladesh, 1982: 810–811 [vol. 6]; Jahan 1972). The language, Bengali, thus became a symbol of a consolidated Bengali identity in opposition to the West Pakistani identity. This symbol was used to "imagine", or construct, a unified Bengali community, using mechanisms such as the use of the printing press in the European context (Anderson 1983).

In Sindh, Balochistan, the N.W.F.P and South Western Punjab the languages used as identity symbols were Sindhi, Balochi, Brahvi, Pashto and

Siraiki. The resulting linguistic mobilization of especially the intelligensia made them powerful ethnic symbols, able to exert political pressure (Rahman 1996). However, Urdu was not resented or opposed much except in Sindh, where there were language riots in January 1971 and July 1972 (Ahmed 1992). But even in Sindh, the crucial issue was that of power. The Mohajirs were dominant in the urban areas and the rising Sindhi salariat resented this. The most evocative symbol with which to mobilize the community was language.

Apart from the riots, the general population's conduct remained pragmatic. The Mohajirs, knowing that they can get by without learning Sindhi, do not learn it except in rural areas where it is essential. The Sindhis, again because they know they cannot get by without learning Urdu, do learn it (Rahman 2002, Chapter 10).

However, if people learn languages for pragmatic reasons (Rahman 2002: 36), they then give less importance to their own languages. This phenomenon, sometimes called "voluntary shift", is not really "voluntary" (see Nettle and Romaine 2000: 94–97, concerning Hawaiian). What happens is that market conditions are such that one's language becomes deficit in terms of what Pierre Bourdieu would call "cultural capital" (Bourdieu 1991: 230–231). Instead of one's language being an asset, it becomes a liability. It prevents one from rising in society. In short, it is ghettoizing. Even if language movements and ethnic pride do not create a sense of shame, minority language speakers might not want to teach their language to their children, because it would overburden the children with far too many languages. For instance, Sahibzada Abdul Qayyum Khan (1864–1937) reported in 1932 that the Pashtuns wanted their children to be instructed in Urdu rather than Pashto (LAD-F 12 October 1932: 132). In 2003, the MMA government chose Urdu, not Pashto, as the language of the power domain in the N.W.F.P. In Baluchistan, too, the same phenomenon was noticed. Balochi, Brahvi, and Pashto were introduced as the compulsory medium of instruction in government schools in 1990 (LAD-B 21 June and 15 April 1990). Language activists enthusiastically prepared instructional material but, on 8 November 1992, these languages were made optional and parents opted for Urdu as the medium of instruction for their children (Rahman 1996: 169). Such decisions negatively influence the survival of minor languages and even somewhat devalue major languages, but this is precisely the kind of policy which has created what is often called "Urdu imperialism" in Pakistan.

In short, the state's use of Urdu as a symbol of national integration has had two consequences. First, it has made Urdu the obvious force to be resisted by other ethnic groups. This resistance makes them strengthen their languages by corpus planning (writing books, dictionaries, grammars, orthographies) and acquisition planning (teaching languages, pressurizing the state to teach them, using them in the media) (for these terms see Cooper 1989). Second, it has jeopardized the additive multilingualism recommended by UNESCO (2003) and others (e.g. Edwards 1994) as the use of Urdu has spread, assisted by the media and urbanization. This adversely affects the other Pakistani languages and threatens linguistic and cultural diversity in the country.

2.2. Status of English in Pakistan

English was supposed to continue as the official language of Pakistan until the time that the national language(s) replaced it. However, this date came and went, as did many other dates before it and English is as firmly entrenched in the domains of power in Pakistan today as it was in 1947. The major reason for this is that this is the *de jure* but not the *de facto* policy of the ruling elite in Pakistan. The *de facto* policy can be understood with reference to the elite's patronage of English in the name of efficiency and modernization.

Initially the Civil Service of Pakistan (CSP) was an Anglicized body of men who had moulded themselves in the tradition of the British. The officer corps of the armed forces, as Stephen P. Cohen suggests, was also Anglicized. It was, in his words, the "British generation" which dominated the army until 1971 (Cohen 1998: 162–163). It is understandable that members of this elite group had a stake in the continuation of English because it differentiated them from the masses. It gave them a competitive edge over those with an Urdu-medium or traditional (*madrassa*) education and, above all, it was the kind of cultural capital which held an elitist position and constituted a class-identity marker. What is less comprehensible is why members of these two elite groups, who now come increasingly from the lower-middle and middle classes who have studied in Urdu-medium schools (or schools which are called English-medium but teach mostly in Urdu), should also want to preserve, and indeed strengthen, the hegemony of English – a language which has always been instrumental in suppressing their own class?

The answer lies in the fact that the elite has invested in a parallel system of elitist schooling of which the defining feature is teaching all subjects other than Urdu through the English medium. This has created new generations of young people who have a direct stake in preserving English. All the arguments which applied to a small Anglicized elite of the early generation of Pakistan now apply to the young aspirants who stand ready to enter the ranks of this elite. Their parents, themselves not at ease in English, have invested far too much in their children's education to seriously consider decreasing the cultural capital of English.

Moreover, most people think in terms of present-day realities which they may be critical of at some level but which they assume as permanent facts of life. This makes them regard all attempts at change as either utopian or as suspiciously radical activities. For the last century and a half, the people of this part of the world have taken the ascendancy of English for granted. In recent years, with more young people from the affluent classes taking the British "O" and "A" level examinations, with the world-wide coverage of the BBC and CNN, with globalization and the presence of English as a world language, with stories of young people emigrating all over the world armed with English, English has become a commodity more in demand than ever before.

The present author carried out a survey of 1085 students from different schools in Pakistan in 1999–2000 to study their attitudes towards English. The results of this survey are presented in Table 2 (Rahman 2002). (The results do not add up to 100 % in some cases because those choosing two or more languages have been ignored.)

Table 2. School-going youngsters' attitude towards English

	Madrassas (N=131)	Sindhi medium schools (N=132)	Urdu medium schools (N=520)	English-medium schools		
				elitist (N=97)	Cadet college (N=86)	Ordinary (N=119)
1. What should be the medium of instruction in schools?						
Urdu	43.51	9.09	62.50	4.12	23.26	24.37
English	0.76	33.33	13.65	79.38	67.44	47.06
Mother tongue	0.76	15.15	0.38	2.06	Nil	1.68
Arabic	25.19	Nil	0.19	Nil	Nil	0.84
No response	16.79	37.88	16.54	5.15	Nil	8.40
2. Do you think higher jobs in Pakistan should be available in English?						
Yes	10.69	30.30	27.69	72.16	70.93	45.38
No	89.31	63.64	71.15	27.84	29.07	53.78
NR	Nil	6.06	1.15	Nil	Nil	0.84
3. Should English-medium schools be abolished?						
Yes	49.62	13.64	20.19	2.06	12.79	5.88
No	49.62	84.09	79.04	97.94	86.05	93.28
NR	0.76	2.27	0.77	Nil	1.16	0.84

The results suggest that sixteen-year-old students of matriculation level or the equivalent in Pakistani schools are not in favour of English as the medium of instruction in schools unless they are already enrolled in English-medium schools. However, as they grow up and enter elitist positions, their investment in English, which then becomes the language of schooling of their children, beocmes apparent. They no longer support policies which would replace English with other languages.

However, paradoxically, even school students in non-English medium schools do not support the abolition of English-medium schools. Perhaps this seems too radical, visionary, and impractical to them. Perhaps they feel that English-medium schools provide good quality education and should remain available for the modernization of the country. Or perhaps they understand that such schools are a ladder out of the ghetto of their socio-economic class into a privileged class, one which their siblings or children might make use of. In short, it is probably because of their pragmatism and a shrewd realization that nothing is going to change that they want the English-medium schools to keep flourishing.

As mentioned earlier, the British colonial government and its successor, the Pakistani government, have rationed out English. Their stated policy was to support Urdu but their underlying aim was perhaps only to create a subordinate bureaucracy at a low cost (vernacular-medium education is less expensive than English-medium education) and to maintain an anti-ethnic and ideological symbol within the country.

The armed forces, which were better organized than any other section of society, created cadet colleges from the 1950s onwards. These schools, run on the lines of the elitist British public schools, were subsidized by the state. In the 1960s, when students from ordinary colleges, who came by and large from vernacular-medium schools, protested against these bastions of privilege, the government appointed a commission to investigate their grievances. The findings of this commission agreed that such schools violated the constitutional assurance that "all citizens are equal before law" (Paragraph 15 under Right No. Vl of the 1962 Constitution). However, the Commission was also convinced that these schools would produce suitable candidates for filling elite positions within the military and the civilian sectors of the country's services (Government of Pakistan 1966: 18). This meant that the concern for equality was merely a legal nicety. This, indeed, is what has happened. Today the public schools are as well-entrenched in the educational system of the country as ever before.

In short, by supporting English through a parallel system of elitist schooling, Pakistan's ruling elite acts as an ally of the forces of globalization at least as far as the hegemony of English is concerned. The major consequence of this policy is the weakening of local languages and the lowering of their status. This, in turns, opposes linguistic and cultural diversity, weakens the "havc-nots" even further, and increases poverty by leaving the best-paid jobs in the hands of the international elite and the English-using elite of the peripheries.

3. Language vitality in Pakistan

The year 2000 saw three excellent books on language death: David Crystal's *Language Death*, Daniel Nettle and Suzanne Romaine's *Vanishing Voices*, and Tove Skutnabb-Kangas's *Linguistic Genocide in Education – or World-wide Diversity and Human Rights*. Works such as these, along with other related efforts, have made linguists conscious that standardization and the increasing dominance of a restricted number of languages is negatively affecting a large number of smaller languages of the world.

In Pakistan, as mentioned earlier, the linguistic hierarchy is as follows: English, Urdu, and local languages. In the N.W.F.P and Sindh, however, Pashto and Sindhi are seen as identity markers and are spoken informally. In Punjab, unfortunately, there is a widespread culture-shame about Punjabi (Mansoor 1993: 132). In all of the elitist English-medium schools the author visited, there were policies forbidding students from speaking Punjabi. If anyone spoke it, s/he was called *Paendu* ('rustic, village yokel') and made fun of. Many educated parents speak Urdu rather than Punjabi with their children.

The children of elitist English-medium schools are indifferent to Urdu and claim to be completely bored by its literature. They are proud to claim their lack of competence in the subject even when they get "A" grades in the "O" and "A" level examinations. They read only English books, not books in Urdu or other Pakistani languages. TV programs in Pakistan use the term "Urdu-medium" to refer to less-sophisticated programs.

Such prevailing attitudes have a negative effect on Pakistani languages. Urdu is secure because of the huge pool of people very proficient in it and especially because it is used in lower level jobs, the media, education, the court system, commerce, and other such domains in Pakistan. Punjabi is a large language and will survive despite culture shame and neglect. It is used in the Indian Punjab in many domains of power and, what is even more significant, it is the language of songs, jokes, intimacy, and informality in both Pakistan and India. This makes it the language of private pleasure and if it continues to be used in this manner, it is in no real danger.

Sindhi and Pashto are both major languages and their speakers have a sense of pride. Sindhi is also used in the domains of power and is the major language of education in rural Sindh. Pashto is not a major language of education, nor is it used in the domains of power in Pakistan. However, its speakers see it as an their identity marker and it is used in some domains of power in Afghanistan. It, too, will survive, although the Pashto variety which is spoken in cities in Pakistan is now adulterated with Urdu words. Educated Pashtuns often code-switch between Pashto and Urdu or English. Thus, the language is under some pressure.

Balochi and Brahvi are small languages under much pressure from Urdu. However, there is awareness among educated Balochs that their languages must be preserved. As they are not used in the domains of power, they will survive as informal languages in the private domain. Nevertheless, the city varieties of these languages will become very "Urdufied".

Over fifty very small languages of Pakistan (see Appendix 1), mostly in Northern Pakistan, are under tremendous pressure. The Karakorum Highway linking these areas to the plains has placed much pressure on these languages. The author visited Gilgit and Hunza in August 2002 and met, among others, local language activists. They all agree that their languages should be preserved, but are so appreciative of the advantages of the highway that they accept the threat to their languages with equanimity. Urdu and English words have already entrenched themselves in Shina and Burushaski and, as people emigrate to the cities, they are shift to Urdu.

In the city of Karachi the Gujrati language is being abandoned, at least in its written form, as young people seek to be literate in Urdu and English – the languages used in the domains of power.

In Sindh there are small languages so lexically close to larger ones that it is difficult to determine whether they are, in fact, varieties of the larger languages or were different languages but are now shifting towards the larger ones under pressure. These languages are described on the authority of other researchers in Appendixes 2 and 3. Observations on possible language shift and vitality have been made, but the author has not done any field work in Sindh, at least as far as language vitality is concerned, and makes no claim to authority in this field. As far as the languages of the Northern areas are concerned, more certainty can be claimed, since some of these questions have been rechecked in the field by the author himself.

The languages of areas outside Sindh which are facing extinction include:

Badeshi

It has ceased to exist now according to field researchers who visited the valley in February and March 2004. The earlier reports about the people in the Chail Valley of Swat speaking what was probably a variety of Persian are wrong although the *Ethnologue* (Gordon 2005) still reports this. This language has died some generations ago (Zaman 2004a).[2]

Chilliso

Spoken by a small number of people on the east bank of the Indus in the district of Kohistan, it is under great pressure by Shina. According to Hallberg, "A point which further underscores the idea that language shift is taking place in this community is the fact that of the thirteen individuals who were asked, four said that they spoke Chilisso in their home as a child but speak Shina in their home today" (Hallberg in SSNP–1: 122–123).

Domaaki

This is the language of the Doma people in Mominabad (Hunza). Backstrom reported only 500 speakers in 1992 (Backstrom in SSNP–2: 82). The present author visited the village in 2002 and estimated only 300.

Gowro

Spoken on the east bank of the Indus in district Kohistan, mainly in the village of Mahrin, by the Gabar Khel class. Hallberg says that "it would seem that the dominance of Shina may be slowly erasing the use of Gowro" (Hallberg in SSNP–1: 131). Baart confirms that only 1000 speakers are left now and it may be dying (Baart 2003).

Ushojo

This is spoken in the Chail Valley of Swat. According to Sandra J. Decker of the SIL, it was spoken by 2000 people in 1990 (Decker in SSNP–1: 66). She also reported that both men and women spoke Pashto with her (Decker in SSNP-1: 76). J. Baart suspects that the language is under great pressure and is moribund (Baart 2003).

The smaller languages of Chitral, too, are about to be lost. The Kalasha community, which follows an ancient religion and lives in the valleys of Chitral, is in danger of losing its languages. Some young people are reported to have left the language when they converted to Islam (Decker in SSNP–5: 112). Other small languages (Yidgha, Phalura and Gawar-bati) are also losing their vitality.

Two small languages which would have been lost otherwise are being documented by local language activists with the help of Baart. The first is Ormuri, the language of the village of Kunigaram in South Waziristan, which was described as "a strong language in that area" by Hallberg in 1992 (Hallberg in SSNP–4: 60). This language is being documented by Rozi Khan Barki, a resident of the village, with the help of J. Baart. The other is Kundal Shahi, which was discovered by Khwaja Abdur Rahman and is spoken in the

Neelam Valley in Azad Kashmir, about 75 miles from Muzaffarabad. This is being preserved by Khwaja Rahman with the help of Baart.

In short, while only the remotest and smallest of the languages of Pakistan are in danger of dying, other languages have decreased in stature. The undue prestige of English and Urdu has made all other languages burdens rather than assets. This is the beginning of language sickness, if not death.

Although very little information is available on the languages of Pakistan, an effort has been made here to make observations about the use and vitality of a large number of these languages (a summary is presented in Appendix 3). The main point is that as small and isolated communities open up to the forces of modernity, their languages come under threat and may disappear if nothing is done to reverse the language shift.

4. Can language shift be reversed?

Awareness of language shift and the need to reverse it came to the attention of linguists through an epoch-making book by Joshua A. Fishman aptly entitled *Reversing Language Shift* (1991). Ten years after the book appeared the question was revisited by another volume edited by Fishman called *Can Threatened Languages be Saved?* (2001). However, these books are not known in Pakistan and the view they support – that language shift ought to be reversed – is seen as fatuous or sentimental nonsense. The indigenous languages are seen as markers of backwardness or symbols of ethnic resistance to the center and are not taken seriously. A few anti-globalization enthusiasts, however, pay some attention to language issues. In February 2004 speakers in a conference on Green Economics (arranged by an NGO called Shirkat Gah) pointed out that varieties of wheat and other agricultural products have decreased in number and that people do not even have names for varieties which existed about thirty years ago. The disappearance of local names is symptomatic of the depletion of local knowledge. Moreover, as people leave their languages, children are alienated from their ancestors, their roots, their culture and their essential self. Unfortunately, very few people in Pakistan think of this as a problem, and there are no policies about preserving the linguistic diversity of the country.

Under such prevailing circumstances can anything be done to preserve the languages of the country? I believe it can be, but that the first step would be to persuade the government to create a new language policy. This new policy would have to go beyond affirming that everyone has the right to preserve their language and culture. In addition to that, the policy would create programmes to teach children through their mother tongues. Primers would have to be produced on the lines of material already produced by language activists and linguists (provided in Appendix 2). As the UNESCO and other NGOs could finance this project, public funds will be saved and may later be used to hire teachers and provide additional assistance.

A crucial aspect of teaching children in their mother tongue is to overcome the cultural shame associated with the traditional indigenous cultures and communities. This can be done by teaching all children, including those from the elite, through their mother tongue. Such teaching will, of course, be a bridge to the languages of wider communication (such as Urdu or the major provincial language).

Three RLS strategies are mentioned by Fishman: "One is 'shoot for the moon!' Another is 'anything is better than nothing'. The third is 'the right step at the right time' " (Fishman 2001: 474). Out of these, the third strategy seems to most fit Pakistan's case. Individuals may be made sensitive to the necessity of using the language in private domains while taking advantage of such governmental interventions in favour of their languages as much as possible. Among these interventions, apart from teaching, should be the radio, TV, and computer programmes aimed at by RLS activists.

These steps may reverse or at least slow down the language shift which is in evidence in Pakistan. Language shift may eventually occur but those conscious of the loss it entails to their identities will at least have the satisfaction of having done something to try to slow it down.

5. Conclusion

We have seen that the language policies of Pakistan, both declared and undeclared, have increased both ethnic and class conflict in the country. Moreover, our Westernized elites, in their own interests, are threatening cultural and linguistic diversity. As a result they are impoverishing the already poor and creating much resentment against the oppression and injustice of the system.

While it may not be possible to reverse language shift, it is possible to promote the concept of additive bilingualism rather than subtractive bilingualism. This means that we should add to our repertoire of languages to gain power while retaining skills and pride in our own languages. In order to do this, the state and our education system should promote the concept of linguistic rights.

There are tolerance-related and promotion-oriented rights. In Pakistan we have the former but not the latter. This means that, while we keep paying lip service to our indigenous languages, we create such market conditions that it becomes impossible to gain power, wealth or prestige in any language except English and, to a lesser extent, Urdu. It is this which must be changed and the change must come by changing the market conditions. This is what they did in the case of Catalan, a language which had been banned by General Franco of Spain, and which has been revived. Since they made Catalan the language of jobs and the government of Catalonia (Hall 2001), it changed the power equation and people started learning Catalan.

What we need in Pakistan are such promotion-oriented rights for our languages. What will go along with such rights is a good but fair system of schooling which will teach the mother tongue, English and Urdu equally to all children, not as it is done now, with English being taught very well to the elite but very badly to all others (for details, see Rahman 2002, Conclusion). Such steps might save us from the more harmful linguistic effects of language policies.

Appendix 1: Minor languages and dialects of Pakistan

The number of languages listed in the *Ethnologue* (Gordon 2005) for Pakistan is 72. This chart however, lists 55 languages and dialects. The major languages (Punjabi, Sindhi, Pashto, Siraiki, Urdu and Balochi) are given elsewhere. The dialects of Pashto (3), Balochi (3), Hindko (3), Greater Punjabi (Pahari, Potohari) are subsumed under the language head itself. English, Sign Language, Badeshi (which is dead) have been excluded. Marwari, mentioned twice, is entered only once here. Kundal Shahi, not mentioned in the *Ethnologue*, is, however, included. Lexical similarity and intelligibility of varieties of a language are given if known. Judgments concerning a form of speech being a language or a dialect are not given.

Language/ dialect	Other names/ lexical similarity to other languages|dialects	Where spoken	Speakers	Source
Aer	None. 78% lexical similarity with Katai Meghwar and Kachi Bhil. 76% with Raburi; 76% with Kachi Koli.	Jikrio Goth around Deh 333, Hyderabad and Jamesabad. Also in Kach Bhuj in Gujrat (India)	200 (1996)	Gordon 2005
Bagri	(Bahgri ; Bagria; Bagris; Baorias; Bauri). Dialect of Rajasthani 74% lexical similarity with Marwari Bhil of Jodhpur; 54% with Jandavra.	Sindh and Punjab (nomadic between India and Pakistan)	200 000 in Pakistan including 100 000 in Sindh	Gordon 2005

Language/ dialect	Other names/ lexical similarity to other languages/dialects	Where spoken	Speakers	Source
Balti	Baltistani, Sbalti	Baltistan; also India	270 000 (Pakistan); 337 000 (World)	SSNP-2: 8; Gordon 2005
Bateri	(Bateri Kohistani; Batera Kohistan; Baterawal; Baterawal Kohistani) 58–61% lexical similarity with Indus Kohistani; 60% with Gurgula.	Indus Kohistan Batera village (East of Indus North of Besham)	28 251 (Pakistan); 29 051 (World)	Breton 1997: 200; Gordon 2005
Bhaya	Lexical similiarity to Marwari sweeper 84% and to Malhi 75%; Bhat 73%; Goaria,72–73%; Sindhi Meghwar 70–73%, Sindhi Bhil 63–71% and Urdu 70%.	Kapri Goth near Khipro Mirpur Khas (Lower Sindh)	70–700 (1998)	Gordon 2005
Brahvi	Brohi, Brahuidi, Kurgalli, Brahuigi, (no similarity with any language in Pakistan but with many loan words from Persian, Balochi and Urdu).	Kalat region and East Balochistan. Also spoken by small communities in Sindh and Iran etc.	2 000 000 (Pakistan); 2 210 000 (World) (1998).	Gordon 2005
Burushaski	Mishaski, Biltum, Werchikwar Khajuna (language isolate with no similarity with any language. Some words borrowed from Urdu, English and Shina).	Hunza, Nagar, Yasin valleys (Northern areas)	87 049 (2000)	SSNP-2: 37; Gordon 2005
Chilisso	(Chiliss, Galos) 70% lexical similarity with Indus Kohistani; 65–68% with Gowro; 50% Bateri; 48–65% with Shina.	Koli, Palas, Jalkot Indus Kohistan	1600–3000 (1992)	Breton 1997: 200; Gordon 2005

Language/ dialect	Other names/ lexical similarity to other languages/dialects	Where spoken	Speakers	Source
Dameli	(Gudoji, Damia, Damedi, Damel) 44% lexical similarity with Gawar-Bati, Savi, and Phalura, 33% with Kamviri, 29% with Kativiri.	Damel Valley (Southern Chitral)	5000 (1992)	SSNP-5: 11; Gordon 2005
Dehwari also see Persian	(Deghwari) Iranian language somewhat close to Persian and influenced by Brahvi.	Kalat, Mastung (Central Balochistan)	13 000 (1998)	Breton 1997: 200; Gordon 2005
Dhatki	(Dhati) Dialects are Eastern, Southern and Central Dhatki, Malhi and Barage. Varies from Northern Marwari but intelligible. 70–83% lexical similarity with Marwari dialects.	Lower Sind in Tharparkar and, Sanghar	131 863 (Pakistan); 148 263 (World)	Gordon 2005
Domaaki	(Domaski, Doma) loan words from Shina and Burushaski but not intelligible to speakers of both.	Mominabad (Hunza & Nagar)	300 plus (2002)	SSNP–2: 79. Author's personal observation in 2002
Gawar-Bati	(Narsati, Nurisati, Gowari, Aranduiwar, Satr, Gowar-bati) 47% lexical similarity with Shumashti, 44% with Dameli, 42% with Savi and Grangali.	Southern Chitral, Arandu, Kunar river along Pakistan-Afghanistan border	1500 (1992)	SSNP-5: 156; Breton 1997: 200; Gordon 2005
Ghera	(Sindhi Ghera, Bara) Quite different grammatically from Gurgula and similar to Urdu. 87% lexical similarity with Gurgula. 70% with Urdu.	Hyderabad Sindh	10 000 (1998)	Gordon 2005

Language/ dialect	Other names/ lexical similarity to other languages/dialects	Where spoken	Speakers	Source
Goaria	75–83% lexical similarity with Jogi; 76–80% with Marwari sweeper; 72–78% with Marwari Meghwar; 70–78% with Loarki.	Cities of Sindh	25 426 (2000)	Gordon 2005
Gowro	(Gabaro, Gabar Khel) 62% lexical similarity with Indus Kohistani; 60% with Bateri; 65–68% with Chilisso; 40–43% with Shina.	Indus Kohistan (on the eastern bank, Kolai Area, Mahrin village)	200 or less (1990)	Breton 1997: 200; Gordon 2005
Gujari	(Gujuri, Gojri, Gogri Kashmir Gujuri, Gujuri Rajasthani) close to Hindko and related varieties of Greater Punjabi. 64–94% lexical similarity among dialects.	Swat, Dir, Northern areas, Azad Kashmir and Punjab	300 000–700 000 plus (1992)	SSNP-3: 96; Gordon 2005
Gujrati	(Gujrati)	Karachi, other parts of Sindh. Major language in India.	45 479 000 (India); 46 100 000 (World); Probably 100 000 in Pakistan.	Gordon 2005
Gurgula	(Marwari, Ghera) 87% Lexical similarity with Ghera	Karachi, cities of Sindh	35 314 (2000)	Gordon 2005
Hazargi	(Hazara, Hezareh, Hezare'i) similar to Persian	Quetta and other cities of Pakistan. Also in Afghanistan.	156 794 (2000)	Gordon 2005
Hindko	(Hazara Hindko, Peshawar Hindko, Hindki) a variety of Greater Punjabi. Intelligible to Punjabi and Siraiki speakers.	Mansehra, Abbottabad, Haripur, Attock Districts. The inner city of Peshawar and Kohat, etc.	3 000 000 in 1993 i.e. 2.4% of the population.	Gordon 2005

Language/ dialect	Other names/ lexical similarity to other languages/dialects	Where spoken	Speakers	Source
Jandavra	(Jhandoria) 74% lexical similarity with Bagri and Katai Meghwar, 68% with Kachi Koli.	Southern Sindh from Hyderabad to Mirpur Khas	5000 (1998)	Gordon 2005
Jatki	(Jatgali, Jadgali, Jat)	Southern Balochistan and Southwest Sindh. Also in Iran.	100 000 in both countries (1998)	Gordon 2005
Kabutra	(Nat, Natra) intelligibility with Sansi and Sochi. 74% lexical similarity with Sochi.	Umarkot, Kunri, Nara Dhoro (Sindh)	1000 (1998)	Gordon 2005
Kachchi	(Cutch, Kachi) similar to Sindhi.	Karachi	50 000 (1998)	Gordon 2005
Kalami	(Bashgharik, Dir Kohistani, Bashkarik, Diri, Kohistani, Dirwali, Kalami Kohistani, Gouri, Kohistani, Bashkari, Gawri, Garwi)	Upper Swat Khoistan from Kalam to upper valleys also in Dir Kohistan	60 000–70 000 (1995)	Baart 1999: 4
Kalasha	(Kalashwar, Urtsuniwar, Kalashamon, Kalash)	Kalash Valleys (Chitral) Southern	5029 (2000)	SSNP-5: 11, 96–114; Gordon 2005
Kalkoti	69% lexical similarity with Kalami but Kalami speakers do not understand Kalkoti.	Dir Kohistan in Kalkot village	6000 (2002)	Breton 1997: 200; Zaman 2002a; Gordon 2005
Kamviri	(Skekhani, Kamdeshi, Lamertiviri, Kamik) There is a variety of Kativiri also called Skekhani.	Chitral (Southern end of Bashgal Valley)	2000 (1992)	SSNP-5: 143; Gordon 2005
Kashmiri	(Keshuri)	The Valley of Kashmir & Diaspora in Pakistan	4 391 000 in India. About 105 000 in Pakistan (1993)	Breton 1997: 200; Gordon 2005
Kativiri	(Bashgali, Kati, Nuristani, Shekhani) Eastern Kativiri in Pakistan.	(Chitral) Gobar Linkah Valleys	3700–5100 (1992)	Gordon 2005

Language/ dialect	Other names/ lexical similarity to other languages/dialects	Where spoken	Speakers	Source
Khetrani	Similar to Siraiki but influence by Balochi	Northeast Balochistan	4000	Gordon 2005
Khowar	(Chitrali, Qashqari, Arniya, Patu, Kohwar, Kashkara)	Chitral, Northern areas, Ushu in northern Swat	222 800 (Pakistan); 242 000 (World)	SSNP-5: 11, 25–42; Breton 1997: 200; Gordon 2005
Kohistani	(Indus Kohistani, Dir Kohistani, Kohiste, Khili, Maiyon, Maiya, Shuthun, Mair)	Indus Kohistan West bank of river	220 000 (1993)	Gordon 2005
Koli Kachi	(Kachi, Koli, Kachi Koli) similar to Sindhi and Gujrati (78% lexical similarity) but influenced more by Sindhi in Pakistan. Its dialects are Rabari, Kachi Bhil, Vagri, Katai Meghwar, Zalavaria Koli and Tharadari Koli.	(Lower Sindh) around Towns of Tando Allahyar & Tando Adam also in India around the Rann of Kach.	170 000 (1998)	Gordon 2005
Koli Parkari	Parkari (Lexical similarity with Marwari Bhil and Tharadari) 77–83% lexical similarity with Marwari Bhil; 83% with Tharadari Koli	Lower Thar Desert Nagar Parkar. Also in India.	250 000 (1995)	Gordon 2005
Koli Wadiyara	(Wadiyara, Wadhiyara) intelligibility with Kachi Koli and its varieties.	Sindh in an area bounded by Hyderabad, Tando Allahyar and Mirpur Khas in the north, and Matli and Jamesabad in the South.	175 000–180 000 in Pakistan). Total in Pakistan and India 360 000 (1998).	Gordon 2005
Kundal Shahi		Neelam Valley, Azad Kashmir	500 (2003)	Baart and Rehman 2003
Lasi	(Lassi) similar to Sindhi but influenced by Balochi.	Las Bela District (south east Balochistan)	15 000 (1998)	Gordon 2005

Language/ dialect	Other names/ lexical similarity to other languages/dialects	Where spoken	Speakers	Source
Loarki	82% lexical similarity with Jogi and 80% with Marwari.	Sindh – various places	21 000 (1998)	Gordon 2005
Marwari	(Rajasthani, Meghwar, Jaiselmer, Marawar, Marwari Bhil) 79–83% lexical similarity with Dhatki; 87% between Southern and Northern Marwari; 78% Marwari Mehwar and Marwari Bhat.	Northern Marwari in South Punjab North of Dadu Nawabshah. Southern Marwari in Tando Mohammad Khan and Tando Ghulam Ali etc.	220 000 (1998)	Gordon 2005
Memoni	Similarities to Sindhi and Gujrati	Karachi	Unknown	Gordon 2005
Od	(Odki) similarity with Marathi with some Gujrati features. Also influenced by Marwari and Punjabi 70–78% lexical similarity with Marwari, Dhatki and Bagri.	Scattered in Sindh & south Punjab	50 000 (1998)	Gordon 2005
Ormuri	(Buraki, Bargista) 25–33% lexical similarity with Pashto.	Kaniguram (south Waziristan) some in Afghanistan	1000 (Pakistan); 1050 (World)	SSNP-4: 54; Gordon 2005
Persian	(Farsi, Madaglashti Persian in Chitral Dari, Tajik, Badakhshi and the dialects mentioned earlier). Dialects of Persian spoken in Pakistan. The standard variety is used for writing.	Balochistan, Shishikoh Valley in Chitral, Quetta, Peshawar, etc.	2000–3000 (1992)	SSNP-5: 11; Gordon 2005

Language or dialect	Other names/ lexical similarity to other languages and dialects	Where spoken	Speakers	Source
Phalura	(Dangarik, Ashreti, Tangiri, Palula, Biyori, Phalulo) 56–58% lexical similarity with Savi; 38–42% with Shina	7 villages near Drosh, Chitral possibly 1 village in Dir Kohistan	8600 (1990)	SSNP-5: 11, 67–95; Gordon 2005
Sansi	(Bhilki) 71% lexical similarity with Urdu; 83% with Sochi.	North-western Sindh	16 200 (2000)	Gordon 2005
Shina	(Sina, Shinaki, Brokpa)	Giligit, Kohistan, Baltistan and Ladakh	300 000 (Pakistan); 321 000 (World)	SSNP-2: 93; Gordon 2005; Kohistani and Schmidt in this volume
Sindhi Bhil	(Bhil) close to Sindhi. Its varieties are Mohrano, Sindhi Meghwar, Badin etc.	Badin, Matli, Thatta (Sindh)	56 502 (2002)	Gordon 2005
Torwali	(Kohistani, Bahrain Kohistani) 44% lexical similarity with Kalkoti and Kalami.	Chail and Bahrain (Swat)	60 000	Breton 1997: 200; Lunsford 2001; Gordon 2005
Ushojo	(Ushoji) 35–50% lexical similarity with varieties of Shina.	Upper part of Bishigram Valley (Chail) in Swat	1000 (2002)	Zaman 2002a; Gordon 2005
Vaghri	(Vaghri Koli) 78% lexical similarity with Wadiyara Koli.	Sindh many places. Also in India.	90 000 (India); 10 000 (Pakistan) (1998)	Gordon 2005
Wakhi	(Kheek, Kheekwar, Wakhani, Wakhigi, Wakhan) some influence from Burushaski.	Northern ends of Hunza & Chitral	9100 (Pakistan); 31 666 (World)	SSNP-2: 61; Gordon 2005
Wanetsi	(Tarino, Chalgari, Wanechi) 71–75% lexical similarity with Southern Pashto.	Harnai (East of Quetta)	95 000 (1998)	SSNP-4: 51 ; Breton 1997: 200; Gordon 2005
Yidgha	(Yidghah, Luthuhwar) 56–80% lexical similarity with Munji in Afghanistan. Also influenced by Khowar.	Upper Lutkoh Valley (Western Chitral)	6145 (2000)	SSNP-5: 11, 43–66; Gordon 2005

Appendix 2: State of the languages of Pakistan

This chart provides information on the availability of written material in a language, particularly that which is suitable for teaching small children or illiterate adults. The names of the writers of a primer are given in the third column. The names of authors of other material have not been given.

Language	**Material available**	**Names of writers of primers**
Aer	—	
Bagri	—	
Balochi	Alphabet book, primers, folktales, health books, phrase book Balochi–Urdu–English dictionary, printed books on Islamic observances, poetry, modern literature, textbooks etc.	Tan et al. 1999; Farrell and Sadiq 1986
Balti	Ancient records (Devanagari based script); Grammar, parables (Roman); verse, folksongs etc (Nastaliq script).	Hussanabadi 1990
Bateri	—	
Bhat	—	
Bhaya	—	
Bhil Sindhi	Material in Sindhi may be used.	Many primers
Brahvi	Alphabet book, primers, folktales, health books, phrase book, Brahvi–Urdu–English dictionary, printed books on Islamic observances, poetry, modern literature, textbooks etc.	Many primers
Burushaski	Transition primer (Urdu to Burushaski), folktales, bilingual vocabulary: Burushaski-English.	Nasir n.d.
Chilisso	—	
Dameli	—	
Dehwari	—	
Dhatki	Alphabet book, primer, transition primer, folktales, stories for children.	Das et al. 1991; Payne 1991; Various authors 1991
Domaaki	—	
Gawarbati	—	
Ghera	—	
Goaria	—	
Gowro	—	
Gujari	Poetry books, short stories, songs etc.	Many primers
Gujrati	Primers, grammars, textbooks, books etc. (in India also in digital form).	Many primers
Gurgula	—	
Hazargi	Alphabet book, folktales, health books, proverbs, stories for children. Material in standard Persian may also be used.	HLA 1997
Hindko	Primers, literature, prose, dictionaries, magazines etc.	Akbar 1994 & other primers
Jandavra	—	

Language	Material available	Names of writers of primers
Jatki	Primers, word lists, grammars. Naskh/Nastaliq.	Baloch 2003
Jogi	—	
Kabutra	—	
Kachchi	Primers of Sindhi may be used.	Many primers
Kachchi (Bhil)	—	
Kachchi (Katiawari)	—	
Kalami	Alphabet book, transition primer, poetry books, collection of texts from Gawri writers' workshop, proverbs, phrase dictionary Gawri–Urdu–English.	KCS 2002; Zaman 2002b; Zaman 2002c; Shaheen 1989
Kalasha	Alphabet book, pre-reader, dictionary.	Akbar 1994
Kalkoti	—	
Kamviri	—	
Kashmiri	Primers, folktales, poetry, textbooks, other books etc. (most of this literature is in India).	Many primers
Kativiri	—	
Khetrani	—	
Khojki (Script not a language)	Ancient records, *Ginans*, old documents, primers, school textbooks, other books.	Ali 1989
Khowar	Primers, grammar, dictionary, folktales, poetry, religious books, other popular books.	Faizi 1987
Kohistani (Indus)	—	
Koli (Kachi)	Alphabet books, folktales, health books, stories for children, primer.	Masih and Woodland 1995
Koli (Parkari)	Alphabet book, primer, folktales, health books, bilingual vocabulary: Parkari-English, stories for children.	A. Hoyle 1996; R. Hoyle 1990; Hoyle and Samson 1985; Hoyle et al. 1990
Koli (Tharadari)	—	
Koli (Wadiyara)	—	
Kundal Shahi	—	
Lasi	—	
Loarki	—	
Marwari	—	
Memoni	Primers of Sindhi may be used.	Many primers
Od	—	
Ormuri	Primer, grammar, word list [Roman] verse, prose, grammar, word list Ormuri (Pashto script)	Barki 1999
Pashto	All types of textbooks and books; also in digital form. (also used in Afghanistan in some domains of power).	Many primers
Persian	All types of books (also in digital form).	Many primers

Language	Material available	Names of writers of primers
Phalura	—	
Punjabi	Books on literature; history; textbooks etc in Nastaliq script. (All types of books in the Gurmukhi script in India).	Many primers
Sansi	—	
Shina	Poetry, grammar, word lists, folktales, songs, religious books etc.	Taj 1989; Zia 1986; Namus 1961; Kohistani and Schmidt 1996
Sindhi	All types of books also in digital form.	Many primers
Sindhi Bhil	—	
Siraiki	Ancient poetry, modern literature, magazines etc.	Mughal 1987 & other primers
Torwali	Lexicographic work using Nastaliq is in progress.	Kareemi 1982
Urdu	All types of books, also in digital form.	Many primers
Vaghri	—	
Wakhi	Primer, word list, folksongs, proverbs, word lists.	Sakhi 2000
Wanetsi	Primer, songs, folktales, word lists Nastaliq (Pashto variant).	Askar 1972
Yidgha	—	

Appendix 3: Domains of use and vitality of the languages of Pakistan

Language	Domains of use	Vitality	Source
Aer	Used in all functions within the group. Worship songs in Gujrati	Women monolingual. Men multilingual, generally in Sindhi. No evidence of language shift but shift possible to Sindhi as children go to school.	Jeffery[3] 1999
Bagri	Used in all functions within the group. Used in weddings, to tell jokes, in songs.	All multilingual, mostly in Sindhi. No evidence of language shift.	Jeffery 1999
Balti	Used in all functions within the group. Used by teachers as informal medium of instruction for small children if they are MT speakers themselves. Also cultivated by language activists and media persons (radio announcers etc).	Some bilingualism in Urdu especially among the educated and the employed. Positive attitude to MT. Desirous of learning to read their language. No evidence of language shift.	Backstrom in SSNP-2: 23–26
Bateri	Used in all functions within the group.	Some multilingualism in Pashto and Urdu, especially among the educated and those who travel on business. Positive attitude towards MT. No evidence of language shift.	Hallberg in SSNP-1: 137–139
Bhaya	Not known	Shifting to Sindhi and related to Marwari dialects.	Gordon 2005; Author's personal information

Language	Domains of use	Vitality	Source
Bhil Sindhi	Used in traditional ceremonies and worship.	Bilingualism in Sindhi.	Jeffery 1999
Burushaski	Used in all functions within the group. Used by teachers as informal medium of instruction. Also cultivated by language activists, media persons etc.	Increasing bilingualism in Urdu and English, However, the language is being maintained. Desirous of learning Urdu and English but expressing positive feelings for MT.	Backstrom in SSNP-2: 52–53
Chilisso	Many speakers do not use the language even at home.	Bilingualism in Shina. Language shift to Shina in progress. People want their children to learn Shina and Urdu.	Hallberg in SSNP-1: 121–122
Dameli	Spoken by older people at home but younger people also use other languages.	Multilingualism in Pashto and Khowar. However, positive attitude to MT is expressed. Possibility of language shift to Pashto.	Decker in SSNP-5: 124–127
Dehwari	Not known	Influenced by Brahvi.	Gordon 2005
Dhatki	Used by the Malhi group for all functions. Urdu and Sindhi used for songs.	Multilingualism in many languages.	Jeffrey 1999
Domaaki	Possibly used by very few elderly people with each other. Most people do not know it.	Language shift to Burushaki is complete with no hope of reversal.	Backstrom in SSNP-2: 81–83
Gawar-Bati	Used for all functions within the group.	Multilingualism in Pashto and to a lesser extent in Khowar. Positive attitude to MT. However, the language is under pressure by Pashto.	Decker in SSNP-5: 161–163
Ghera	Used for all functions within the group.	Multilingualism in Sindhi and Urdu. Being influenced by both.	Jeffrey 1999
Goaria	Used for all functions within the group. Hindi used in worship. Children use Sindhi and Urdu.	Multilingualism in many languages. Children use Sindhi or Urdu with outsiders.	Jeffery 1999
Gowro	Still spoken by older people, but younger people mix it with Shina and sometimes speak only Shina.	Bilingualism in Shina. Language shift to Shina in progress.	Hallberg in SSNP-1: 129–132; Zaman 2004b
Gujari	Used in some communities but not among in Gujars settled in the Punjab and Azad Kashmir. Language activists are creating literature in the language. Songs and music are broadcast from the radio and there is a TV programme from India.	Multilingualism in many languages, especially Urdu among the educated. In the NWFP, Northern areas and parts of Azad Kashmir, the language is maintained. In the Punjab and near Muzaffarabad and Mirpur, there is language shift to the local languages. Educated people use Urdu.	Hallberg and O'Leary in SSNP-3: 100

Language	Domains of use	Vitality	Source
Gujrati	Used for conversation within the family but younger people are switching to Urdu or English (depending on socio-economic class). All types of literature exist. Used in the media and in the state of Gujrat in India.	Multilingualism in Urdu and English as well as other languages. Language shift to Urdu and English is in progress at least in Pakistan.	Author's field research in Karachi
Gurgula	Language used within community is strong.	Multilingual in many languages.	Jeffery 1999
Hazargi	Used in the group for all functions.	Multilingualism with Pashto, Balochi and Persian. Language is under pressure.	
Jandavra	Private.	People proud of their language.	Jeffery 1999
Jatki	Not known	Not known	—
Kabutra	Used in the group for all functions.	Multilingual in many languages. Positive attitude and pride in language. No shift.	Jeffery 1999
Kachchi (Bhil)	Used in the group for all functions.	Bilingualism in Sindhi. Being rural it is maintained at present. Shift to Sindhi going on.	Jeffery 1999
Kachchi (Katiawari)	Used by older people in some domains.	Shift to Sindhi ongoing.	Jeffery 1999
Kalami	Used for all functions within the group.	Widespread bilingualism in Pashto. Educated people also know Urdu. Attitude towards MT positive and no language shift is observed.	Rensch in SSNP-1: 57–61
Kalasha	Used for all functions within the group.	Positive attitude to MT but those who convert to Islam shift to Khowar or the language of their spouse. Some multilingualism in Khowar and Urdu because of tourism and education. The language is under pressure and there is a possibility of language shift.	Decker in SSNP-5: 107–113
Kalkoti	—	Kalami used is a second language. Most people also speak Pashto.	Gordon 2005
Kamviri	Used for all function within the group.	Multilingualism in Pashto and surrounding languages. Positive attitude to MT but under pressure by Pashto.	Decker in SSNP-5: 146–147
Kashmiri	Small diaspora in Pakistan but used for all functions within the Valley of Kashmir held by India. All kinds of literature available. Used in media and in teaching etc. Also taught at university level.	Multilingualism with Urdu and the local languages. Language shift in progress in Pakistan but is maintained in India.	Aziz 1983; Bukhari 1986

Language	Domains of use	Vitality	Source
Kativiri	Used in all functions within the group.	Positive attitude towards the MT but men multilingual in Pashto and surrounding languages. Difficult to predict language shift.	Decker in SSNP-5: 144–147
Khetrani	—	—	—
Khowar	Used in all domains in the group. Used by teachers as informal medium of instruction for small children if they are MT speakers themselves. Also cultivated by language activists, media persons (radio, TV announcers etc).	Some bilingualism in Pashto, local languages and Urdu, the last especially among the educated and the employed. Positive attitude to MT. Desirous of learning to read their language. No language shift observed.	Decker in SSNP-5: 39–42
Kohistani (Indus)	Used for all functions within the group.	Multilingualism in Pashto and Shina is not common even among them. Positive attitude towards MT. People want it as a medium of instruction for small children. No language shift is observed.	Hallberg in SSNP-1: 110–113
Koli (Kachi)	Probably used in the group	Bilingualism in Sindhi.	Jeffrey 1999; Gordon 2005
Koli Kachi	Used for all functions within the group.	Multilingualism in Sindhi but language being maintained.	Grainger and Grainger 1980: 42
Koli Parkari	Used for all functions within the group.	Multilingualism in Sindhi but language being maintained.	Grainger and Grainger 1980: 42
Koli Parkari	Not known	Bilingualism in Sindhi but language being maintained.	Gordon 2005
Koli Tharadari	Used for all functions within the group.	Men multilingual in many languages. Women and children maintain the language.	Jeffery 1999
Koli Wadiyara	Used for all functions within the group.	Multilingualism in Sindhi but language being maintained.	Jeffery 1999
Kundal Shahi	Used only by the elderly in the family. No longer used by children.	Language shift to local language and Urdu in progress.	Baart and Rehman 2003
Lasi	Not known	Not known	—
Loarki	Used for all functions within the Loar group	Multilingualism in Sindhi and some knowledge of Urdu.	Jeffery 1999
Marwari (Southern)	Used in all domains of the group.	Multilingualism in Sindhi.	—
Memoni	Probably used by older speakers in the group as spoken language.	Most speakers are educated and multilingual in Sindhi, Urdu and Gujrati. The language is shifting to these three languages.	Gordon 2005
Od	Used in some Od communities while others use local languages.	Multilingualism in surrounding languages. Language shift in progress in this iterant community.	Grainger and Grainger 1980: 31

Language	Domains of use	Vitality	Source
Ormuri	Used for most functions in the Kaniguram area. Words of Pashto are common among young people.	Bilingualism with Pashto. Though positive attitude to MT is expressed, language shift to Pashto is visible.	Hallberg in SSNP-4; Barki 1999; Barki n.d.
Phalura	Used at home. Used informally by teachers.	Multilingualism in Khowar, Pashto and Urdu. Language shift to Khowar in evidence. However, ethnic Kalasha have shifted to Phalura in some areas. Vitality picture mixed.	Decker in SSNP-5: 92–94
Rabari	Used in all domains of the group.	Being maintained.	Jeffery 1999
Sansi	Used for worship and weddings.	Multilingualism in Sindhi and slightly in Urdu and Siraiki. No language shift observed.	Jeffery 1999
Shina	Used in all domains in the group. Used by teachers as informal medium of instruction for small children if they are MT speakers themselves. Also cultivated by language activists, media persons (radio announcers etc).	Considerable bilingualism in Urdu especially among the educated and the employed. Positive attitude to MT. Ambivalent about learning to read their language. No language shift observed. However, there is pressure of Urdu.	Backstrom in SSNP-2: 173; Kohistani and Schmidt in this volume
Sochi	Used in singing, weddings and telling stories.	Multilingualism in Sindhi and slightly in Urdu.	Jeffery 1999
Torwali	Not known	Men bilingual in Pashto but language being maintained.	Gordon 2005
Ushojo (Ushuji)	Used at home at least by the older speakers. There is much mixing of Pashto.	Multilingualism in Pashto and Torwali but educated people know Urdu. Young people who know the MT use Pashto in some areas. Language is under threat from Pashto. Language vitality is varied and mixed.	Decker in SSNP-1: 75–79
Vaghri	Used in private domains.	Bilingualism in Sindhi. Positive attitude to the language in spite of pressures.	Jeffery 1999
Wakhi	Used in all domains of the group. Language activists and radio broadcasters also cultivate it.	Bilingualism with Urdu among younger, educated people. Also knowledge of Burushaski. Positive attitude towards MT. Desirous of learning the written language in school. However, the language is under pressure from Urdu.	Backstrom in SSNP-2: 70–73
Wanetsi (Waneci)	Used in private domains but those who live in cities do not use it.	Bilingualism with Pashto. Positive attitude towards MT. However, under pressure from Pashto.	Hallberg in SSNP-4; Askar 1972
Yidgha	Used for in group functions. Used informally by teachers and for explaining religious texts.	Multilingualism in Khowar and sometimes Urdu, Persian and Bashgali. Language shift to Khowar in evidence.	Decker in SSNP-5: 56–57

Notes

1. Such protests remind one of the works of linguists who oppose the arrogance of monolingual English speakers (see, for example, Skutnabb-Kangas 2000; Crystal 2000: 84–88; Nettle and Romaine 2000).
2. I am grateful to the authors for providing me access to the manuscript.
3. Quoted by the kind permission of the author.

References

Abdullah, Syed
1976 *Pakistan mein urdu ka masla* [The Status of Urdu in Pakistan]. Lahore: Maktaba Khayaban-e-Adab.
Ahmed, Feroze
1992 The language question in Sind. In *Regional Imbalances and the Regional Question in Pakistan*, Akbar S. Zaidi (ed.), 139–155. Lahore: Vanguard Books.
Akbar, Mujahid
1994 *Hindko qaida* [Hindko Primer]. Peshawar: Maktaba Hindko Zaban.
Ali, Mumtaz Tajddin
1989 *Khojki: Self Instructor*. Karachi: Privately printed.
Alavi, Hamza
1988 Pakistan and Islam: Ethnicity and ideology. In *State and Ideology in the Middle East and Pakistan,* Fred Halliday, and Hamza Alavi (eds.), 64–111. London: Macmillans.
Anderson, Benedict
1983 *Imagined Communities: Reflections on the Origin and Spread of Nationalism*. London: Verso.
Askar, Umar Gul
1972 *Wanetshi*. Quetta: Balochi Academy.
Aziz, Mir Abdul
1983 The Kashmiri language in Azad Kashmir and Pakistan. *Pakistan Times*, 20 June.
Baart, Joan L. G.
1999 *A Sketch of Kalam Kohistani Grammar*. Islamabad: National Institute of Pakistan Studies and Summer Institute of Linguistics.
2003 Interview by the present author. 10 August, Islamabad.
Baart, Joan, and Khwaja A. Rehman
2003 The language of the Kandal Shahi Qureshis in Azad Kashmir, Unpublished manuscript.
Backstrom, Peter C., and Carla F. Radloff (eds.)
1992 *Sociolinguistic Survey of Northern Pakistan. Volume 2: Languages of Northern Areas* (SSNP-2). Islamabad: National Institute of Pakistan Studies and Summer Institute of Linguistics.
Baloch, Nabi Baksh
2003 *Jatki boli* [Jatki Grammar, Wordlist]. Hyderabad: Sindhi Language Authority.

Barki, Rozi Khan
1999 *Makh akhui zaban ta gor ghagu zar zeen* [Ormuri Phonetics, Wordlist, Stories, Poetry]. Islamabad: Privately published.
n.d Dying languages with special focus on Ormuri language. Typescript.
Bourdieu, Pierre
1991 *Language and Symbolic Power*. Cambridge: Polity Press.
Breton, Roland J. L.
1997 *Atlas of the Languages and Ethnic Communities of South Asia*. New Delhi: Sage Publications.
Bukhari, M. Yusuf
1986 *Kashmiri aur urdu zaban ka taqabli muta'ala* [The Comparative Study of Urdu and Kashmiri]. Lahore: Markazi Urdu Board.
Census
2001 *1998 Census Report of Pakistan*. Islamabad: Population Census Organization Statistics Division, Government of Pakistan.
Cohen, Stephen P.
1998 *The Pakistan Army*. 2nd edition. Karachi: Oxford University Press.
Cooper, Robert L
1989 *Language Planning and Social Change*. Cambridge: Cambridge University Press.
Crystal, David
2000 *Language Death*. Cambridge: Cambridge University Press.
Das, Seval, Mike Payne, et al.
1991 *DhaaTkii akhar aan phooTuu* [Dhatki Words and Pictures]. Hyderabad: New Foundations.
Decker, Kendall D. (ed.)
1992 *Sociolinguistic Survey of Northern Pakistan. Volume 5: Languages of Chitral* (SSNP-5). Islamabad: National Institute of Pakistan Studies and Summer Institute of Linguistics.
Edwards, John
1994 *Multilingualism*. London: Routledge.
Faizi, Inyatullah
1987 *Khowar bol chal* [Khowar Primer, Grammar, Wordlist, Conversation]. Chitral: Anjuman Taraqqi-e-Khowar.
Farrel, Timothy, and Abdul Haleem Sadiq
1986 *Bunii kitaab* [Basic Book, Alphabet Book]. Quetta: Shal Association.
Fishman, Joshua A.
1991 *Reversing Language Shift*. Clevedon: Multilingual Matters.
2001 *Can Threatened Languages be Saved?* Clevedon: Multilingual Matters.
Government of Bangladesh
1982 *History of Bangladesh War of Independence*, Vol. 6. Dhaka: Government of Bangladesh, Ministry of Information.
Government of Pakistan
1966 *Report of the Commission on Student's Problems and Welfare and Problems*. Islamabad: Ministry of Education, Government of Pakistan.
Gordon, Raymond G. Jr. (ed.)
2005 *Ethnologue: Languages of the World*. 15th edition. Dallas: SIL International. Online version: <http://www.ethnologue.com/>.

Grainger, Peter S., and Nita C. Grainger
1980 A preliminary survey of the languages of Sind, Pakistan. Summer Institute of Linguistics Report.
Hall, Jacqueline
2001 *Convivencia in Catalonia: Languages Living together* Barcelona: Fundcio Jaume Bofill.
Hallberg, D. G. (ed.)
1992 *Sociolinguistic Survey of Northern Pakistan. Volume 4: Pashto, Wanechi, Ormuri* (SSNP-4). Islamabad: National Institute of Pakistan Studies and Summer Institute of Linguistics.
HLA
1997 *Beeid kee alif bee yaad bigiirii* [Let's Learn the ABC's]. Quetta: Hazaragi Literacy Association.
Hoyle, Anne
1986 *Achoota paRhoon* [Let's Read]. Hyderabad: Church of Pakistan.
Hoyle, Richard
1990 *Paarkari bhaNoo* [Read Parkari]. Pakistan: Parkari Language Committee.
Hoyle, Richard, Viro Goel, Malo Samson, Naru, and Meghraj
1990 *Paarkari aakhar an phooTuu* [Parkari Letters and Pictures]. Hyderabad: Parkari Language Committee.
Hoyle, Richard, and Malo Samson
1985 *Parkari bhaNyaa roo kitaabb* [Parkari reading book]. Hyderabad: Parkari Language Committee.
Hussanabadi, M. Yusuf
1990 *Balti zaban* [Balti Language]. Skardu: Privately published.
Jahan, Rounaq
1972 *Pakistan: Failure in National Integration.* New York: Columbia University Press.
Jeffery, David
1999 Sindh survey month November 1996. Unpublished report.
Kalam Cultural Society
2002 *Gawri alif be* [Gawri Primer]. Kalam: Kalam Cultural Society.
Kareemi, Abdul Hameed
1982 *Urdu kohistani bol chall* [Conversation in Urdu and Kohistani]. Swat: Kohistan Adab Academy.
Kohistani, Razwal, with Ruth Laila Schmidt
1996 *Shina qaida* [Shina Environmental Primer]. Islamabad: Himalayan Jungle Project.
LAD-B = Legislative Assembly Debates of Baluchistan (dates and other details follow in the text).
LAD-F = Legislative Assembly Debates of the North-West Frontier Province (dates and other details follow in the text).
Lunsford, Wayne A.
2001 An overview of linguistic structures in Torwali: A language of northern Pakistan. M. A. Thesis, University of Texas, Arlington.
Mansoor, Sabiha
1993 *Punjabi, Urdu, English in Pakistan: A Sociolinguistic Study.* Lahore: Vanguard.

Masih, Mavo, and Andy Woodland
1995 *Kachhii akhar anee phooTuu* [Kachi Letters and Pictures]. Hyderabad: New Foundations.
Mughal, Shaukat
1987 *Siraiki qaida* [Siraiki Primer]. Multan: Siraiki Majlis Adab.
Namus, M. Shuja
1961 *Gilgit aur shina zaban* [Gilgit and Shina Languages]. Bahawalpur: Urdu Academy.
Nasir, Nasiruddin
n.d. *Buruso birkis* [Buruso Primer]. Hunza: Burushaski Research Academy.
Nettle, Daniel, and Suzanne Romaine
2000 *Vanishing Voices: The Extinction of the World's Languages* New York: Oxford University Press.
Payne, Joan
1991 *DhaaTkii paRhoo! Pahlkoo kitaab* [Read Dhatki, Book 1]. Hyderabad: New Foundations.
Rahman, Tariq
1996 *Language and Politics in Pakistan.* Karachi: Oxford University Press.
2002 *Language, Ideology and Power. Language Learning among the Muslims of Pakistan and North India.* Karachi: Oxford University Press.
Rensch, Calvin R., Sandra J. Decker, and Daniel G. Hallberg (eds.)
1992 *Sociolinguistic Survey of Northern Pakistan. Volume 1: Languages of Kohistan* (SSNP-1). Islamabad: National Institute of Pakistan Studies and Summer Institute of Linguistics.
Rensch, Calvin R., C. E. Hallberg, and Clare F. O'Leary (eds.)
1992 *Sociolinguistic Survey of Northern Pakistan. Volume 3: Hindko and Gujari* (SSNP-3). Islamabad: National Institute of Pakistan Studies and Summer Institute of Linguistics.
Sakhi, Ahmad Jami
2000 *Wakhi zaban tarikh ke aene men ma qaida* [Wakhi Primer]. Gilgit: Privately published.
Shaheen, M. Parvesh
1989 *Kalam kohistan: Log aur zaban* [Kalam Kohistan: Its People and Language]. Mingora: Academy of Swat Culture.
Skutnabb-Kangas, Tove
2000 *Linguistic Genocide in Education – or Worldwide Diversity and Human Rights.* London: Lawrence Erlbaum.
SSNP
1992 *Sociolinguistic Survey of Northern Pakistan* 5 Vols. Islamabad: National Institute of Pakistan Studies and Summer Institute of Linguistics.
SSNP-1 = Rensch, Decker, and Hallberg 1992.
SSNP-2 = Backstrom and Radloff 1992.
SSNP-3 = Rensch, Hallberg, and O'Leary 1992.
SSNP-4 = Hallberg 1992.
SSNP-5 = Decker 1992.
Taj, Abdul Khaliq
1989 *Shina qaida* [Shina Primer]. Rawalpindi: Privately published.

Tan, Eunice, M. Gulrang Lal, and Nazia Gul Mohammad
1999 *Baloochii qaaedaa* [Baluchi Primer]. Karachi: Azat Jamaldini Academy.
Trail, Ronald L., and Gregory R. Cooper
1999 *Kalasha Dictionary – with English and Urdu.* Islamabad: National Institute of Pakistan Studies and Summer Institute of Linguistics.
UNESCO
2003 *Education in a Multilingual World.* Paris: UNESCO
Various authors
1991 *Awhii sindhii paRhee dhaaTkii paRhoo!* [You read Sindhi, read Dhatki!]. Hyderabad: New Foundations.
Zaman, Muhammad
2002a Report on language survey trip to the Bishigram Valley. <http://www.geocities.com/kcs_kalam/ushoji.html>.
2002b *Gawri, urdu, angrezi bol chal* [Gawri, Urdu, English Conversation]. Kalam: Kalam Cultural Society.
2002c *Aao gawri parhen* [Come Let's Read Gawri]. Kalam: Kalam Cultural Society.
2004a The Badeshi people of Bishigram and Tirat Valley, Madyan, Swat, Surveyed by Shamshi Khan and Muhammad Zaman. Unpublished report.
2004b Interview with Muhammad Zaman by Tariq Rahman, 28 January 2004.
Zia, Mohammad Amin
1986 *Shina qaida aur grammar* [Shina Primer and Grammar]. Gilgit: Zia Publications.

Lesser-known language communities of South Asia: Linguistic and sociolinguistic case studies

Vanishing voices: A typological sketch of Great Andamanese

Anvita Abbi

1. Introduction

The Andaman Islands consist of a long chain of approximately 250 islands situated in the southeastern region of the Indian sub-continent in the Bay of Bengal. The chain of islands runs north to south, and is spread over an area of 6430 sq. km. The Ten Degree channel in the south separates these islands from the Nicobar Islands. The capital city of the Andaman Islands is Port Blair, which is situated in the southern most part of the Islands, at a distance of 1255 km. from Kolkata.

Various linguistic and genetic studies in the past, including Basu (1952), Burenhult (1996), Portman ([1898] 1992), suggest that the Andamanese languages might be the last remnants of pre-Neolithic Southeast Asia. They possibly represent modern humans' initial settlement (Hagelberg et al. 2003). In their article in *Current Biology*, Hagelberg and her team compared various genetic markers seen in present day members of the Onge, Jarawa and the Great Andamanese tribes. Their conclusion was that Andamanese have closer affinities to Asian than to African populations and that they are descendants of the early Palaeolithic (old-stone age) colonizers of Southeast Asia.

Genetic and epigenetic data (Endicott et al. 2003) suggest a long-term isolation of the Andamanese for a substantial period of time, an extensive population substructure, and/or two temporally distinct settlements. Geographical isolation, scientists believe, probably aided in the survival of ancient human lineages in the Andamanese. Some recent studies by geneticists indicate that the Andamanese are possibly related to the Negritos of the Malay peninsula and in the Philippines despite the differences in blood type frequencies (the Andaman Association 1995–2002). If modern linguistics can shed any light on early human prehistory, Andamanese languages warrant an in-depth study before they disappear.

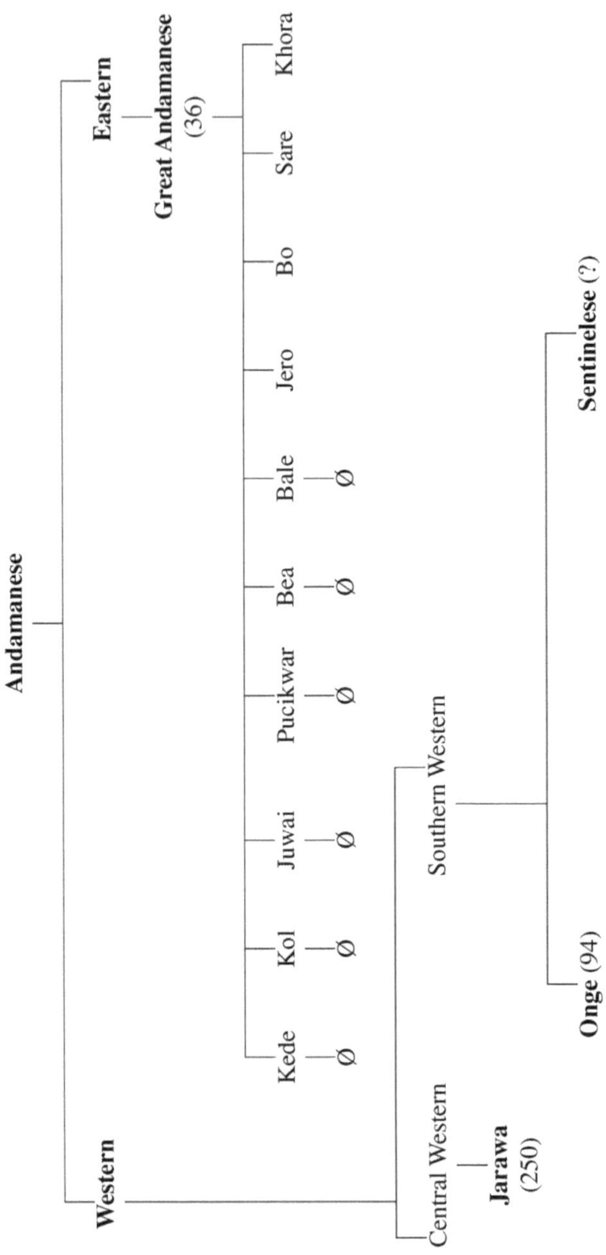

Figure 1. Present state of the Andamanese languages, number of speakers in parentheses

1.1. The present state

Living Andamanese tribes can be grouped into four major groups, i.e. the Great Andamanese, the Jarawa, the Onge and the Sentinelese. Barring the Sentinelese, the other tribes have come into contact with the mainlanders. The present paper focuses only on the linguistic structure of the Great Andamanese language. The demographic scale of these islanders is inversely related to the amount of contact with mainlanders: the broader the contact, the smaller the population. The map of the territory occupied by the Great Andamanese in the nineteenth century as opposed to the present map issued by the Andaman Association charts an inevitable journey towards the extinction.

The estimated population of 5000 to 8000 Great Andamanese prior to the establishment of the Penal settlement in Port Blair in 1858 was reported as being reduced to 625 in 1901 (Annamalai and Gnanasundaram 2001; Awaradi 1990; Weber 1998). Since then, the population has declined drastically, to only 36 (19 males and 17 females).[1] Table 1 illustrates the population decline of the Great Andamanese over the last seventy years.

Table 1. Population of the Great Andamanese over the last seventy years

1901	*1911*	*1921*	*1931*	*1951*	*1961*	*1971*	*1981*	*1991*	*1998*	*2002*
625	453	209	90	23	19	24	28	33	40	36

Source: The Billboard of the office of the Anthropological Survey of India, Port Blair.
Note: Census figures for 1941 are not available.

"Great Andamanese" is a broad term that has been used to refer to ten disparate groups of the tribe. These groups once inhabited the entire region of the Andaman Islands, but have now settled on Strait Island. Our recent fieldwork[2] could only verify four of these ten tribes, as seen in Figure 1, where the number of existing tribes is given in parentheses and the number of extinct tribes is indicated by the symbol Ø.

Table 2 categorically charts the vanishing sub-tribes/ethnic groups within the Great Andamanese family.

Major factors contributing to the diminishing population of the Great Andamanese include environmental "disturbances", contagious diseases as a result of contact with city dwellers, and a high mortality rate, assisted by addictions to alchohol, tobacco and opium.

The tribes are for the most part hunter-gatherers, and in the case of Great Andamanese and the Onge, food from the city is distributed on a regular monthly basis. Despite the distribution of food by government officials, males prefer to hunt in the sea and in the forest, and females prefer to gather roots and vegetables from the jungles. We noticed that the love for hunting was so

Table 2. Decline in the ethnic groups over the last hundred years

Ethnic groups	Adult	Child	Total 1901	Total 1975	Total 2002
Sare	31	8	39	4	
Khora	63	33	96	1	
Bo	31	17	48	6	
Jeru	178	40	218	11	
Kede	54	5	59	0	
Kol	8	3	11		
Juwoi	40	8	48	0	
Pucikwar	45	5	50	0	
Bale	15	4	19	0	
Bea	30	7	37	1	
Total	495	130	625	23	36

Source: Figures for the period 1901 are from the census taken in India. Figures for the period 1975 are from Annamalai and Gnanasundaram (2001). The figure for 2002 is based on the current fieldwork. Information on the distribution of ethnic groups was not available.

great that the Great Andamanese would describe their hunt for turtles even if the activity was undertaken several months ago.

The hierarchical society seen in India is not present here. However, the leader of the tribal group is selected by government officials, and serves as a representative of the community during official communications. Generally a person with a functional knowledge of Hindi tends to be nominated as this representative leader.

In 1968 the Andaman Government, acting on the recommendation of the anthropologist T.N. Pandit, resettled the surviving tribes on Strait Island, located about 68 nautical miles from Port Blair. As of today, these tribes frequently visit Port Blair to receive monthly allowances, medical care and other necessary aid from the government. Out of the remaining 36 Great Andamanese, 3 women and 2 men are employed with various local governmental organizations. When the need arises, elderly people do not hesitate to be hospitalised or receive modern medical help, even though this involves frequent trips from the Strait Island to Port Blair or extended stays in the city hospital. Great Andamanese are totally acquainted with metropolitan ways of life, and aspire to adopt these ways. We met 2 Great Andamanese women working for the police force who were completely immersed in the metropolitan city culture, serving us tea and snacks from a kitchen fully equipped with modern utensils and appliances.

1.2. The linguistic configuration

The Great Andamanese, especially those who frequently visit Port Blair, have a functional knowledge of Hindi and Bangla (the Indo-Aryan languages). Some of them also understand a few words of spoken English. They speak the contact language used in the Islands, i.e., Andamani Hindi, a variety of Hindi that is similar to a pidgin, lacking all agreement features. As of today, the Great Andamanese speak a mixed version of two or three of the original ten language varieties within the same language family. It would not be an exaggeration to say that Great Andamanese is an amalgam of ten different but mutually intelligible languages once spoken on the mainland of the Andaman Islands. No two speakers speak the same language, as each speaker is descended from mixed marriages between different tribes within the same language family. However, mutual intelligibility between these various language varieties assists basic communication. Our main speaker, Lico, a female in her early forties, spoke a mixed language of Khora (her grandparents' language), Sare (the language of her adopted mother) and Jero (her father's language). Of the 36 living Great Andamaneses at least those who visit Port Blair frequently use Hindi to speak to their children. Some of the children we interviewed could not create simple sentences such as 'I am hungry' in Great Andamanese. Middle-aged and older people, however, use the language in some domains.

The lack of young boys and girls eligible for marriage is a serious problem, and may eventually force the members of the society to marry outside the tribe. We were acquainted with one such member of the family who had recently married a non-tribal. In such an environment we should not be surprised if the language under consideration eventually borrows features from non-tribal languages spoken in the Islands, and thus becomes more complex than what it is at present.

1.3. Perils of urbanization

Though some tribals who are employed by the Indian government earn wages, they still have a general feeling that the life in the jungle is far better than that of the city, as there are no restrictions of time and place in the former. In interviews with the tribals, they would say that they felt like free birds in the jungle. They marvelled at us being able to sit in the same chair in one place for a long period of time in an office doing "nothing". This perhaps was the reason why government officials constantly complained about the Great Andamanese "escaping from their duties" and running away to the jungle as and when it suited them. One of the male tribals I interviewed directly indicated to me that the tribals were against the literacy and education program run by the government. He saw that the program as making the tribes subservient to the locals. He asked me "What will you give me after I get educated? Most likely I will be serving as a peon in an office and the boss will expect me to get vegetables from the market for his family. I'd rather be naked in the jungle and roam free and be my own master". The more I saw of the semi-educated and semi-literate Great Andamanese the more sorry I felt for them. All attempts to bring the tribe into the mainstream have created havoc in their lives and have disturbed the social, economic and ecological balance they held three hundred years ago.

2. A grammatical sketch

In the following, I sketch the major typological features of Great Andamanese based on our pilot survey. The results cannot be considered conclusive due to the sparseness of data. The language is characterized by variation at the phonological and lexical levels as a result of the linguistic makeup discussed earlier.

2.1. The sound system

Though the languages of the Andamanese family have been removed from the Indian areal pressures, we find, surprisingly, an abundance of aspirated and retroflex consonants, the characteristic phonological features of the Indian languages spoken on the mainland. Great Andamanese offers four way phonemic contrasts in nasals, while aspiration contrast is limited to voiceless sounds. Thus, /p/ and /p^h/, /ʈ/ and /$ʈ^h$ /, /k/ and /k^h/ contrasts exist. The palatal /č/ has no aspirated counterpart. A striking feature is the absence of glottal fricative [h] and velar plosive [g]. The former is now incorporated as a borrowed sound from Hindi, especially in the use of the auxiliary [hɛ] 'to be'. Plosives are unreleased in the word-final position. The discovery of the occurrence of voiced and voiceless bilabial fricatives [β] and [ɸ] was amazing, as these sounds are not known to exist in any other language of the Andaman nor in any other Indian language. The following sound pairs are in free variation at the level of intra-community, i.e., within the same clan.

[ɸ ~ p^h ~ f]
[β ~ l ~ w]
[k^h ~ x]

The intra-community variation, thus, renders a large number of sound inventories as shown in Table 3.

Table 3. Consonant sounds of Great Andamanese

	Bilabial	Labio-dental	Dental	Alveolar	Retroflex	Palatal	Velar
Plosive	p b		t d		ʈ ɖ	č ǰ	k
	p^h				ʈ^h		k^h
Nasal	m			n		ɲ	ŋ
Trill				r	ɽ		
Fricative	ɸ β	(f)		s		ʃ	(x)
Lateral				l			
Approximant	w					y	

The Great Andamanese is an eight-vowel system, as can be seen in Table 4, and offers very large combination possibilities in the area of diphthongs, as represented in Table 5.

Table 4. Vowels sounds of Great Andamanese

	Front	Central	Back
High	i		u
Higher Mid	e		o
Mean Mid		ə	
Lower Mid	ɛ		ɔ
Low			a

Table 5. Diphthongs in Great Andamanese

Front	Central	Back
ia, iu, ie, i:o, i:e, io,		ua, uo, uə,
ei, eo,	əu,	oa, oi, oe, o: ɔ, o:a
ɛo,		ɔi,
		ao, a:e, ai

Length is phonemic, /bočo/ 'peel' but /o:čɔ/ 'net', however, long and short /u/ varies freely before a final vowel in a diphthong situation. Thus, speakers varied between two renderings for 'my ear' /ʈ^{h}ɛr-bu:o/ and /ʈ^{h}ɛr-buo/. Back vowels [u] and [o] as well as front vowels [e] and [ɛ] varied freely in the word final position in inter-community situation, i.e., across the members of different clans.

2.2. The lexicon

Inalienable possessions, such as names for various body parts and kinship terms, can be classified into different classes according to the phonetic shape of the *personal prefix* just preceding the root. The personal prefix is constituted of two parts: the pronominal clitic indicating the possessor and the body part classifying prefix (which serves as a host to the clitic). Human body parts in Andamanese in general and the Great Andamanese in specific are classified in several classes according to the division of the body made by the native speaker. The schema of the personal prefix can be presented as:

ʈʰ + V + (C) = Noun

The initial sound of the personal prefix is a pronominal clitic meaning 'self', followed by various body part-classifying genitive affixes, each represented by a distinct prefix. These prefixes are obligatory and align to the pronominal prefix. As indicated earlier, they suggest the ways the human body is classified. We could identify four distinct types of these genitive prefixes though there is a variation within a speaker. Among the speakers of different clans we could identify seven distinct types of the genitive prefix. The genitive prefix varies in its thematic vowel and an optional consonant, according to the nature of the area/part of the body that is referred to [possessed]. Thus /ʈʰɛr-/ refers to head and individual parts of the face, such as mouth, eye, lips, neck, tooth etc. /ʈʰut- ~ ʈʰot/ refers to the hairy part of the body, and /ʈʰum ~ ʈʰom ~ ʈʰoŋ ~ ʈʰun/ refers to the extremities of the body such as finger, nails, wrist, ankle, foot, toes etc. and /ʈʰa-/ refers to tongue and kinship terminology. Consistency is lacking, as the language contains several varieties of past and present dialects. In general, a speaker has two or more forms in the verbal repertoire that s/he can vary freely in all contexts (see Table 6).

For kinship terms, there are only three types of personal prefixes, again constituted of a pronominal clitic and a monosyllabic open genitive affix, e. g., /ʈʰa-/, /ʈʰu-/, and /ʈʰɛ-/, with its thematic vowel varying according to the hierarchy of generation regarding the referent and the ego. Thus, /ʈʰa-/ refers to one generation higher than the ego or to the same generation', e.g. /ʈʰa-mimi toc-tue/ 'mother's brother' and /ʈʰa-ra sulu thui ka:ʈa/ 'younger sister'; /ʈʰu-/ refers to one generation lower than the ego', e.g. /ʈʰu ʈhirɛ/ 'daughter' and /ʈʰe-/ refers to an affinal relationship, e.g. /ʈʰɛ-boi/ 'husband/wife', 'spouse'.

Physical ailments are treated as inalienable possessions, e.g. /ʈʰɛr-βuc'/ 'my cold', /ʈʰɛr-cot'/ 'my cough'.[3]

Table 6. Personal prefixes with noun roots for kinship terms and body parts terms

Personal prefix	Kinship	Body parts
/ʈʰ-ɛr-/ 1st SG		/ʈʰɛr-co/ 'my head' /ʈʰɛr-be:ŋ/ 'my forehead' /ʈʰɛr-kɔtʰo/ 'my nose' /tʰɛr-ɸoŋ/ 'my mouth' /tʰɛr-ulu/ 'my eye'
/neli-ɛr-/ 2nd SG HON		/neli-ɛr-ulu/ [nelirulu] 'your (HON) eye'
/ʈʰ-oŋ-/ 1st SG		/ʈʰoŋ-kara/ 'my nails'
/ʈʰ-un-/ 1st SG		/ʈʰ-un-ʈo:/ 'my wrist'
/ʈʰ-um-/ 1st SG		/tʰum-rono/ 'my heel' /ʈʰum-ɔʈo/ 'my foot'
/ʈʰ-ot'-/ 1st SG		/ʈʰot'-bo/ 'my back'
/ʈʰ-ut'-/ 1st SG		/ʈʰut'-bec'/ 'my hair'
/ʈʰ-a-/ 1st SG	/ʈʰa-mai/ 'my father' /ʈʰa-mimi/ 'my mother' /ʈʰa-mimi-toc'-tue/ 'my mother's elder sister/brother	/ʈʰa-tat'/ 'my tongue'
/ak-a-/ 3rd SG	/aka-mimi-tara-tob'/ 'his grand-mother' /aka-mai-tara-tob'/ 'his grand-father' /aka-mai/ 'his father'	
/ʈʰ-u-/ 1st SG	/ʈʰu-ʈʰirɛ/ 'my daughter', 'my son'	
/ʈʰ-e-/ 1st SG	/ʈʰe-boi/ 'my husband', 'my wife' /ʈʰe-boi-toc'-tʰu/ 'my husband's brother'	

Alienable possessions are case marked to the possessor nouns/pronouns.

(1) *Peje* *ni-sɔ-imu* *enɔl* *hɛ*
Peje 2nd-GEN-cap beautiful AUX [<Hindi]
'Peje, your cap is beautiful'

2.3. Pronouns and deixis

The language is immensely rich in showing various deictic relations. First person plural has the distinction of exclusive/inclusive, as in /miyo/ and /mer-enla/ respectively, i.e., addressee component may or may not be added in the pronoun.

A four-way distinction in third person pronouns was observed. Thus, proximate, distant [visible], remote [invisible] and a reference to someone/something in an intermediate position between proximate and very proximate in pronouns such as /k^{h}udi/ vs. /k^{h}idi/ in Table 6 are maintained. Further, proximate place deixis has two-way distinction 'close' and 'very close' and distant place deixis has three way distinction, signifying a generic distance and distance measured against the vertical and horizontal axis. See Table 7. There are some gaps in the table necessitating more intensive research on the language.

The animate/non-animate distinction is not maintained in interrogative pronouns, e.g. /acaʃyu/ 'which', 'who', unlike most of the languages of India.

Second person and third person plural forms can be used as honorific. Except in the language of one informant (Nao II) as exemplified below, we did not find any reflexive pronoun.

/ʈhiu-ʈhɔt'/	/ɲiyu-ɔʈ'/	/nili-ut'/
'I-myself'	'you-yourself'	'you all-yourself'

Table 7. Pronominals and place deixis in Great Andamanese

Pronouns	Singular	Plural
Person		
1st person [exclusive]	ʈ^{h}u	miyo
1st person [inclusive]	ʈ^{h}ənɛ	merenla
2nd person	ni	niliya
3rd person [distant, visible]	ɖu	ɲɔt
3rd person [distant, invisible]	u	niyo
3rd person [proximate, intermediate]	k^{h}udi	—
3rd person [proximate, very close]	k^{h}idi	—
Place		
Proximate 'here'	k^{h}urɔ	—
Proximate 'here, very close'	ʈ^{h}idi	—
Distant 'there'	k^{h}uʈol	—
Distant in height	tuluk^{h}ui	—
Distant in length	tara-ʈ^{h}uro	—

2.4. Case

Four distinct case markings /-bi ~ -i ~ -bɛt / are for accusative and dative, /-ʈa / for instrumental and ablative, /tɔʈ/ for locative and /-sɔ/ for genitive are suffixed to the appropriate nouns. Nominative and agentive are unmarked.

2.5. Verbs

Verbs are classified in different classes based on the nature of the initial consonant[3] of the verbal suffix. We only found six such suffixes, each identified by a specific but different consonant, with a vowel attached to the verb root. Let us term this consonant a "thematic consonant" because this determines the verb class. Verb stem, thus, could have any of the following CV endings, where a verb may belong to a *b* class, or a *l* class or to a *k* class, etc.:

–bV or –lV or –kV or –rV or -p^{h}V or, -mV

The vowel of the verb stem indicates aspect or mood. The verb schema could be presented as below. No other Indian language shows even a slight resemblance to these verb structures.

Verb root + Cons Class	+ Vowel	+ Consonant/zero
	[Aspect/Mood]	[Tense]

The vowel of the verbal suffix may represent the following aspectual and mood categories.

/-o ~ -ɔ/	indicative
/-e ~ -ɛ/	imperative
/-i/	perfective

The presence or absence of the final consonants of the verb schema represents tense. Great Andamanese maintains a two-way distinction in tense, past and non-past. Past tense is unmarked.

/-m/	non past [present or future]
/-Ø/	past

(2)

k^hudi-o	*ceo-ʈa*	*cokbi-bi*	*ekp^{h}uti-k-ɔ-m*
3rd SG	knife-INSTR	turtle-ACC	cut-CL-IND-NON.PAST

'He cuts a turtle with a knife'

(3)

k^hudi-o	*ceo-ʈa*	*cokbi-bi*	*ekp^{h}uti-k-ɔ-Ø*
3rd SG	knife-INSTR	turtle-ACC	cut-CL-IND-PAST

'He cut a turtle with a knife.'

(4) *ɛk^{h}u-l-ɛ*
lift-CL-IMP
'Lift up in lap'

(5)

i:no-bi	*k^hu-k-ɛ*
water-ACC	drink-CL-IMP

'You drink water.'

In addition, the suffix /-bɛ ~ -bi ~ -be/ occurs in existential stative verb sentences as well as with predicative adjectives.

(6) *t̪ʰɔ* *t̪ɔp-be*
1st SG bathe-BE
'I bathe.'

(7) *o* *cakʰəm-be*
3rd SG old-BE
' That person is old.'

(8) *myɔ-tot'* *tara* *ra:tʰomo-be*
rock-LOC leg/behind pig-flesh-BE
'There is pig's flesh on the rock.'
('(Meat of) a leg of a pig is on the rock')

(9) *peǰe* *ɲyo-be*
Peǰe home-BE
'Peje is at home.'

2.6. Word order

Great Andamanese is a head final language. The order of the constituents is: S + DO + IO + Verb. The modifier follows the modified (see examples 10, 11, 12) and the intensifier follows the modifier (see example 15) but the genitive precedes the governing noun both in alienable and inalienable possessions (for example, 11, 13 and 14).

(10) *kʰudie* *čaythomo-bi* *ǰi-k-o-Ø* *araliso*
3rd SG meat-ACC eat-CL-IND-PAST all
'He ate all the meat.'

(11) *nu* *t̪ʰirɛ* *ərkʰole* *tara* *ka-m-o-Ø*
your child playful lot play-CL-IND-PAST
'Your child was very playful', or
'Your playful child played a lot.'

(12) *kʰidɛr* *fun* *bot̪o* *kiya* [Hindi kiya 'did']
coconut ripe fall PAST
'The ripe coconut fell down'

(13) *aka* *mimi* *vɛsɛrɛ* *b-o-m*
his mother hits CL-IND-NON.PAST
'(child's) mother hits'

(14) *aka* *mimi* *ata-ʈo:p* *b-o-m*
his mother TR-bath CL-IND-NON.PAST
'His mother bathes (the child).'

(15) *ʈ*^{h}*ire-ka* *leko* *elevka* *əka:lɛ-k-ϕ*
child-3 SG GEN small superlative die-CL-INDI PAST
'(his) smallest / youngest child died'.

The negative marker is /-p^{h}o ~ -fu/ and occurs post verbally before the markers for Tense, Aspect and Mood. In negative existential sentences it may occur as a free morpheme.

(16) *k*^{h}*riŋkoso* *uč-čone-fu-b-i-l-o-Ø*
Strait Island why-go-NEG-CL-PERF.EPENTH-IND-PAST
'Why had he not gone to the Strait Island?'

(17) *ino-bi* *k*^{h}*u-fu-b-e*
water-ACC drink-NEG-CL-IMP
'Do not drink water!'

(18) *k*^{h}*idi* *enɔl-ʈɔʈceo* *p*^{h}*o*
3rd SG good-person NEG
'This person [proximate] is not good.' or 'This is not a good person.'

The typological characteristic features of Great Andamanese are summarized in Table 8.

Considering the nature of the language, especially in the realm of the verb morphology, Great Andamanese appears to be very different from other languages of the Andaman tribes, i.e., Jarawa and Onge (Abbi 2003, forthcoming). Great Andamanese also differs from these languages in the absence of any cognates in the basic vocabulary, kinship terms or body parts terms. While Onge and Jarawa provide enough pieces of evidence to establish a genealogical relation between them, Great Andamanese appears to be a family apart. If Andamanese is considered a fifth language family (Manoharan 1986) then perhaps Great Andamanese is either a distant relation of this fam-

Table 8. Typological features of Great Andamanese

Index of grammatical features	Types of grammatical features available
Unmarked clause order	SOV
Adjective	Head-modifier
Possession system	Dependent marking
– Alienable	Possessor-GEN-possessed
– Inalienable	Personal prefixes
Case markings	Postpositions
Pronoun system	Inclusive vs. exclusive
Deictic system	More than three-way distinction
Verb system	
– Distinct classes	Defined by the suffixed consonant
– Agreement	No agreement with S/P
– TAM	Shown as A/M T
– Clause linkage	Parataxis
– Tense distinction	Past/Non past
Reduplication	Absence
Compounds	Endocentric
Kinship and body parts by personal	Yes
prefix	Pronominal clitic + V (C)
Dependent marking	Yes
Serial verbs	Yes

ily or may even constitute the sixth language family of the Indian subcontinent. An intensive study of the lexicon and the structure of Great Andamanese is warranted before it is too late.

3. Conclusion

Our general impression after meeting the tribals was an unhappy one. They are the lost tribe who do not know what is good for them and those who do know, such as the old people, do not know how to go back to the balanced and equilibrium stage they once enjoyed. They have reached a no return point and suffer from various diseases contaminated by the city dwellers. According to Awaradi (1990: 258) the immediate problem is of marriage and procreation. The marriage alliances with non-autochthons lead to exploitation and loss of identity. The self-reliant economy that they enjoyed once has turned into complete dependence on the local government. As they are few in number and largely alienated from their native culture, there is very little hope that they will either retain whatever is left of the language and culture or increase in the number of population. Their way of living is in a flux and it will be sometime till they reach any comfortable equilibrium stage. We should not forget that they are only thirty-six in number and no matter how fast they run to catch this stage they are bound to miss it.

Appendix

Table 9. Tribal settlement areas

Name of the Island	Tribal Settlement	Area
Strait Island	Great Andamanese	3.11 sq km

Source: Anthropological Survey of India information board, Port Blair, 2001.

Table 10. Population sketch based on the pilot survey[1]

Great Andamanese (2001–2002)

Total Population	=	**36**
Male	=	**19**
Female	=	**17**

Family-wise Description

1.	Jirake	M	65
	Surmai	F	(second wife)
Their children			
	Meo	M	
	Ilφe	M	
	Renge	F	21
	Buro	F	
	Tango	F	
	Reya	F	18
	Nyaramo	M	
	Kanmo	M	
	Buluba	M	
	Dec	M	
	Lico	F	33 (married to Golat)
2.	Boa	F	(lives alone in a separate hut)
3.	Boro	F	74 (widow of late Ilφe)
Her children			
	Golat	M	(his first wife died and he then married Lico)
	Loka	F	(married to non tribal, see 9 below)

	Sulu	M	
	Boa	F	29 (married to Nao Jr.)
4.	Nao Junior	M	55 (first wife died, now married to Boa)
	Bea (son)	M	
5.	Peje	M	60
	Neo	F	
Their children			
	Jo	M	25
	Tong	F	22
	Kaba	F	
	Irep	M	
	Phoro	M	
	Lephai	M	
6.	Children of (Lico + Golat)		
	Kobo	F	
	Moroko	M	
	Buli	M	
	Lephe	F	
7.	Children of (Tong + Ilɸe)		
	Bui	M	
	---	F	(exact name is not known)
8.	Tango's children		
	Beno	F	
	Belei	F	
9.	Loka married a non-tribal girl Ms. Prema and has two children		
	Gulab	F	08
	Sharda	F	05

Notes

1. *Addendum*: In early 2006, as this volume was going through the final stages of editing, I made a field trip to the Andaman Islands, where I learned that the Great Andamanese population had increased to 51 individuals since 2002 when we completed the pilot survey (see note 2).
2. The present essay is based on first-hand data from a pilot survey of the languages of the Andaman Islands personally conducted by two of my students, Shailendra Mohan and Pramod Kumar, and myself, during 2001 and 2002. We are thankful to the Director of the Tribal Welfare Department (*Adim Adivasi Janjati Vikas*

Samiti AAJVS in short) of the Andaman and Nicobar Islands, Dr. S.A. Awaradi, for providing invaluable help and for making it possible to collect data on the language. The pilot survey was supported by the Max Planck Institute, Leipzig in Germany.

3. The abbreviations used in this article are as follows: ACC = accusative; AUX = auxiliary; CL = class; EPENTH = epenthetic; GEN = genitive; HON = honorific; IMP = imperative; IND = indicative; INSTR = instrumental; LOC = locative; NEG = negative; PERF = perfective; PRS = present; SG = singular.
4. Manoharan (1989) reports seventeen distinct classes.

References

Abbi, Anvita

2003 Vanishing voices of the languages of the Andaman Islands. Paper presented at the Max Planck Institute, Leipzig.

forthc. *Endangered Languages of the Andaman Islands.* Munich: Lincom Europa.

The Andaman Association. Switzerland. <http://www.andaman.org>.

Annamalai, E., and V. Gnanasundaram

2001 Andamanese: Biological challenge for language reversal. In *Can Threatened Languages be Saved? Reversing Language Shift, Revisited. A 21st Century Perspective*, Joshua Fishman (ed.), 309–322. Clevedon: Multilingual Matters.

Awaradi, S. A.

1990 *Computerized Master Plan 1991–2021.* Port Blair: Andaman and Nicobar Administration.

Basu, D. N.

1952 A linguistic introduction to the Andamanese. *Bulletin of the Department of Anthropology*, vol. 1–2: 55–70.

Burenhult, Niclas

1996 Deep linguistic prehistory with particular reference to Andamanese. *Working Papers in Linguistics,* 45: 5–24. Lund University, Department of Linguistics.

Hagelberg, Erika, Lalji Singh, K. Thangaraj, A. G. Reddy, V. R. Rao, S. C. Sehgal, P. A. Underhill, M. Pierson, and I. G. Frame

2003 Genetic affinities of the Andaman Islanders. A vanishing human population. *Current Biology* 13: 86–93.

Manoharan, S.

1986 Linguistic peculiarities of Andamanese family of languages. *Indian Linguistics* 47 (1–4): 25–32.

1989 *A Descriptive and Comparative Study of the Andamanese Language.* Calcutta. Anthropological Survey of India.

Portman. M. V.

1898 *Manual of the Andamanese languages.* Repr. Delhi: Manas Publications, 1992.

Weber, George

1998 *The Andamanese Language Family. A Contribution to the Centenary of M. V. Portman's Work.* <http://www.andaman.org>.

Lisu orthographies and email[1]

David Bradley

1. Lisu orthographies

Lisu is a language with nearly a million speakers spanning four countries: China, Burma, Thailand, and India. For further information, see Bradley (1999, 2003). Lisu has various competing orthographies. Four of these represent very similar western dialects; the most widespread one was devised by Protestant missionaries between 1914 and 1916 and is usually called the Fraser script after James Outram Fraser, the leading figure in its creation and spread.[2] Others include one devised by a Lisu traditional priest, Wa Renbo, in about 1925 for his own northeastern variety of western Lisu; one devised by linguists in China from 1956 to 1958 for northwestern Lisu, called New Lisu in China, and finally, one devised by a Protestant missionary in Thailand for the southern subvariety of western Lisu. The earliest orthography for a closely-related language was devised by China Inland Mission members A. G. Nicholls and G. E. Metcalf in central Yunnan in about 1910, for what they called Eastern Lisu but which is now called Lipo. It represents a distinct language which is partly intelligible, with difficulty, to Western Lisu speakers.

The Eastern Lisu or Lipo script is one of a number of so-called Pollard scripts, the first of which was devised by the British Methodist missionary Samuel Pollard starting in October 1904 for a variety of Miao in Guizhou Province. It uses large symbols for initial consonants and smaller symbols for rhymes. The relative position of the rhyme symbol and the initial consonant symbol indicates the tone: above for high tone, top right for mid tone, bottom right for low tone and so on. For a detailed discussion of the Pollard scripts as used for Miao languages, see Enwall (1994). A Pollard script is still in use among the Christian Lipo. A simplified version of a Pollard script with the rhyme always at the bottom right of the initial consonant and the tone indicated by an additional following symbol has been used for some Christian and other publications in Miao in areas near the Lipo in Yunnan Province. There is a computer font, however, for Lipo, which was created in the late 1980s by American missionaries from the Morse family, Robert and David, who work mainly with the Western Lisu. Various Biblical texts were printed in this orthography from 1912 to 1951 and revised versions in corrected Lipo are under preparation.[3]

The Wa Renbo script (also sometimes called the Wang Renpo script from another Chinese form of the inventor's name) is a 1030-item syllabary. Not all canonical syllables have a distinct symbol, so sometimes a near-homophonous syllable with a different tone or a similar initial or rhyme is used instead. Very few people ever used this script. Those who did were all traditional (non-Christian) religious practitioners in Weixi County in northwestern Yunnan. It was mostly written on paper or on ritual funeral boards, with some texts printed from woodblocks. The sole publication is Wa (1999), printed in an edition of only 500 copies, but there is a great deal of manuscript material available.

The New Lisu script was intended to follow the principles of the *pinyin* romanisation of Chinese. It went through at least two draft stages, the first with some Cyrillic letters in 1957 and the second with Chinese loanwords represented exactly as in *pinyin* in early 1958. The final implemented version, later on in 1958, used loanwords in their integrated Lisu form. This version uses postscript consonants to represent the tones of Lisu. For a detailed discussion and comparison of Fraser Lisu and New Lisu, see Bradley and Kane 1981. Although New Lisu was vigourously promoted by the government, and was in sole use for publishing from 1957 to 1979, with thousands of copies of books, it was never well accepted by most Lisu. Textbooks and literature started to appear again in China in Fraser Lisu in 1980. In 1984 the Nujiang Lisu Autonomous Prefecture government officially decided to use only Old Lisu, that is, the Fraser script. After this, the number of books printed in New Lisu gradually decreased; the last textbook was published in 1990. From the mid-1980s, the print runs of the Fraser Lisu textbooks were already much larger than those for New Lisu. All recent textbooks in China are only in Fraser Lisu. However, some publications, notably modern literature written by non-Christian Lisu and some translations of government material from Chinese, still do appear in New Lisu.

The Thai-based script was prepared by the Overseas Mission Fellowship missionary linguist Edward R. Hope, and was intended to comply with Thai government policy that orthographies for minority groups in Thailand should be based on Thai; see Hope (1976). The Thai government has never enforced its policy. This orthography was not accepted by the Lisu in Thailand, and has never been used.

2. The Fraser Lisu orthography and book Lisu

It is most interesting that the variety of Lisu used in Biblical texts, and most widely written in other contexts, represents no one dialect. Everyone has to compromise. Written Fraser Lisu is dramatically different from the southern variety as spoken in Thailand and parts of the Shan State in Burma, and also quite distinct from the northern variety as spoken in northwestern Yunnan, northern Burma and India. It is closest to the northeastern subvariety of the central variety, as spoken where Fraser initially worked in Tengchong County,[4] but contains some phonological and lexical elements of other central subvarieties and the northern variety. It also contains a reduced inventory of grammatical markers. This was almost certainly a conscious choice by the missionaries, so that they could reach as many Lisu as possible. This kind of Einbau rather than Ausbau[5] is unusual, but not without more recent parallels – for example, the case of Romansch in Switzerland or Reformed Yunnan Yi.

At the time of the first contact between central variety evangelists and northern variety speakers in the early 1930s, considerable difficulty was reported in communication. Again, when the first central and northern variety speakers arrived in Thailand in the early 1970s, they had extreme difficulty with the southern variety, and southern speakers could not understand them. After a while, speakers learned to recognize and even accommodate other varieties. However, the written language provides a neutral alternative, called ꓕꓳ: ꓶ: ꓥꓳ: /t^{h}o^{21} ɣɯ^{21} ŋo^{21}/ 'book language', which is used in formal contexts such as church services. Thus the missionaries have created a new alternative dialect. Many literate Lisu happily continue to speak in their own variety most of the time, but write and occasionally speak in the book language.

The structure of the Fraser Lisu script is outlined in detail in Bradley (1979: 57–60). Its most distinctive characteristic is that all letters are upper case, including 15 upper case letters used in inverted form. Another of its unusual features is that an initial consonant with no vowel written after it has an inherent vowel /ɑ/. This is the case in most Indic-derived scripts but is rather unusual for a romanisation. Tones are represented by punctuation marks after each syllable. Presumably the idea for tonal representation came from Ba Thaw, a Karen whose own language, like Burmese, uses some postscript marks that look like alphabetic punctuation to mark tones.[6] Table 1 shows the Fraser Lisu alphabet in its normal order, with phonetic values.

The orthographic representations of the six tones of Fraser Lisu and other types of Lisu are given in Table 2 below. The exact phonetic values of various letters differ allophonically as well as dialectally; see Bradley et al. (2006).

Table 1. Fraser Lisu Alphabet

ꓐ	ꓑ	ꓒ	ꓓ	ꓔ	ꓕ	ꓖ	ꓗ	ꓘ	ꓙ	ꓚ	ꓛ	ꓜ	ꓝ	ꓞ	ꓟ	ꓠ	ꓡ	ꓢ	ꓣ	ꓤ
b	p	pʰ	d	t	tʰ	g	k	kʰ	dʐ	tʂ	tʂʰ	dz	ts	tsʰ	m	n	l	s	ʐ	z
ꓥ	ꓦ	ꓧ	ꓨ	ꓩ	ꓪ	ꓫ	ꓬ	ꓮ	ꓯ	ꓰ	ꓱ	ꓲ	ꓳ	ꓴ	ꓵ	ꓶ	ꓷ	ꓭ		
ŋ	ɦ	x	h	f	w	ʂ	j	ɑ	æ	e	ø	i	o	u	y	ɯ	ɤ	ɣ		

There are also various graphic sequences with ꓬ as second letter representing single segments: ꓠꓬ for /ɲ/ and ꓙꓬ ꓚꓬ ꓛꓬ ꓫꓬ for /dʑ tɕ tɕʰ ɕ/.

In the first draft of the Fraser script, prepared on Christmas 1914 in Myitkyina, Burma by the Scotsman Fraser, the Karen Baptist evangelist Ba Thaw and the American Baptist Mission member J. G. Geis, there were no inverted letters. Rather, a number of diacritics and digraphs were used. The only publication in this draft script is Ba Thaw (1915). The following year, Fraser and Ba Thaw met again in Tengchong, China and came to the decision to use inverted letters instead of digraphs and diacritics. The first textbook, which combined Christian teaching and literacy, was eventually published in Rangoon, anonymous (1922). This book remains in print, with many revised editions in Burma and Thailand over the years.

Printing in Fraser Lisu using movable type was done by inserting the necessary letters upside down, which sometimes led to alignment problems. The first Biblical text translation was the Gospel of Mark, printed in 1921 by the British and Foreign Bible Society in Shanghai. The full New Testament appeared in 1938 and the entire Bible in 1968, with many subsequent revised editions of both. From the 1920s onward, a number of typewriters were converted by welding inverted upper case letters onto existing keys. Some of the typewriters are still in use.

This orthography caused problems and embarrassment for many printers. One entire large edition of the Bible was printed in Hong Kong in 1981 with the outside cover and spine having the gold embossing with inverted letters wrongly printed right side up (*WU-S SI XY L OT*). This was fixed by covering the front cover and spine with red plastic inserts having the correct Lisu (ꓪꓴ-ꓢ ꓢꓲ ꓫꓬ ꓕꓳ ꓶ or /wu^{33} sɑ^{44} sɿ^{21} ɕɑ^{33} tʰo^{21} ɣɯ^{21}/ 'God's holy book'). Note that all of the tones have been omitted, as is often the case on book covers and spines in Fraser Lisu. For further discussion, see below.

3. Fonts and the problem of email

There are three main fonts used for Fraser Lisu: one created by a Chinese software organization in Shandong, one created by David Morse in Thailand, and one created by Andy Thomson of the Overseas Mission Fellowship in Thailand. Some other individuals have also created their own fonts. The Chinese font is used by various publishing organizations in China, but is not commercially available. The Morse font, called LISU_FA, is widely used in Thailand and Burma, and is freely available. The Thomson font, called LEDZA, is also freely available and widely used in Thailand and Burma. The Morse font is probably the most widely used, but the LEDZA font is a close second.

The key mapping principles of these three fonts are quite different. The LEDZA font uses the Lisu alphabetical order, mapping the first letter, ʙ, onto the key for @, the second letter, ᴘ, onto the key for A, the third letter, ᴅ, onto the key for B and so on. This means that material in the LEDZA font can be converted to Lisu alphabetic order automatically using standard software, a quite useful feature for lexicography. The other two fonts use a rather different principle, with the inverted letter on the same key as the upright letter. This is mnemonically more convenient, but the exact mapping differs between the two. All three fonts are for use on PC.

There are PC macros and keymaps available from David Morse to convert LEDZA to his font and vice versa. There is also a Macintosh version of the LEDZA font and a Macintosh macro to convert LEDZA into phonetic transcription available from the author of this chapter. A request for a Unicode standard for Lisu is underway.

Clearly it is very difficult for Lisu computer users to exchange files and communicate electronically, given that the three main fonts use different keymapping and that most users have access to only one font. Also, electronic communication for most users is over slow modem connections on often unreliable lines, which make the transmission of email or attachments (such as fonts, large files with font specification or PDF files) extremely problematic.

4. The problem of tones

It is well-known that orthographies which use diacritics to represent tones often cause problems for users. When the diacritics are not readily available, they cause problems for users of old technology such as typewriters, typesetting, and so on. In the case of Fraser Lisu, the tone diacritics are existing punctuation marks following the relevant syllable, so there is no underlying problem here – the problem is with the inverted letters instead.

However, it is also the case that diacritics representing tones are very often omitted by users of such orthographies. In Lisu this is a matter of orthographic convention. For Fraser Lisu, it is generally regarded as childish or bad style to put in all the punctuation marks representing tones. Depending on the formality of the context, the degree of ambiguity, the relative frequency of the word, whether the word has recently appeared in the context and so on, tone marking is very often omitted. In Biblical text materials, where tones are more frequently marked than in less formal materials, the rate of tone marking varies between about 20% and 40%. This is because some Bible translation committees had a bias for less tone marking, while others wanted more. In informal personal letters, this can be much lower. Newsletters, the Lisu magazine printed bimonthly in Burma, and so on show intermediate rates. On the other hand, some individuals prefer to use very full or even complete tone marking.

Lisu children's teaching materials also tend to use much fuller tone marking. However, most omit marking one of the six tones. In China and in some materials prepared in Burma and Thailand, it is the mid creaky tone (Lisu tone 3 or MY.. CY. marked with two full stops) which is always (or almost always) omitted. However, in other materials prepared in Thailand and Burma, it is the mid noncreaky tone (Lisu tone 4 or MY.. PO., marked with full stop plus comma) which is omitted. Adult literacy materials often have far fewer tone marks, especially in the earliest stages, presumably reflecting a judgement that the tone marking is a problem for learners.

Overall, the lowest rate of tone marking is on the covers and spines of certain Biblical texts: the 1981 Hong Kong printing of the Bible with no tones at all, the 1999 red letter edition of the New Testament with one tone (but more in its less redundant subtitle), the 1978 New Testament with one tone in the title and three in a subtitle, and so on. This may be the case partly because the title is highly redundant. A large bound book in Lisu is almost certain to be something Christian, and the word for Bible in Lisu has been the same since the 1930s.

Even in New Lisu, there is a strong tendency to omit some tones. The unmarked tone in New Lisu is the mid creaky tone (Fraser Lisu Tone 3 or ..), while the mid noncreaky tone (Fraser Lisu Tone 4 or .,) is marked by a postscript *-x*. However, the postscript *-x* is very often omitted in normal New Lisu text (as opposed to materials prepared to teach children). This of course means that the distinction between mid creaky and noncreaky tone is not consistently maintained. The phonetic details of the realisation of these two tones, and their lexical distribution, differ somewhat between varieties of Western Lisu.

5. The development of Advanced Lisu

In Morse and Tehan (2000: 55), David Morse recounts the story of how in August 1968 the letter "M" broke off the only available Lisu typewriter in the remote valley in Burma where his family and a large number of Lisu were living in isolation. Upon his arrival to Thailand soon afterwards, he began working on what he calls Advanced Lisu. This is a series of modified versions of the Fraser script which eliminate inverted letters. Morse and Tehan (2000) outline five stages of this process. In stage 1, all letters are still upper case, but digraphs are used in place of inverted letters. In stage 2, upper and lower case letters are both used. In stage 3, script is used rather than just single letters. Stage 4 is the elimination of spaces between syllables within words, capitalisation of the first letter of proper names, and revisions to punctuation; and stage 5 is the use of postscript consonants to represent tones, instead of postscript punctuation marks.

A primer to teach Advanced Lisu (Morse (2000), including stages 1 to 5 together) was prepared and has gone through many printings. A sample Biblical text in stage 2 Advanced Lisu (plus the capitalisation of proper names from stage 4) was published as anonymous (2000). In Morse and Tehan (2000: 58) it is reported that while older people have negative feelings, younger people find Advanced Lisu easy to learn and easier to write than traditional Fraser Lisu.

The various stages of Advanced Lisu could have caused confusion; but in fact almost no one stays at stage 1, few if any use stage 3 running script, most have adopted the capitalization of the first letter in proper names but not the rest of stage 4, and stage 5 is not yet in use. As in Fraser Lisu, there is a very strong tendency to omit tone marking in Advanced Lisu.

Curiously, the postscript tone markings chosen for stage 5 Advanced Lisu are different from the tone indications in New Lisu. It would have been easy to align the two systems more closely, and perhaps even achieve an integration, but this is not to be. The following table compares the two systems.

Table 2. Fraser Lisu, Advanced Lisu and New Lisu tones

Fraser Lisu number	marking	name	Advanced Lisu stage 5	New Lisu	pitch
1	.	MY.. TI.	*-t*	*-l*	[55]
2	,	N. PO.,	*-c*	*-q*	[35]
3	..	MY.. CY.	*-s*	(unmarked)	[44], creaky
4	.,	MY.. PO.,	*-r*	*-x*	[33]
5	:	MY.. JƎ.,	*-p*	*-t*	[21]
6	;	MY.. N.	*-q*	*-r*	[21], creaky

Two other abbreviation devices are built into Advanced Lisu stage 5 which also cause problems for the integration of the Advanced and New systems. The very frequent verbal postposition /ɑ^{44}/, written with a postscript underline in Fraser Lisu, is written with postscript *-x* in stage 5. Also, a reduplicated syllable is written with postscript *-z* in stage 5. For example, 'happy' ꓘ, ꓛꓲ: ꓛꓲ:_ [kʰɑ³⁵ tɕʰi²¹ tɕʰi²¹ ɑ⁴⁴] comes out as *kha, chi:zx* in stage 2 and would be *khacchipzx* in stage 5; the corresponding New Lisu would be *kaqqitqit a.*[7] It is clear enough why stage 5 is not very widely accepted.

Advanced Lisu stage 5 also provides a postscript letter for each sequence of two tones. There are eight such sequences, any of the four nonlow tones can be followed by either of the two low tones. Most monomorphemic words containing such sequences are final clause markers or locationals with one of the low vowels /ɑ/ or /æ/, such as /bɑɑ̰²¹/ 'maybe' and /nææ̰²¹/ 'inside'; Table 3 shows how the various orthographies treat these two examples.

Table 3. Sequences of two tones

Fraser Lisu	Advanced Lisu	New Lisu	phonetic form	gloss
ꓐ.:	*bam*	*bbal' at*	/bɑɑ̰²¹/	maybe
ꓠꓯ.;	*naej*	*nail' air*	/nææ̰²¹/	inside

The proposed postscript letters for combined tones are *-m* for high plus low, *-g* for rising plus low, *-v* for mid creaky plus low, *-b* for mid noncreaky plus low, *-j* for high plus low creaky, *-k* for rising plus low creaky, *-f* for mid creaky plus low creaky, and *-d* for mid noncreaky plus low creaky. Thus, overall, stage 5 Advanced Lisu uses fourteen different postscript letters for different purposes, and sequences of up to three are possible (tone plus reduplication plus final /ɑ⁴⁴/). Loanwords with final /n/ or /ŋ/, written in Advanced Lisu with final *-n* and *-ng*, are mainly nouns or classifiers, and so are not often reduplicated or followed by the verbal marker /ɑ⁴⁴/.

Advanced Lisu makes some Lisu happy because it is not just upper-case letters, and therefore looks less childish to them. Morse and Tehan (2000: 58) also report the feeling that "At last, we can write using any typing machine, and also use longhand – we have full freedom in expressing ourselves."

6. Solutions

While the full implementation of stage 5 Advanced Lisu seems unlikely, one area where at least stage 2 has taken off is in email. Given the font and data flow rate problems discussed above, a substantial proportion of Lisu email traffic is now using it. Unless two people both happen to know that they use the same font, it is now fairly normal to use Advanced Lisu for this purpose. One estimate[8] is that there are about 800 users, representing less than 0.1% of the Lisu population. Overall, nearly two-thirds of the Lisu are Christian, and so have varying degrees of familiarity with the traditional Fraser script.

As in normal Fraser Lisu, it is possible to omit tone marking in Advanced Lisu stage 2. However, from email data it seems that there is rather less omission than in Fraser Lisu of various genres. This may be because many of the email users are younger literacy or publishing workers who are particularly language-aware. Some of them are among those who prefer to mark tones consistently, apart from leaving one tone unmarked.

It may seem ironic that it is exactly when the wider availability of computers ought to have made it less cumbersome to use an unusual script like Fraser Lisu that those most literate and technically advanced among the speech community are instead starting to use Advanced Lisu. However, that is the situation. It overcomes the technical problems of using the Fraser script for email in the four nations where most Lisu live, and links the Lisu in remote areas of India, Burma, China, and Thailand.

It is still the case that all Lisu first become literate in Fraser Lisu, and as far as I know there is no desire to reprint any Biblical texts in Advanced Lisu. In any case, this would separate the Lisu Christian commnuity from its ninety-year tradition of using Fraser script, and most senior pastors are opposed to it. There is already a great deal of material published in Fraser Lisu. For a partial list, see Morse and Tehan (2000: 61–62). Since many non-Christian Lisu associate this script with Christianity, it may be that the Advanced Lisu system will provide a bridge which will allow literacy to spread more widely without conversion, but this has not yet happened. So far, Advanced Lisu remains an auxiliary device for representing Fraser Lisu over email.

Notes

1. I am pleased to acknowledge the funding support of the Australian Research Council (A59701122, A00001357) and the assistance of many Lisu and other friends and colleagues over the years, particularly members of the Morse family including David, Steve, and others.
2. In China, from 1958 onwards, the Fraser script has also been called Old Lisu.
3. Much of the Lipo material published up to 1951 uses Chinese-influenced syntax; this is being fixed in the new materials. The Lipo Christians are about 60 000 of the total Lipo population of 250 000.
4. Then called T'engyueh, or in *pinyin* Tengyue.
5. Einbau is the creation of a new compromise dialect from elements of various

existing dialects, as opposed to Ausbau, the codification and development of an existing dialect and its spread as a standard for use by speakers of other dialects.

6. Karen uses the Indic *visarga* as a tonal mark; this is similar to the alphabetic colon. Burmese, the script which formed the basic source for the Karen scripts created in the early 19th century, uses a postscript lowered circle, similar to a full stop, for another tone.
7. The northern variety, as represented by the New Lisu orthography, has velars rather than alveopalatals in some words before /i/, so the actual New Lisu form is *kaqkitkit a* [$k^h\text{ɑ}^3$8 k^hi^{21} k^hi^{21} ɑ^{44}].
8. Ahdi Mark (personal communication, Chiangmai 2003).

References

anonymous
1922 M I MI ꓕO ꓷ [M I (spells) MI book] [Lisu Catechism and Hymn Book]. Rangoon: American Baptist Mission Press.

anonymous
2000 G-L-ꓕI-Y SU Tꓯ FI DU ꓒO-ꓕO LE BO ꓶ:_ M ꓕO: ꓶ: Lꓯ; XU/Ga-La-Thi-Ya su tae fi du Pho-Lo le bo gheu:x ma tho: gheu lae; xu [Galatians, a Letter from Paul to Others]. Chiangmai: Christian Literacy Fellowship.

Ba Thaw
1915 *wu-sa t'o-rẽ kw n nyi ā m ba p'o ga ā m jy-k'a* [Catechism in Lisu]. Rangoon: American Baptist Mission Press.

Bradley, David
1979 *Proto-Loloish*. Scandinavian Institute of Asian Studies Monograph Series No. 39. London & Malmö: Curzon Press.
1999 Lisu. In *Hill Tribes Phrasebook: Hill tribes of South-East Asia*, David Bradley, Paul W. Lewis, Nerida Jarkey, and Christopher Court (eds.), 75–98. Melbourne: Lonely Planet.
2003 Lisu. In *The Sino-Tibetan Languages*, Graham Thurgood, and Randy J. LaPolla (eds.), 222–235. London: Routledge.

Bradley, David, with Edward R. Hope, James Fish, and Maya Bradley
2006 *Southern Lisu Dictionary*. STEDT Monograph Series, No. 4. Berkeley: STEDT.

Bradley, David, and Daniel Kane
1981 Lisu orthographies. *Working Papers in Linguistics, University of Melbourne*, 23–38. University of Melbourne.

Enwall, Joakim
1994 *A Myth Become Reality: History and Development of the Miao Written Language*. Stockholm East Asian Monographs no. 5. Stockholm: Institute of Oriental Languages, Stockholm University.

Hope, Edward R.
1976 Lisu. In *Phonemes and Orthography: Language Planning in Ten Minority Languages of Thailand*, William A. Smalley (ed.), 125–148. Pacific Linguistics C-43. Canberra: Department of Linguistics, Research School of Pacific Studies, Australian National University.

Morse, David L.
2000 LI-SU ꓕO ꓶ MU.. F. ꓛO, ꓒꓶ.. M. DU [Advanced Lisu Writing Teacher]. Chiangmai: Christian Literacy Fellowship.

Morse, David L., and Thomas M. Tehan
2000 How do you write Lisu? In *Endangered Languages and Literacy. Proceedings of the Fourth FEL Conference, University of North Carolina Charlotte, 21–24 September 2000*, Nicholas Ostler, and Blair Rudes (eds.), 53–62. Bath: Foundation for Endangered Languages.

Wa Renbo (ed. by Guang Naba, compiled by Mu Yuzhang, Han Gang, and Xu Hongde)
1999 *Jitianguge* [Ancient Songs of Heavenly Worship], vol. 2. Kunming: Yunnan Nationalities Press.

Shina in contemporary Pakistan

Razwal Kohistani and Ruth Laila Schmidt

1. Distribution and genetic affiliation of Shina

Shina is an Indo-Aryan language of the Dardic group, spoken in the Karakorams and the western Himalayas: Gilgit, Hunza, the Astor Valley, the Tangir-Darel Valleys, Chilas, and Indus Kohistan, as well as in the upper Neelam Valley and Dras. Outliers of Shina are found in Ladakh (Brokskat), Chitral (Palula and Sawi), Swat (Ushojo; Bashir 2003: 878) and Dir (Kalkoti).[1]

Bailey (1924: xiii) divides Shina into three dialect groups: (a) Gilgit, (b) Astor, Gures and Dras and (c) Kohistan and Chilas. Two scholars, Strand (2001b) and Radloff (1992: 122–132) identify two main dialect centres for Shina. One is in the area around Gilgit, and the other is in the area around, east, and south of Chilas.

The population figures presented in Table 1 have been obtained from: (a) Kohistan: *District Census Report of Kohistan 1998* (Government of Pakistan 1998), (b) Diamer, Gilgit and Baltistan: *Population and Housing Census of the Northern Areas 1998* (Government of Pakistan 1998), (c) Palula (Henrik Liljegren, Frontier Language Institute), (d) Kalkoti (Muhammad Zaman, Summer Institute of Linguistics) and (e) Shina speakers who have migrated (Kohistani 1998).[2]

In the nineteenth and twentieth centuries, many Shina-speaking communities were conquered by the Maharaja of Kashmir, and came indirectly under British influence. Today, Pakistan administers the Shina-speaking communities of the Gilgit, lower Hunza, Tangir-Darel, Astor and Chilas valleys, as well as those in Indus Kohistan, while India administers the Shina-speaking communities in the Neelam (Kishenganga) drainage, the Gures and Tilel valleys, the Dras plain and Ladakh. In modern times, eastern Shina dialects are in contrast with Balti and Kashmiri, the Gilgiti dialect with Burushaski and Khowar, and the Kohistani dialect with Pashto and Indus Kohistani, all in addition to Urdu (Bashir 2003: 878).

The aim of this paper is to provide information about the use of Shina in its two main centres: Gilgit and Chilas/Kohistan. This focus on Gilgit and Chilas/Kohistan not only reflects the local perception of cultural and linguistic patterning in the Shina-speaking zone but is also governed by pragmatic constraints. Due to adverse weather conditions, geographical inaccessibility, and time constraints, we were not able to visit Astor and Baltistan.[3] When providing statistical data, however, there is a need to separate the Shina-speaking areas of the Northern Areas (comprising five districts, of which Gilgit and Diamer contain the largest number of Shina speakers) from those of district Kohistan, even though parts of Diamer, with its headquarters at Chilas, belong linguistically and culturally to the Chilas/Kohistan dialect area. This is because census and other statistical data for the Northern Areas and district Kohistan are published separately. In describing language use, it is more useful, when possible, to group Chilas, Darel and Tangir with Kohistan.

Our central hypothesis is that to the extent that modern education, communications and technology have reached the Shina-speaking areas of Pakistan, the use of Shina has become restricted to the private and local spheres of life. Communication in public fora increasingly takes place in Urdu (the official language of Pakistan), English, or Pashto (the regional language of the North-West Frontier Province in which Kohistan is included). But there are forces working against this: committed Shina-speaking intellectuals in Gilgit and in the Kohistan and Diamer districts, as well as Islamic missionaries whose priority is communicating with individuals. We shall first make an overview of Gilgit, as an appropriate example of language use in the north, then contrast this with Kohistan (and to the extent possible, Chilas, Tangir, and Darel), where conditions are quite different.

2. Shina-speaking districts in Pakistan

2.1. Gilgit

Gilgit is the center for trade and government in Pakistan's Northern Areas, and a commercial centre for trade between China and Pakistan over the Khunjerab pass. Educational facilities include high schools, degree colleges, and a post-graduate college. It is also to a certain extent a center for communications, both telecommunications and radio (Radio Pakistan broadcasts daily from Gilgit, reaching areas which are unable to receive transmissions from Islamabad).

Although trade is increasingly important in Gilgit town, the economy of the Gilgit valley, as well as of the Shina-speaking areas of the Hunza river valley to the north, is still based on farming and herding. The opening of the Karakoram Highway has improved the marketability of the fruit and nuts which are a specialty of these valleys, and Shina speakers are being integrated into wider economic networks (Schmidt 1984: 681).

Table 1. Shina speaking population in Pakistan

District	**Sub-division**	**Population**
Kohistan, NWFP	Palas	165 613
	Dassu	184 746
	Pattan (estimated Kohistani speakers who can speak Shina in Pattan, Jijial, Duber, Kiyaal, Seo and Khandia valley)	20 000
(Chilas)	Astor	71 666
	Chilas	72 723
	Darel/Tangir	59 193
Gilgit	Gilgit	145 272
	Nagir-II	16 562
	Punial	37 773
	Ishkoman	18 406
	Gupis	29 648
Baltistan	Rondo	11 458
	Gultari	10 000
	Kharmang	12 000
Chitral	Drosh Tehsil (Ashret and Byori valley)	10 000
Dir Kohistan	Upper Dir Tehsil (Kalkoti Shina)	5 000
Swat	Swat Tehsil (Ushojo Shina)	1 200
Total		**871 260**

Table 2. In other districts/areas where Kohistani Shina speakers have migrated

District	**Population**
Batgram	20 000
Mansehra	25 000
Abbottabad	12 000
Haripur	8 300
Rawalpindi	9 930
Islamabad	2 200
Swat	390
Mardan	325
Hazro/Attock	9 250
Muzaffarabad	8 900
Rawalakot	4 680
Lahore	430
Karachi	860
Hyderabad	349
Estimated Shina speaker of NAs of Pakistan in lowland cities	150 000
Total	**252 614**

(Source: Kohistani 1998: 11)

In Gilgit, Shina is the language of the home and the local marketplace, wherever communication with non-Shina speakers does not require the use of the lingua franca, Urdu. But it is not the dominant language of public fora, although some of the speeches in the local elections of 1994 were in Shina (Radloff 1992: 163, 173; Rahman 1996: 218).

Radio Pakistan began broadcasting in Shina from Rawalpindi in 1949 and from Gilgit in 1979. The standard chosen is the Gilgiti dialect. The Gilgiti dialect would thus appear to be the most appropriate focus of efforts to promulgate a standard medium for communication throughout the region, but unfortunately, the percent of lexical similarity between Gilgiti and Kohistani Shina is only about 63–65%.[4] There are also substantial cultural differences between Gilgit and Kohistan/Chilas. The population of Gilgit is predominantly Shia, and has strong cultural links to Iran. The populations of Kohistan and Chilas are overwhelmingly Sunnis of the Deobandi sect, with strong affiliations to Swat, and more recently, to Raiwind in the Punjab. These religious and cultural affiliations often overshadow linguistic and ethnic relationships.

The dialect of Gilgit has a writing system, which is not yet fully standardized, and members of a literary organization based in Gilgit, the Karakoram Writers Forum, attempt to publish in it. M.S. Namus, Amin Zia (1986) and Abdul Khaliq Taj (1990) have developed orthographies by modifying the Persian alphabet (Rahman 1996: 218, quoting Ali Usman 1991).

2.2. Kohistan, Chilas, Tangir and Darel

In Kohistan, Shina is spoken along the left bank of the Indus river in Kolai, Palas, Jalkot, Shatial and Sazin,[5] not only by ethnic Shins, but also by the pastoral Gujars and agricultural Sarkhali population. A closely related form of Shina is spoken in the Chilas, Astor, Darel and Tangir valleys. In Duber and Patan on the Indus right bank, Shina is spoken as a second language, for purposes of trade within Kohistan.

In Kohistan, Chilas and its side valleys, Tangir and Darel, Shina is the means of communication for all purposes, wherever communication with outsiders does not necessitate communication in Urdu or Pashto. The government has established courts in Kohistan, Chilas, Darel and Tangir, but it has only limited means of enforcing the Pakistani legal code in such remote and rugged areas, and customary law continues to be observed throughout the region. In a government court, arguments and verdicts might well be presented in the national language, Urdu, or the provincial language, Pashto; but in a tribal *jirga* (council) of ethnic Shins, arguments continue to be presented in Shina.

In the bazaars of Kohistan, Chilas, Tangir and Darel, non-Shina speakers conduct business in Shina if they can speak it, even if they speak another language among themselves. The rate of bilingualism is low, and among bilinguals, the second language is often a pidgin of Urdu and Pashto, or Urdu and Hindko, which is effective only for basic transactions. Among women, the rate of bilingualism is virtually nil.[6]

2.3. Shina in Chitral, Dir Kohistan and Swat

Before the modern dialects of Shina underwent the processes of grammatical change that resulted in the complex verb tenses and noun inflections which we find today, groups of Shina-speakers migrated from Chilas to Chitral, Dir Kohistan, and Swat. In Chitral and Dir Kohistan we find an archaic variety of Shina, Palula, along with the related dialects of Kalkoti and Sawi. This migration took place according to Morgenstierne (1941: 8) and Cacopardo (1995) around two hundred years ago.

According to Bashir (2003: 882) Palula is spoken by 7000–9000 people in the villages of Ashret, Byori, Ghos and Purigal in southern Chitral. The Palula speakers of Byori say that they have come from the Tangir valley via Swat and Dir; the people of Ashret have a tradition that they migrated from Chilas via Laspur and belong to the Bota tribes of the Chilas area.[7] Working with genealogies and calculating 20 years to a generation, Strand (2000a) places the time of migration earlier than Morgenstierne's and Cacopardo's calculations, reckoning that the people of Ashret must have left Chilas in ca. 1640.

The speakers of Kalkoti, spoken in Kalkot in upper Dir Kohistan, share the same oral history as the people of Ashret. Sawi was spoken originally in the Sau valley south of Arandu (where the Mastuj river flows out of Chitral into Afghanistan), having been transplanted there from Chitral (Buddruss 1967). Although heavily influenced by Gawarbati, it is still mutually intelligible with Palula. Decker (1992: 80) finds 56–58% lexical similarity between Sawi and Palula. Since the 1979–1987 war in Afghanistan, Sawi speakers have been dispersed in refugee camps in Chitral and Dir, and their number is not known (Bashir 2003: 887). Decker (1992: 77–78) estimates the number of speakers at 8000–12 000.

Palula is a marginalized language. Palula speakers are not comfortable speaking their own language in the Drosh bazaar. They either remain silent or speak Khowar, the principal speech of Chitral, which is a distinct language.

Another outlier of Shina is Ushojo, which is spoken in the Bishigram side valley of the Swat river by 1200–2000 people whose oral history says that they migrated there from Kolai in Indus Kohistan several hundred years ago.[8] Lexical research by Decker (1992: 66, 72) shows that Ushojo is most closely related to Kohistani Shina. Many Ushojo speakers also speak Torwali and Pashto.

2.4. Shina outside Shina-speaking areas

Some Shina speaking families have migrated, either temporarily or permanently, out of the Kohistan and Chilas districts to the districts of Batagram, Mansehra, Abbottabad, Haripur, Hazro, Mingora, Rawalpindi, Islamabad, Muzaffarabad, Rawalakot, Lahore, Hyderabad and Karachi. In the Pulandran area of the Kaghan Valley, speakers of Kohistani Shina come from the Hazara District and set up small summer settlements along the main road. Whatever their origin, migrant Shina-speaking families maintain strong links with their native areas, both with their patrilineage ("blood relationships") and their mother's family ("milk relationships") through the exercise of land ownership rights, economic exchange and arrangement of marriages. In Rawalpindi and Mansehra, Shina speakers have established their own *mahallas* (neighborhoods), which range in size from 40 to 100 households.

The settlement of Shina-speakers in lower-lying areas of the Northwest Frontier Province (such as Oghi, Abbottabad, Qalandarabad, Havelian, Haripur and Hazro) where Hindko and Pashto are spoken has lead to an unprecedented level of linguistic interaction. In these areas, Shina-speakers live side by side with speakers of Urdu, Punjabi, Hindko and Pashto, and members of the younger generation often speak several of these languages fluently. In Rawalpindi, the Shina-speaking population interacts with speakers of Kashmiri, Hindko and Punjabi, leading to bi- and multilingualism in those languages. Speakers of Kashmiri, Hindko and Punjabi are also beginning to learn Shina.

In urban areas, Shina-speakers deal daily with the use of Urdu and Punjabi in the marketplace. Education, print and electronic media are in Urdu and English. As a language with few written texts, Shina tends to be restricted to the home. Traditional lore is being lost, much vocabulary is borrowed from Urdu and Punjabi, and grammatical changes are taking place.

3. Language use today

An examination of the use of Shina in private and public domains shows that the language has the most restricted role precisely in the place from which a literary standard might be expected to emerge: in its urban centre, Gilgit. With the exception of a few committed intellectuals, the educated elite consider it an inferior language, which they are ashamed to speak in public. Our data support the rule of thumb for language use given by the senior librarian of Gilgit Municipal Library, Sherbaz Ali Khan Barcha:[9]

Shina is used for:	speaking with parents and family
Urdu is used for:	communication in public, outside the family
English is used for:	employment in private and government sectors
Arabic is used for:	religious rituals

Even though Gilgit is the largest Shina-speaking town and the centre of Shina intellectual life, the lexical difference between the Gilgiti and other dialects, reinforced by religious differences between Shias and Sunnis, prevents it from emerging as an engine for the development of literary and cultural activities in Shina, and hinders efforts to develop a standard written form of Shina. Intellectual life is closely connected with religion, and in this domain, people look outward for inspiration. The Shias of Gilgit continue to look to Iran; the Sunnis of Chilas and Kohistan look southward to Raiwind in the Punjab (the centre of the Tablighi Jamaat).[10] Under the government of Zia ul-Haq, religious antagonism escalated into open conflict, and a large village in Gilgit, Jalalabad, was set on fire by Sunnis from Kohistan, Chilas, Darel and Tangir.

Shina is strongest in rural areas, where modern education has only recently become available. (Consequently bilingualism in Urdu is rare.) These areas are so geographically vast, and the terrain is so difficult, that there is little chance of Shina suffering language death in the near future. But this relegation of Shina to rural and domestic spheres hinders it from developing a single standardized form, or more than a rudimentary literature.

The Astor valley lies on the Astor river, which is a tributary of the Gilgit river, but is separated both from the Gilgit valley and from Chilas by the Nanga Parbat massif. Data for Astor provided by Radloff (1992: 163) suggest that language use in Astor is similar to that of Gilgit.

3.1. Language use in the home, *baithak*s and *hujra*s

A *baithak* (Shina *byak*) is a room or rooms where men sit and where male guests are entertained. A *hujra* is a guest house for men, where guests are housed and fed for free (a custom borrowed from Swat). It serves as a venue for negotiating social and economic exchange. Only well-to-do men can maintain *hujras*. *Baithaks* are found all over the Shina-speaking region (sometimes called *drawing rooms* in Gilgit). *Hujras* are a typical feature of tribal Kohistan (where the ethnic Shin community is composed of lineages descended from a common ancestor, holding at least in theory equal status, sharing all resources equally and recognizing no overlord).[11] They are found only in the valleys of Indus Kohistan, Chilas, Tangir, and Darel. Gilgit is not tribal but instead a stratified society with a long history as a state headed by a monarch, and a *hujra* would have no function there.

Shina is the language of the home throughout the Shina-speaking areas. But as increasingly more new products, tools, technology, and travelers appear, loan words from Urdu and English continue to increase, especially in commercial hubs like Gilgit town or in cities outside the Shina-speaking areas. Indigenous vocabulary decreases correspondingly. Construction of the Karakoram Highway (which runs from Besham along the Indus and Gilgit rivers, through Gilgit and Hunza to China over the Khunjerab Pass) has resulted in a quantum increase in trade, tourism, communication with lowland Pakistan, and government activities. The result is that Gilgit town is more tightly linked with the rest of Pakistan, and the influence of Urdu and English has greatly increased. In Chilas, Darel, Tangir and Indus Kohistan, the pace of change is much slower, but even there, change is visible.

Folklore and folk knowledge are being displaced by information broadcast through television and radio. As discussion in the home increasingly revolves around television programs and newscasts, traditional topics of discussion are abandoned. In a panel interview, Amin Zia and Abdul Khaliq Taj reported that they are not in a position to adequately transmit the folklore and folk knowledge they learned from their parents and grandparents to their children.[12]

Shina-speaking children whose families have migrated to cities and towns in the lowlands suffer the most language loss, and are often chided for their broken Shina when they visit their places of origin. Grammatical changes have been observed among children who live in other urban areas of Punjab and the North West Frontier Province. To give two examples: the feminine plural forms of verbs are being lost, and the oblique stem *aso-* of the 1st pl. pronoun *bes* is now used as the nominative stem, both with system-wide effects. In Rawalpindi, where Shina speakers live in four *mahallas* or neighborhoods, Shina-speaking children interact with Punjabi and Hindko-speaking children on the street, in the marketplace and in schools. Viewing of television is widespread. In contrast to older migrants who left Kohistan between 1940–1970, the younger generation is rapidly replacing Shina vocabulary with Punjabi and Urdu. This will have profound effects on urban Shina when the older generation dies. One wonders to what extent the links with rural Shina-speaking areas can be maintained under these circumstances.

Rural areas are far less affected by modernization, and here both the Shina language and traditional lore are robustly maintained. These areas include rural parts of the Gilgit valley (Punial, Jaglot, Nagir, Ishkoman, Gupis, Astor, Hanzil, Gulapur, Sherot and Bagrot) and all the Shina-speaking areas of the Indus valley and its side valleys (Darel, Tangir, Chilas and Indus Kohistan). In Kohistan, even Shina speakers who know Urdu object to being addressed in Urdu, or for that matter Punjabi or Pashto. Kohistani migrants to the lowland continue to speak Shina at home (a necessity until such time as women's education becomes widespread), and to correct their children's speech. In the Kohistan and Chilas districts, language use in the home has hardly changed for the past three decades despite the introduction of government schools. But here, also, new vocabulary from Urdu, Punjabi and Pashto is introduced by Tablighi missionaries, and by girls from urban areas who are married into rural families.

Under the influence of Deobandi Islam, women have less opportunities to meet and share experiences. In Kohistan, for example, women used to collect wood and wild vegetables, and to participate in collective agricultural tasks (for example, the collective harvesting known as *hashar*). They tend to be more more restricted to the home now than before, although they still meet one another in the annual migration to the summer pastures.

Interaction among boys and men is also somewhat decreased in Kohistan, as the custom of males sleeping in the collective *hujra* at night is being given up in favor of individual *baithak*s attached to homes. In the *baithak*s, conversation is turning from traditional topics such as agricultural issues and feuds, to local politics, elections, and administrative and developmental affairs.

When the men sleep at home, not only do they have less opportunity for conversational exchange, but the women are inhibited from talking among themselves. There is, however, one positive impact of this change: children now have more opportunity to interact with their parents. In former times, Kohistani children used to have little contact with their fathers in their early years.

3.2. Language use in the market place

In the Indus valley (Chilas, Darel, Tangir, Shatial, Dassu and Komila) and in Astor and Shina-speaking parts of Nagir, Shina is spoken outside the home for local communication and in small local marketplaces. However, if a shopkeeper is addressed in Urdu or (in the Indus valley) in Pashto, he responds in that language.[13]

In Gilgit town, uneducated local residents speak Shina in public fora, but educated local residents prefer to speak Urdu or English. In the Northern Light Infantry[14] Market, the shopkeepers (the majority of whom are Pashtuns) prefer to speak Pashto, although many of them also know Urdu and Shina. Shopkeepers in smaller neighborhood markets (the majority of whom are local people, as are their customers) speak Shina. When leading intellectual figures were interviewed, however, they reported an increase in the use of Urdu. Sher Baz Ali Khan Barcha reports that literate Shina speakers feel "cultural shame" at speaking Shina in the marketplace or official workplaces.[15] Amin Zia (poet and scholar), Abdul Khaliq Taj (poet, writer, and magistrate), Jamshed Khan Dukhi (poet and writer) and Tariq (radio and television reporter) report that the use of Urdu is increasing in Gilgit town.[16]

In the major Indus valley towns (Chilas, Darel and Tangir), Shina is the language of the marketplace for local people, although shopkeepers readily speak Urdu or Pashto with outsiders. In the Kohistan district (Shatial, Dassu, Komila, Sumar Nala etc., where there are no major towns) all business is transacted in Kohistani dialects of Shina; likewise in the smaller bazaars of Darel and Tangir.

Several small bazaar towns are located on the west bank of the Indus (Patan, Jijal and Duber), where non-Shina Dardic languages referred to under the umbrella term "Indus Kohistani" are spoken. Here, Shina serves as a second language and as a lingua franca among people who are neither ethnic Shins nor Indus Kohistanis (Syeds, Pashtuns, Sarkhali, Gujars, and Shamogas[17]).

Bazaars are, like anywhere a major entry point for new technologies, products, and tools. The Urdu or English names for these products and techologies naturally follow. Abdul Khaliq Taj reports that his children (like many others) have lost the words for traditional agricultural tools and technology, because the tractor has replaced the plough, manufactured cloth is replacing handwoven wool, and so on. When the old technology is cast away, the younger generation does not learn the vocabulary which referred to it.

3.3. Language use in religious rituals

In bilingual areas, Shina tends not to be used in ceremonial speech on formal religious occasions, such as the Friday sermon, Eid sermon, or Muharram recitations. Apart from the Arabic portions, these are usually in Urdu or Pashto. Ordinary religious instruction is, however, either given in Shina or translated from Arabic into Shina. Difficulties of translation have led to some attempts to write and publish religious tracts in Shina.

In the predominantly Shia rural areas of Gilgit, Shia *zakrins* (Shia mullahs) address their congregations in Shina during Muharram,[18] but in Gilgit town, they use Urdu in Muharram recitations. Both Shia and Sunni mullahs address their congregations in Urdu in Friday prayers, as well as on the occasion of the two major religious festivals or *eids*. But practical religious instruction is given in Arabic, and explained in Shina; and *du'a* (non-obligatory personal prayer) is said in Shina by practically everybody.

Tariq of Astor reports that in Astor, the medium of Tablighi preachings and religious lessons (*dars*) held in the mosque is Shina.

In Kohistan, the Friday sermon and Eid sermons may be in Urdu or Pashto, but ordinary religious instruction is in Shina. Throughout the Indus valley proper (Kolai, Palas, Jalkot, Harban and Shatial, Chilas, Darel and Tangir), the majority of Kohistani and Chilasi men belong to the Tablighi Jamaat, and may leave Kohistan to spend fixed periods of three days, forty days, four months, six months, or one year in study groups held in any of the four provinces of Pakistan. Whatever they learn, they communicate to their family and relatives in Shina. Religious instruction is given to children by their parents and grandparents, and such instruction is always in Shina. The Tablighi *nisab* or syllabus (published in Urdu) is read in Urdu, along with a direct translation into Shina, which is a long-established practice. The more remote the area, the greater the use of Shina as the medium of instruction.

In Kohistan, there are many small *madrasas* (religious schools) where children study Islamic books written in Arabic, Persian, or Pashto, which the teacher then explains in Shina. Local mullahs give religious instruction in Shina, and there are two or three places in Kohistan where girls are able to learn the Quran through Shina translation.

The need to reach out to people in the local language has resulted in attempts at publishing religious works in Shina. Hussain Akbar has published a biography of the prophet, *Somolo Rasul* [The Holy Prophet], in 1992. Nasiruddin Chilasi [no date] published a 254-page work in Chilasi Shina, *Zad-e-Safar* [Provisions for the Journey], and Abdul Rauf of Palas is preparing a Tablighi primer in Shina, with publication expected shortly. The target audience of all these works is local people, especially women, who are not literate in Urdu or Pashto. Other Tablighis are preparing a tract in Shina on the six principles of Tabligh.[19] Despite their high motivation, these authors are hindered by the lack of a standard orthography, especially for the Kohistani dialect.

3.4. Language use in education

3.4.1. The school system

Before independence, the Shina-speaking areas of Gilgit, Astor, and Dras were located in the princely state of Jammu and Kashmir. This geographical area, broken into isolated valleys by the mountain ranges of the Himalayas and Karakorams and the river systems of the Jhelum and Indus, is one of the most polyglot regions of the world, and seems always to have used some administrative language of non-Kashmiri, non-Dardic origin: Sanskrit in pre-Islamic times and Persian under Mughal rule. Urdu superseded Persian as the administrative language in the second decade of the twentieth century (Sufi 1974: 812–813). After independence, Gilgit and Astor came under Pakistani administration. The Government of Pakistan is the largest provider of educational services, and since its beginnings, has worked to make Urdu a symbol of Pakistani national identity, of Islam and of the values of Mughal culture (Rahman 1999: 67–68). The Northern Areas and Kohistan follow the broad outlines of the curriculum defined by the Federal Ministry of Education, and the textbooks in effect determine the curriculum (World Bank 1997: 7–8), in which Shina finds no place, either as a subject of study or as a medium of instruction; nor can it be studied as a subject at university level. (In primary grades, of course, teachers may use Shina to explain the textbooks to the students.) Shina-speaking children use textbooks written in Urdu and English. In middle schools, they usually also learn Arabic or Persian. In the Indus valley, which lies in the North West Frontier Province, they are also supposed to learn the provincial language, Pashto. To teach all these languages effectively would require a much better teaching standard than exists in most schools. But one message does, however, get through: a language is not a language (*zaban*) unless it has a written literature. An unwritten language like Shina is only a dialect (*boli*), and not worthy of study.

Except in Gilgit town, the impression one gets includes educational facilities in poor condition, often poorly trained or untrained teachers with little supervision or academic support and limited and/or inappropriate instructional materials. In Kohistan, many schools were observed to be closed, or were used to store hay.

In *madrasas* (religious schools), Arabic is taught, in many cases alongside Persian (Rahman 1999: 106–107, 115–116).

The Northern Areas has received substantial educational benefits from non-governmental donors. The second largest provider of educational services in the Northern Areas is the Agha Khan Rural Support Program, with

250 schools established in cooperation with village organizations across the five districts of the Northern Areas. This program has also actively recruited female teachers. Even so, only 44% of primary teachers hold the Primary Teaching Certificate, and 11% have not completed the matric, or grade 10 (World Bank 1997: 6). Low educational standards do not negatively affect Shina, however. In Gilgit town, where educational facilities are best, Urdu and English have taken over more domains from Shina than in rural areas, and Shina suffers from greater stigma.

In Kohistan (Kolai, Palas and Jalkot) there were no schools until 1984; the only education available was religious instruction. Government schools have been in existence for only about two decades. In these schools, the medium of instruction is Urdu except in one or two English-medium private schools. Although the majority of the teachers here are Shina speakers, this does not result in better instruction. Pedagogical techniques are usually limited to reciting and translating textbooks written in Urdu, and monolingual Shina-speaking children make slow progress.

In Kohistan, children accompany their families on the seasonal migration to alpine pastures in the summer, where one may occasionally find schools, or occasionally find teachers, but seldom both together. Even if this problem could be solved, children may be put to collecting medicinal plants and mushrooms which are sold for cash income, and so would be unable to attend school. By contrast, among the Kohistani Shina-speakers who have emigrated to urban areas of Pakistan, an estimated 80% of families educate their children, both boys and girls. In Rawalpindi and some other highly urbanized areas, the figure approaches 100%. This reflects the fact that a primary reason for emigration is to take advantage of the better educational facilities in cities. Boys are educated to the highest level the family can afford, usually to the F.A. (12th class) or B.A., and girls generally up to the matric (10th class) or F.A. About 50% of these educated girls are married back into families residing in Kohistan.

In the Gilgit district, educational facilities (both Urdu and English medium) are much better than in the Kohistan district. In addition to government schools, there are four good "public schools" (i.e., private or semiprivate schools, not government schools) offering education to both boys and girls up to the best Pakistani standard: (a) Army Public School, (b) Al-Hayaat Public School, (c) Fauji Foundation School and (d) Al-Azhar School & College.

Foreign aided educational projects to raise the literacy rate have been carried out throughout the region both by the Government of Pakistan and by non-governmental organizations (namely, the International Union for Conservation of Nature, Agha Khan Rural Support Program, World Wildlife Fund-

Pakistan). These attempts have been effective in Gilgit town and its side valleys, but not in the Indus valley.

In Kohistan there is an intermediate college, which is not functional at present. In Chilas and Gilgit there are both colleges and a university (the International Karakoram University, run by the Government of Pakistan in Gilgit). Gilgiti intellectuals favor the inclusion of Shina in the school curriculum, and Amin Zia, Jamshed, Tariq and Abdul Khaliq Taj advocate the opening of a Shina section in the Karakoram University, but theirs are minority voices, and no concrete steps are being taken.

Education of girls is a problematical issue in some places. In Gilgit, there is little opposition to it, and in both Gilgit and Chilas town, girls attend school. In Kohistan, the number of girls attending school is miniscule, and in Darel several girls' schools were recently set on fire as a protest against the education of women. By contrast, in lowland urban areas, Shina-speaking girls are not kept out of school, and many who return to Kohistan, Chilas or Astor find employment as schoolteachers.

To the degree that educational services are provided, they directly result in a lôss of the mother tongue, Shina. Modern educational policies and practices have simply continued the ancient tradition of adopting and promoting a non-indigenous administrative language.

3.4.2. Literacy and enrollment

The literacy rate for the Northern Areas as per the 1998 Census has not been released by the Government of Pakistan. The literacy rate reported by the 1981 Census was 14.7% (World Bank 1997: 2). The literacy rate in Kohistan district among the population aged ten years and above is 11.08%, an increase of 9.68% since 1981.[20] These figures are skewed by the Census' definition of literacy: the ability to read a newspaper and write a simple letter. In Kohistan, one must know Urdu, Pashto or English to possess these skills. We have observed that many people who are not literate by this definition are still able to read the Quran in Arabic; and following the publication of a Shina primer in Arabic characters (Kohistani and Schmidt 1996) we observed so-called "illiterates" who were able to read the Shina texts in the primer.

According to the World Bank (1997: 51), just over half of school-age children of primary school age in the Northern Areas were enrolled in school in 1994. In the Diamer district, enrollment of girls was much lower, only 14% of the school-age population, reflecting a local bias against the education of females.

Table 3. Literacy rate, enrollment in schools and number of schools

	Both sexes	Male	Female
Literacy rate in the Northern Areas (1981)	14.7%	24%	3%
Enrollment rate at primary school level (1994)	N.A.	74%	35%
Enrollment rate at middle school level (1994)	N.A.	54%	18%
Literacy rate in Kohistan:	11.08%	17.23%	2.95%
Educated Persons by sex in Kohistan:[24]	10.19%	16.15%	2.32%
Enrollment rate in Kohistan[25]	6.89%	10.60%	1.34%
Schools in Gilgit District, Northern Areas		**Male**	**Female**
High schools		26	8
Middle schools		27	10
Primary schools		58	25
Schools in Diamer District, Northern Areas			
High schools		14	1
Middle schools		28	6
Primary schools		127	22
Schools in Shina-speaking areas of District Kohistan			
High schools		3	nil
Middle schools		18	nil
Primary schools		310	60

3.5. Language use in public administration

As already observed, Urdu and English are the languages of public administration (police stations, hospitals, and courts) in Gilgit. In the Indus valley, the languages of public administration are Shina with Shina-speaking personnel, and Urdu and Pashto with outsiders. Unfortunately, Shina-speaking policemen, magistrates or doctors are exceptions worth mentioning, rather than the rule, and in the absence of statistics we must restrict ourselves to on-the-spot observations. In Kohistan, where the rate of bilingualism is low, local people communicate in Shina wherever possible, but this is only the case where administrative personnel also speak Shina. Given the low levels of education in Kohistan, many posts are held by outsiders who do not speak Shina, but there are exceptions, especially in the Department of Education. In Kohistan District headquarters at Dassu, local employees and local people speak Shina in offices, hospitals, and the courts, a situation which has been observed for the past 12 years on official visits to these places. The same was observed in Chilas on the occasion of employment interviews at Diamer district headquarters in March 2004. In addition to the hundreds of young people from Astor,

Darel, Tangir and Chilas who had come for the interview, there was a rush of onlookers from the bazaar. The Deputy Superintendent of Police, who along with his policemen was controlling the crowds, was observed speaking Shina with them.

Because women are the least educated segment of the population, Shina is more likely to be used in hospitals than in any other sector of public administration. Again, this is only possible when there is medical staff who know Shina. Such personnel are the exception rather than the rule, but may provide role models for younger Shina-speakers. Dr. Hafiza Bano of Gilgit (a Kashmiri woman who now speaks Shina at home) was observed speaking Shina when treating patients. For many years, there was a German health project in Kohistan (unfortunately now closed) which organized monthly health clinics at Pattan and Dassu. The women doctors in this project realized that the local women could not communicate properly in Urdu or Pashto, and so learned Shina and Indus Kohistani to gain a better understanding of their medical problems, with good results. In remote areas (like Sheryal and Paro in the Palas valley) where no female doctors have been posted, male doctors or dispensers treat patients, and some have been observed speaking Shina with both male and female patients. However a male doctor may not examine a woman to diagnose complaints of the reproductive system, and if the woman must rely on a translator (who is usually a male, since a woman must be accompanied by a male relative when she travels), social norms make it difficult for the woman to discuss intimate problems. Junior level women health workers (such as midwives) who do speak Shina may serve as an interface between patient and doctor in these situations. In lowland urban areas, most Shina-speaking women know enough Urdu, Hindko, Punjabi, or Pashto to report their symptoms to doctors in these languages.

4. Effects of the media

4.1. Radio

News reports in Shina are broadcast from Islamabad Radio Station. In the past Islamabad also used to broadcast recitations of the Quran with Shina translations, folk songs, letters etc. Gilgit Radio Station broadcasts folk songs, *ghazals*,[21] dramas, commentaries, and stories. A number of pre-eminent writers of Shina in Gilgit (Amin Zia, Abdul Khaliq Taj, Jamshed Khan Dukhi, and Tariq) are dissatisfied with current radio programming, commenting that the broadcasting staff have low levels of competence and creativity, and receive little technical support in writing. They also maintain that the news broadcasts in Shina are perfunctorily translated from Urdu, and are full of linguistic errors which degrade the status of Shina rather than promoting its use. To illustrate these opinions, they produced many examples from radio programming.[22] Sher Baz Ali Khan Barcha, however, reports that the impact of radio broadcasting in Shina is very positive, resulting in an increased local interest.[23]

The staff of the Islamabad Radio Station get little support for their efforts. Most work on an hourly basis, and have other full time jobs, so have little time for improving their knowledge of Shina. They have also failed to adopt a standardized writing system, so manuscripts are produced with no systematic rendering of Shina phonology and vocabulary. At the Gilgit Radio Station, the day-to-day interface with Shina-speaking writers and scholars has led to an improvement in language use, and script writers do try to standardize the spelling of Shina to be consistent with Shina phonology. This represents the beginning of a standardization process, though only in Gilgit. In Kohistan, there is little interest in the Shina broadcasts, which are not easily understood due to the low level of lexical similarity as mentioned in section 2.1 above.

Radio programming does have an impact in the listening audience, as it introduces new ideas, international news, national and local political coverage, information about the economy, religious ideas, and so on. To properly assess the impact, a listener survey would be necessary, which was beyond the resources of the authors.

4.2. Television

The introduction of the satellite dish has brought not only national television programming to Pakistan's Northern Areas and Kohistan, but also international programming. English and Urdu films can be viewed by means of satellite dishes, video cassette recorders/players, compact disks, and internet services, wherever the facilities and resources are available (not only in Gilgit town but also in remote villages).

In areas where television broadcasts are accessible, evening hours at home are often spent watching them. This is traditionally the time when parents and children interact, so television hence reduces contact between children and parents. A cultural impact can also be observed. In the bazaars of Gilgit, Chilas and Kohistan, one can now buy products ranging from soap, shampoo and lipstick to video players and motorcycles. In households where dried fruit and nuts would have once been served to guests, western-style biscuits are now offered. Carpets and cushions are being supplemented by chairs and sofas. Chemical fertilizer is replacing manure. People are adopting the fashions and hairstyles of urban cities in the lowlands, as well as western furniture. Films have introduced the concept of modern romance, even though it remains taboo to a certain extent. The *hujra* and *baithak* are becoming *drawing rooms*, and *babu* and *aji* are becoming *mummy* and *papa*.

A concern among Shina-speakers in Gilgit is that the demand for products produced in lowland Pakistan takes money out of the local sector and puts it into the hands of the producers and their middlemen (who are often also non-local). Half the business income in areas such as Gilgit goes to non-local people. But rather than acting as a brake on consumption, this encourages migration to the lowlands to seek better-paid employment.

There is no television programming whatsoever in the Shina language, although there is programming in some regional languages such as Gujari, Kashmiri and Hindko. Some efforts have been made to videofilm dramas in Shina, both in the Gilgit district and in the Ghizar District. As the number of television channels, both government and private, expands, demand will also increase, and it is hoped that the barriers presented by the lack of a standardized script can be overcome through television and radio broadcasting.

4.3. Print media

There are no newspapers of any sort published in Shina. In Gilgit, Chilas and Kohistan, a wide range of Urdu and English daily newspapers are available.

While a great deal of information about the Northern Areas, and even about Kohistan, is available in European languages, published works in Shina are far more limited. We were able to obtain details on the following published books (although the list is not complete, since some publications were mentioned to us, but could not be traced). Many works do not furnish information about the publisher or the date of publication. The process of publication has become slightly easier with the introduction of Urdu fonts on the computer, but Shina characters which do not exist in Urdu must still be added by hand.

In Shina

Akbar, Hussain Akbar
1992 *Somolo Rasul* [The Holy Prophet]. Islamabad: Modern Book Depot Islamabad.

Chilasi, Nasiruddin
[N.D.] *Zad-e-safar: Kalam Baba Chilasi* [Provisions for the Journey: Collected Poems of Baba Chilasi].

Kohistani, Razwal, with Ruth Laila Schmidt
1996 *Shina Qaida* [Shina Environmental Primer]. Islamabad: Himalayan Jungle Project.

Kohistani, Razwal
1997 *Kostan da Adoyati Goryo Riwaiti Istimal* [Indigenous Knowledge of Medicinal Plants in Indus Kohistan]. Rawalpindi: Shina Research Forum-Karakorum.
Kohistani, Razwal
1999 *Bunyadi Shina Urdu Lughat* [Basic Shina-Urdu Dictionary]. Rawalpindi, S.T. Printers.
Mursaleen, Maulvi Zia ul
[N.D.] *Qaida Kohistani Masdar Haroof* [Primer of Kohistani Shina].
Taj, Abdul Khaliq
1989 *Shina Qaida* [Shina Primer].
Zia, Mohammad Amin
1974 *San* [The Wine-Pit]. [Collected Poems, in Shina]. Rawalpindi, S.T. Printers.
Zia, Mohammad Amin
1978 *Soweno Moriye* [Words of the Wise]. Shina Proverbs, in Shina. Islamabad: Folk Heritage Institute.

Noteworthy works on Shina and Shina-speakers in Urdu
Kohistani, Razwal
1998 *Indus Kohistan* [Indus Kohistan (ethnography and history)]. Rawalpindi: Shina Research Forum-Karakorum.
Namus, Dr. M. Shuja
1961 *Gilgit aur Shina Zaban* [Gilgit and the Shina Language]. Bahawalpur: Urdu Academy.
Zia, Mohammad Amin
1986 *Shina Qaida aur Graimar* [Shina Primer and Grammar]. Gilgit: Zia Publications.

5. Impact of tourism

Tourism, both international and national, has been a major source of income for Pakistan's Northern Areas, including Gilgit and Hunza, but following the attacks of September 11, international tourism has dramatically dropped and hotels are virtually empty. (The Government of Pakistan does not allow tourism in Indus Kohistan.) Everywhere that tourists go, they promote the learning of English and Urdu, so although they provide much-needed economic resources, they contribute nothing towards strengthening the use of Shina.

6. Conclusions

The vast area and rugged geography of the Shina language area ensure that the language is in little danger of suffering language death in the near future.

But there is equally little chance of it developing a single standardized form, or more than a rudimentary written literature. Modernization, including modern education, reinforces the time-honoured policy of adopting and promoting a non-indigenous language: Urdu, and more recently, English. This inculcates in educated Shina speakers the attitude that their mother tongue is inferior, and makes them ashamed to speak it in public.

In Gilgit, the interaction between script writers at the Gilgit Radio Station and Shina-speaking writers and scholars has led to the the beginning of a standardization process for that dialect. However, in Kohistan, where the percent of lexical similarity with Gilgiti Shina is only about 65%, there is little interest in the Shina broadcasts from Gilgit, or in adopting a Gilgiti standard variety. Lexical divergence between Gilgiti and Kohistani is reinforced by a sectarian split between Shias and Sunnis, which prevents the largest Shina-speaking town, Gilgit, from emerging as an engine for development of literature and cultural activities in Shina, and hinders the development of a standard written form of the language. The Shina outliers (Palula, Kalkoti, Sawi and Ushojo) spoken by small, isolated populations in areas dominated by other languages are in the weakest position, and there is a real risk that some of them may die out. The dominant languages in public administration are Urdu and English in the Gilgit District, and Urdu and Pashto in the Diamer and Kohistan Districts. Shina-speaking public servants are the exception rather than the rule. Educational services throughout the region are substandard, except in Gilgit town, but improvement of the educational standard results in the loss of Shina linguistic domains to Urdu and English. Nowhere is there any formal instruction in Shina, nor is there any initiative from the government to introduce it, or much public support for its introduction. Islamic preachers, whose work challenges many traditional customs, actually promote the use of Shina in both oral and written forms by preaching in it and, to a lesser extent, using it to write religious tracts.

Thus, in the Shina-speaking areas of Pakistan, we see a typical South Asian pattern in miniature: the imposition of non-indigenous, standard languages on regions without a written language of their own, the imposition of high-culture values and norms on indigenous cultures, and an attempt to extend national legal and educational systems into these areas. We also see, at the same time, Shina being vitalized through the activities of religious preachers who are mobile throughout this region, and by the efforts by intellectuals to produce regional written literatures.

Notes

1. Strand 2001a: 254; Strand 2001b; Kohistani, interview with Muhammad Zaman, Besham, 27th March 2004.
2. In addition to the rather sparse published information about language use in the Shina-speaking area, data has also been collected through participant observation by Razwal Kohistani in the Diamer and Kohistan districts, and in lowlands communities to which Shina-speakers have migrated. Data for Gilgit and Astor was collected from a panel interview of four leading intellectuals (Amin Zia, Abdul Khaliq Taj, Jamshed Khan Dukhi and Tariq) in Gilgit on 8 March 2004, from an interview with Sherbaz Ali Khan Barcha in Gilgit on 8 March 2004 and through participant observation in the marketplace. Henrik Liljegren of the Frontier Language Institute and Muhammad Zaman of the Summer Institute of Linguistics provided information about dialects of Shina in Chitral, in an interview held in Besham on 27 March 2004. Where no other source is cited, the data is taken from Razwal Kohistani's notes.
3. Information provided in this paper on the Shina-speaking population of Astor and the use of Shina in that valley is based on the already published material and the information provided by Tariq from Astor in the panel interview held in Gilgit, 8 March 2004.
4. Radloff (1992: 122–132) finds that in areas at the northern and southern extremes of the Shina-speaking zone, lexical similarity is quite low. 11 of 12 respondents from Kohistan said that "their language" was not broadcast on Radio Pakistan.
5. For a description of the ethnic makeup of Kohistan, see Zarin and Schmidt (1984).
6. Monolingualism among women has been a characteristic feature of rural areas of the Northwest Frontier Province. However, a new generation of educated girls is being married back into Kohistan, and this may bring about change.
7. Interviews with Muhammad Zaman and Henrik Liljegren, 27 March 2004. See also Cacopardo and Cacopardo (2001: 86–87).
8. Bashir 2003: 878; Interview with Muhammad Zaman of the Summer Institute of Linguistics, 27 March 2004.
9. Interview, 8 March 2004.
10. The Tablighi Jamaat is an Islamic missionary movement which works to invigorate faith at an individual level. Some sources link it to the Deobandi sect, but the Tablighis themselves insist that they remain aloof from all sectarian or political activity.
11. Although the ethnic Shins of Kohistan cherish an egalitarian tradition among themselves, non-Shin communities such as Yeshkuns, Kamins and Doms have a lower status than Shins.
12. Interview with Amin Zia, Abdul Khaliq Taj, Jamshed Khan Dukhi and Tariq, 8 March 2004.
13. Shopkeeping is a profession requiring more language skills than others. In the 1970s, Schmidt observed that the shopkeepers of Rawalpindi Bazaar had learned rudimentary Italian, as a result of Italian housewives from the Tarbela Dam project shopping there.
14. The Northern Light Infantry are the Pakistani government troops posted to the Northern Areas.

15. Interview with Sher Baz Ali Khan Barcha, 8 March 2004.
16. Panel interview with Amin Zia, Abdul Khaliq Taj, Jamshed Khan Dukhi and Tariq, 8 March 2004.
17. Shamogas have migrated from the Banihal area of Kashmir. Their language is reported as Pahari, but this just means "of the hills" (as does "Kohistani"). Their language may in fact be a dialect of Poguli (see Schmidt and Koul 1983: 11), but the authors have not been able to confirm this.
18. Muharram is the Shia month of mourning, during which the martyrdom of the Prophet's grandson Husain is remembered through recitation of the sacred stories and ritual mourning.
19. The six principles, or "numbers" of the Tablighis are: (1) the *kalima*, or declaration of the faith, (2) the five daily prayers, (3) knowledge of and memorization about God (Allah), (4) benevolence to Muslims, (5) purity of intention and (6) spending time and money in spreading the word of God.
20. Government of Pakistan 1998: 23.
21. A ghazal is a lyrical poem in the Mughal tradition.
22. Panel interview with Amin Zia, Abdul Khaliq Taj, Jamshed Khan Dukhi and Tariq, 8 March 2004.
23. Interview with Sher Baz Ali Khan Barcha, 8 March 2004.
24. Government of Pakistan 1998: 25.
25. Government of Pakistan 1998: 25.

References

Ali, Usman

1991 *Shinalogy*. Gilgit: Usmani Kutab Khana [in Urdu].

Bailey, T. Grahame

1924 *Grammar of the Shina Language*. London: The Royal Asiatic Society.

Bashir, Elena

2003 Dardic. In *The Indo-Aryan Languages*, George Cardona, and Dhanesh Jain (eds.), 818–894. London: Routledge.

Buddruss, Georg

1967 *Die sprache von Sau in Ostafghanistan. Beiträge zur Kenntnis des Dardischen Phalura*. München: J. Kitzinger.

Cacopardo, Alberto M., and Augusto S. Cacopardo

1995 Peoples of Southern Chitral: A survey of some so-far unstudied ethnic groups of northern Pakistan. Part II: The Palulo. Paper presented at the third international Hindu Kush conference in Chitral, September 1995.

Cacopardo, Alberto M., and Augusto S. Cacopardo (eds.)

2001 *Gates of Peristan. History, Religion and Society in the Hindu Kush*. Rome: Istituto Italiano per L'Africa e L'Oriente.

Decker, Kendall D.

1992 *Languages of Chitral. Sociolinguistic Survey of Northern Pakistan*, vol. 5. Islamabad: Quaid-e-Azam University and Summer Institute of Linguistics.

Government of Pakistan
1998 *Kohistan District Census Report; Population and Housing Census of the Northern Areas.* Islamabad: Population Census Organization, Statistics Division. [This document was not made available in its entirety, and so page numbers cannot always be quoted.]

Kohistani, Razwal
1998 *Indus Kohistan.* Rawalpindi: Shina Research Forum-Karakorum [in Urdu].

Kohistani, Razwal, with Ruth Laila Schmidt
1996 *Shina Qaida* [Shina Environmental Primer]. Islamabad: Himalayan Jungle Project.

Morgenstierne, Georg
1941 *Notes on Phalura. Skrifter utgitt av Det norske videnskaps-akademi i Oslo.* No. 5.

Radloff, Carla F.
1992 The Dialects of Shina. In *Languages of Northern Areas. Sociolinguistic Survey of Northern Pakistan*, vol. 2, P. Backstrom, and C. Radloff (eds.), 89–203. Islamabad: Quaid-e-Azam University and Summer Institute of Linguistics.

Rahman, Tariq
1996 *Language and Politics in Pakistan.* Karachi: Oxford University Press.
1999 *Language, Education and Culture.* Karachi: Oxford University Press.

Schmidt, Ruth Laila
1984 The Shina speakers of Pakistan and India. In *Muslim Peoples: A World Ethnographic Survey*, R. V. Weekes (ed.), 678–684. Westport, Connecticut: Greenwood Press.

Schmidt, Ruth Laila, and Onkar N. Koul
1983 Dardistan revisited: An examination of the relationship between Kashmiri and Shina. In *Aspects of Kashmiri Linguistics*, Onkar N. Koul, and Peter E. Hook (eds.), 1–26. New Delhi: Bahri Publications.

Strand, Richard
2000a <http://users.sedona.net/~strand/IndoAryan/Indus/Atsaret/AtsaretCulture/ Atsaretgen.html> Posted 27 June 2000.
2000b <http://users.sedona.net/~strand/IndoAryan/Indus/Atsaret/AtsaretTexts/ AtsaretHistory.html> Posted 14 April 2000.
2001a The tongues of Peristan. In Cacopardo and Cacopardo (eds.), 251–259.
2001b <http://users.sedona.net/~strand/IndoAryan/Indo Aryas.html> Posted 12 June 2001.

Sufi, Al Haji Dr. G. M. D.
1948–1949. *Kashir, A History of Kashmir*, vol. 1–2. Lahore: Punjab University. Repr. New Delhi: Light and Life Publishers, 1974.

World Bank
1997 *World Bank Educational Appraisal, Report No. 158340-PAK. Staff Appraisal Report Pakistan, Northern Education Project.* <http://www-wds.worldbank.org/servlet/WDSContentServer/WDSP/IB/1997/09/30/000009265_3980420170649/Rendered/PDF/multi_page.pdf> Accessed 27 April 2004.

Zarin, M. M., and Ruth Laila Schmidt
1984 *Discussions with Hariq: Land Tenure and Transhumance in Indus Kohistan*. University of California, Center for South Asia Studies. Working Papers.

The rise of ethnic consciousness and the politicization of language in west-central Nepal

Michael Noonan

> The paradox is this: if economic rationality tells us that the next century will be the age of global integration of the world's economies, cultural "irrationality" steps in to inform us that it will also be the century of ethnic demands and revived nationalisms.
>
> *Carlos Fuentes (1997)*

1. Introduction

A while ago, my friend and colleague, Ram Prasad Bhulanja, told me that he had received an email from his cousin, a man I know, who had journeyed from Beni, in the Myagdi District of west-central Nepal, to the capital Kathmandu. After discussing the content of the letter, the man's situation and so on, I asked him what language the email was written in. Ram and his cousin are native speakers of Chantyal, a Tibeto-Burman language; both are also fluent speakers of Nepali, the national language. He replied that the email was in Nepali, and on further questioning he revealed that he had never received an email or conventional letter in Chantyal, even from close relatives, though phone conversations with these same people are always in Chantyal.

It's worth noting in this context that the email was written in an internet shop in Kathmandu, and that in such shops the keyboards are configured for English and the screens display only roman characters, not the Devanagari used to write Nepali. Nepali was thus de-privileged in this context, and Ram's cousin had to transliterate Nepali into roman characters in order to send Ram a message. Presumably, it would have been just as easy to render Chantyal in the roman alphabet, but this wasn't done, and indeed is seldom, if ever, done.

A further consideration worth noting is the fact Ram's cousin has been involved in the 'Nepal Chhantyal Association' [Nepāl Chhantyāl Saŋgha[1]], the organization of Nepal's Chantyal ethnic group.[2] He has had a hand in producing some of the publications of this group, including a book titled *Chhantyāl bhāṣākā kehī śabda jān, vyākarañ paddhati ra sādhārañ kurākānī* ['Some vocabulary, the grammatical system and general conversation of the Chantyal language'] which contains wordlists with Nepali and English translations, some grammatical discussion, and even some 'practical' dialogue with translations in Nepali and English. So, he has actually helped produce a publication which included written Chantyal, but it never occurred to him to send his cousin a written message in that language.

In this paper I would like to provide a bit of background for this state of affairs, and in the process describe the state of play between the rise of ethnic consciousness, attitudes toward language, and the state of language endangerment of some Tibeto-Burman speaking peoples of west-central Nepal. The main points to be discussed are that language issues have not yet been pushed to the forefront of the political agenda in this part of Nepal, that ethnic organizations are still at the beginning stages of dealing with matters of documentation, standardization, and orthography of their ancestral languages, and that primary education in minority languages is vital for language preservation. One consequence of all this is that the uses of print and electronic media that have been important in language preservation efforts of minority languages in other parts of the world are having little effect in this region of Nepal. After a general discussion of the state of minority languages in Nepal, I'll illustrate these matters with examples taken from the experiences of the Chantyal, Gurung, Magar, and Tamang communities, speakers of Tibeto-Burman languages spoken primarily in west-central Nepal.[3]

2. Ethnic identity and ethnic consciousness

In traditional cultures ethnic identity, as a component of one's social identity, may be asserted to a greater or lesser extent dependent on contingent factors, in particular its instrumental value in gaining some economic or political advantage. It is, moreover, a much more flexible concept than modern-day nationalists and ethnic activists would like to admit, with boundaries that may be fluid rather than static, and where ethnic identities may be multiple or overlapping.[4] Further, the language one speaks may or may not be a determinant of, or even a major component of, one's ethnic identity. Indeed language shifts are a perennial feature of human affairs, as the examination of the history of virtually any inhabited region of the planet will demonstrate.

Over the last two centuries, we have witnessed a phenomenon, most prominently in the West and then increasingly in the rest of the planet, of politicizing ethnicity, thereby transforming ethnic identity into something I will refer to as *ethnic consciousness*. Ethnic consciousness manifests itself in attempts to 'define' the ethnic group, establishing what it means to be a member of the group; in this way, ideas like language, dress, religion, history are used to 'define' the group, and thus become both conscious and politicized – subject to debate both within the community itself and in the larger political arena. The ideology that underlies this rise in ethnic consciousness accords language a central role: a proper ethnic group, this ideology maintains, should have its own language, and the group should have rights with regard to its language.

In traditional cultures, shift in language (which may result in language death when whole ethnicities are involved in the shift) is commented on and may be the source of regret, but is seldom the launchpad for political or social action with the aim of preserving the language, although socio-political states of affairs that have the effect of inducing language loss may be the object of political or social action.[5] With the ideology that underlies modern ethnic self-consciousness comes the potential for politicizing language issues. In what follows we will examine some ways in which the rise in ethnic self-consciousness has been played out among certain ethnic groups in west-central Nepal, in particular about how issues relating to language have become politicized.

3. Language and the Nepalese state

Nepal is an undeveloped country which has experienced relatively little investment in either infrastructure (communications, transportation, etc.) or education. It has also experienced despotic regimes which have actively discouraged expressions of ethnic identity that were at variance with the official state promotion of Nepalese nationalism and the Hindu religion. One result of all this was that ethnic, religious, cultural, or political activism for much of the twentieth century was mostly confined to exiles. The lack of infrastructural and educational development served the nationalist and Hindu-supremacist interests of the state since it helped maintain a population uninfluenced by outside – and potentially subversive – ideas.

The constitution established in the aftermath of the 1990 people's movement reaffirmed the status of Nepal as a Hindu state and Nepali as the sole national language, but legalized the establishment of ethnic organizations and the use of indigenous languages other than Nepali in the schools. (English and Sanskrit had been long established in education, though in different spheres.) Prior to the establishment of the new constitution, indigenous languages other than Nepali were effectively banished from the public sphere, and only Nepali was permitted in education, in broadcasting and, to a significant degree, in the print media. The new constitution recognizes "the right of every citizen to develop and promote their languages, script and culture" (Article 18), and the government has recognized twelve minority languages to be used on a regular basis on the national broadcast media. Other than these radio and television broadcasts, however, very few concrete steps have been taken to promote the use of minority languages in the country. For example, there has been almost no progress in the development of minority languages in primary education, and recent court decisions have prohibited the use of indigenous languages at the local government level, while at the same time, the government continues to subsidize Sanskrit at all levels of education (Gurung 2003).

4. Ethnic consciousness and language

Against this background, it is not surprising that ethnic consciousness (the politicization of ethnic identity) has been slow to develop in rural west-central Nepal. Moreover, it has developed in a climate of increasing economic and cultural pressures. Demographic pressures reduced productivity of the land and accelerated a process of ecological decline that is rapidly rendering subsistence farming in the hills unsustainable. The search for wage employment, once confined to a few men only, is now the norm among younger adults, with many moving to the cities and foreign countries in search of work. This, combined with fear of induction into the Maoist army, has led to an outmigration of young people, leaving many villages without young adults. Additional cultural pressures include the spread of Nepali-medium schooling, Nepali-medium radio and television, consumer capitalism, foreign tourism, and the desire to participate in Nepalese national Hindu culture.

While the effects of the cultural pressures have been profound, the effects of the economic pressures are easier to understand and articulate, and so the political demands of the ethnic organizations have mostly been in the form of what might be called 'civil ethnicity': the demand for a more equitable allocation of resources within the Nepalese state. As a result, ethnic political activism has seldom placed language issues at the forefront for several reasons:

- economic grievances are considerable and well grounded in the experience of ordinary people;
- most people don't see a link between their economic grievances and the official status of their mother tongues;
- bilingualism in Nepali is sufficiently widespread among ethnics so that few could claim lack of access to economic resources on the basis of language alone;
- there has been no tradition of privileging local languages other than Nepali on a national level, and few traditions (even "remembered" traditions) of privileging these languages on a local level *vis-à-vis* Nepali; that is, for the

last 200 years, no indigenous language has been in a position to challenge the dominance of Nepali, not even locally;
- and ethnic consciousness translatable into political action has not penetrated deeply into Nepalese society, despite the strength of ethnic identity.[6]

As a result, language issues have been largely backgrounded, despite the attempts by some activists to foreground them.

Despite the fact that language issues have seldom been foregrounded on the national level, ethnic organizations have devoted considerable attention to them. In general, however, the reach of the ethnic organizations is not great; that is, the activities of the ethnic organizations, with few exceptions, have little impact, intellectually or otherwise, on the populations they claim to represent. Language issues are largely the concern of educated elites, but this is not to say that ordinary people are insensitive to these issues. A recent poll reveals that there is considerable support for government efforts to preserve indigenous languages, though there is much less support for using them as the medium of education.[7] Further, the rise in percentages of ethnics claiming their ancestral language as their mother tongue has increased, as documented by the latest census figures, attesting to the success of the ethnic organizations in promoting ethnic consciousness.[8]

But the poll failed to address the question of how important the promotion of indigenous languages was compared to other issues, e.g. economic issues. All the evidence suggests that the promotion and use of indigenous languages does not rank high in importance compared to other issues, mostly because the case has not been made that the use of Nepali seriously disadvantages minority ethnic populations. The only language issue to surface at the national level in recent times has been the teaching of Sanskrit, which had been compulsory from the fourth grade but was recently made an optional subject.[9] Sanskrit was seen to privilege unfairly certain castes and ethnicities. It is unlikely that language issues will assume a high degree of importance to ethnic minorities until one of two contradictory situations occurs: the state substantially improves the lot of ethnic minorities so that economic issues no longer dominate the list of grievances, or the state is perceived as failing utterly to provide peace and economic stability and there are political movements aiming to break up the state.[10]

5. The state of the minority languages

Finally, there is the state of the minority languages themselves. Despite an increase in the last census in the percentages of ethnics reporting their ancestral languages as their mother tongues, the true state of affairs with regard to the Tibeto-Burman languages of west-central Nepal is that the percentage of fluent speakers is declining rapidly as young people either fail to learn the language or learn it only imperfectly. What the census numbers reveal is an increase in ethnic consciousness, while at the same time hiding the true condition of the languages.[11] Further, these languages have no tradition of literacy and have only recently come to be written, and even then not often, and when they are written they are written so as to provide examples of writing in the language rather than as media of communication. Most of the existing documentation comes from foreign linguists, though there are hopeful signs that this may change.

6. Progress in efforts at promotion and standardization

I'll now briefly report on the progress of the efforts of the Chantyal, Gurung, Magar, and Tamang communities to standardize and promote their languages.

Magar

In the 2001 census, 1 622 421 people claimed to be Magars, while 770 116 claimed the language as their mother tongue. Less has been done to standardize and promote Magar than any of the other languages considered in this study. This no doubt is in large part a reflection of the fact that a considerable percentage of those now considered ethnic Magars do not speak Magar as well as the fact that those who do may speak dialects which are very different and not fully mutually intelligible.[12]

It should be noted that until quite recently many non-Magars have claimed Magar status. These groups, which include the Chantyal, Kham, Kaike, Kusunda, Raute, and Raji ethnic groups, were all too small or remote to be classified in the Muluki Ain of 1854, the national legal code which classified Nepalis according to a single caste hierarchy. These groups *claimed* to be Magars because Magars were officially classified as "clean" (in the Hindu sense) and "unenslavable", and because the British were interested in hiring Magars for their Gurkha regiments. These people *had license* to call themselves Magars because until recently there was little sense of a larger Magar ethnicity and hence no core Magar community which could challenge these claims.[13] The fact that these people had their own languages whose relationships to Magar is not obvious to non-linguists (and non-existent in the case of Kusunda) apparently did not affect their claims, though their distinct languages were important later for claims to separate ethnic identities.

The Nepal Magar association conducts meetings in the national language Nepali, and the *Magar Studies Center* website[14] is entirely in English. There has been very little publication in Magar and the ethnic organization has not broached the subjects of orthography and standardization to any significant extent.

Gurung

In the 2001 census, there were 543 571 Gurungs, of whom 338 925 claimed to have Gurung as their mother tongue. Like the Magars, the Gurungs were also classified in the Muluki Ain as clean and unenslavable, and also like the Magars, Gurungs served as Gurkha soldiers. As a result, the Gurungs, too, were an attractive group for others to claim association with, and even today ethnic groups which now claim separate status still use "Gurung" as a surname: these groups include Ghales (I refer here to those who speak the Ghale language, which is only distantly related to Gurung), Mananges, and Nar-Phus, the last two speaking languages which are related to Gurung within the Tamangic family of languages though not especially close to Gurung within that family.

The Gurungs are much in advance of the Magars as regards publication in their language and attention paid to orthography and standardization, though many issues have been contentious and few issues have been resolved. There have been advocates for four distinct scripts in recent years (Glover 2002): the Tibetan script, particularly among Gurung Buddhists; the Khemaa lipi खेमाँ लिपि, an indigenous adaption of the Devanagari script; the roman script, advocated mostly by veterans of the British and Indian Gurkha regiments; and the Devanagari script. While the other scripts still have their adherents, Devanagari is clearly the script of choice for publication in Gurung.

There have been a significant number of publications in Gurung,[15] at least by the standards of minority languages of Nepal that have recently come to be written. This literary activity is at least in part owing to the fact that a number of Gurungs have achieved literacy and some degree of economic well-being as a result of their service in the Gurkha regiments; it is also a result of the geographic location of some key Gurung settlements in and around the tourist resort of Pokhara and the popular trekking district around Annapurna. Within the general conventions imposed by the Nepalese version of the Devanagari script, each author has felt free to invent his own spellings. Glover (2002) proposes a number of spelling conventions for Gurung publications, in particular those that would be used for a Gurung–Nepali–English dictionary. It remains to be seen if these spelling conventions will be generally adopted.

Chantyal

In the 2001 census, 9814 people claimed to be Chantyal and 5912 claim the language as their mother tongue. The last figure cannot possibly be correct since the language is only spoken in a few villages whose combined population does not exceed 2000. The figure attests, however, to the dramatic rise in ethnic consciousness that has occurred over the last couple of decades.[16]

Because only about one in five Chantyals speak the language, meetings of the ethnic organization are always conducted in Nepali. There were no publications using the Chantyal language until 1987 (Chhantyal 2044 [1987]), and indeed there have been no publications exclusively or even primarily in Chantyal save Bhulanja and Noonan (1995), the only publication in Chantyal and the only one not somehow associated with the Nepal Chhantyal Association. The first publication in 1987 and the few publications by the Chantyal community which have appeared subsequently have all used the Devanagari script: unlike all the other languages discussed in this paper, there has been no discussion about the use of any other script for this language. Bhulanja and Noonan proposed Devanagari spelling conventions for Chantyal, appended to a collection of children's stories that were distributed to schools in Chantyal-speaking villages. These conventions have not been observed in subsequent publications.

Chantyals have engaged in an interesting kind of language planning by inventing out of whole cloth a set of numbers and a set of calendar terms, with the intention, one presumes, of replacing the terms long ago borrowed from Nepali.[17] This kind of language planning is unique in this part of Nepal, but it remains to be seen if these terms will actually be used. No doubt, the motivation for this was a realization of the extent of borrowing of vocabulary from Nepali (80–85% of the total) that became obvious when Chantyal phrases where provided Nepali translations, an important feature of all but one of the publications that have so far appeared.

Tamang

Tamang has the largest number of speakers of any Tibeto-Burman language in Nepal: in the 2001 census, there were 1 282 304 Tamangs, of whom 1 179 145 claimed Tamang as their mother tongue. Tamangs are, even by the standards of Nepalese peasants, economically depressed and have never had good relations with the Nepalese government. In the Muluki Ain of 1854, Tamangs were classified as clean but enslavable. There was a Tamang peasant rebellion as late as 1951 (Gurung 2003: 14), and recent ethnic activism had taken on a much more defiant tone than other groups considered here, even before the recent upsurge in Maoist activity.

Despite a common sense of grievance, Tamangs have had until recently very little sense of shared community. Indeed, Sonntag has referred to Tamang as "a language in search of an ethnic group" (1995: 113). Macdonald (1989:176) has maintained that "Tamang identity insofar as it can be said to exist is a Nepalese administrative invention and a concept formulated by non-Nepalese researchers to facilitate written communications between themselves. There does not seem to be much evidence to show that isolated Tamang villagers are conscious of belonging to a pan-Tamang social identity."[18] While much the same could be said of other Nepalese ethnicities, it appears to be true of the Tamang to an unusual degree. This state of affairs is partly due to the fact that Tamang settlements are widely scattered throughout the "middle hills" of Nepal (*i.e.* the area lying between the flat, malarial Terai and the high Himalaya), and partly due to the fact that Tamangs lacked the shared experience of serving in the Gurkha regiments, which has had a formative influence on the Gurungs' sense of common ethnicity.

Still, there is the Tamang language: though riddled with deep dialect divisions, Tamang is still recognizably a single language, and it is this sense of shared ancestral language which can, and now does, serve to bring Tamangs together. Tamang has come to be written only recently. Choice of writing system often divides activists. For example, while almost all publication in Tamang has thus far been in the Devanagari script,[19] there are vocal advocates in the Tamang community for the Tibetan script, including a simplified version of the script referred to as *Tamyig*. As with the other languages considered here, problems of orthography and standardization have not been addressed, which means that written Tamang isn't used for real communicative purposes among Tamangs: there are several Tamang websites, a couple of which post announcements of community activities. The language used in these websites is primarily English, with translations of some, but not all material in Nepali. Where Tamang is present, it is either the subject of discussion or a translation of something displayed more prominently elsewhere. The basic function of Tamang on these websites is simply to display the language, not to use it as a medium of communication.

One stark fact which unites all these languages is this: while speakers know these languages *can* be written (*i.e.* literates have typically seen examples of them in writing), none of them is *used* in writing to any significant degree, even in correspondence between native speakers. This follows from a situation widely observed throughout the world: people ordinarily write the languages they were taught to write in school unless there is a strong local tradition of literacy in another language. If the school system ignores minority languages, as it does in Nepal, minority languages are unlikely to be used in writing, even between coethnics. I personally have observed similar situations elsewhere in the world: people I know in the Lango community of Uganda write letters home to their relatives in English, even though their elderly relatives do not speak English and require someone to translate the letters for them. These letter writers know how to write their native language (they have seen it written and can write it on request), but have had little experience doing so and prefer to write in the language in which they had received explicit instruction in writing. In Nepal, since virtually all adults know Nepali, the disincentives for writing in Nepali are even fewer than those for the Langi writing in English.

Literacy in minority language communities is not necessary for language documentation efforts since documentation can be carried out by outsiders using conventional fieldwork techniques, recordings, and so on. It is necessary, however, for standardization if the minority communities are to carry out this work themselves, and for language preservation efforts. This suggests that educational reforms promoting the use of minority languages in the early years of schooling are the best way to encourage literacy in the minority languages. Literacy in these languages would, in turn, encourage the use of minority languages in writing, and this could have a major impact on language preservation efforts.

The use of minority languages in the schools involves a kind of chicken-and-egg problem: languages can't be used in the schools until there is an agreed-upon orthography and a standardized form that can be used in textbooks and other educational materials, and so far language activists have failed to provide such standard forms. Were there the political will, the government could step in. Given the current political climate in Nepal, the will is lacking. We can hope that a future government will make such a move a priority.

In sum, language activists have not succeeded in creating a political situation that would encourage language preservation. Their failure is understandable, given all the obstacles that they have had to overcome in the larger political arena. But even within their own communities, they have barely begun to address issues of orthography and standardization and have, in general, ignored suggestions by foreign linguists regarding them (Noonan 2005). Further, for many of these languages (Gurung, Magar, and Tamang), there are significant differences between dialects which will have to be resolved before a standard can emerge. If these languages are ever to serve as vehicles of education or administration, these problems will have to be resolved. Ethnic organizations, for the most part, have not begun to address these issues seriously, and most language activists of my acquaintance do not seem to grasp the fact that these issues must be dealt with before their languages can achieve the official status they desire for them.

7. Conclusion: Globalization, media, and language preservation in west-central Nepal

In this paper I've tried to sketch the political, economic, and cultural pressures faced by some ethnic communities in west-central Nepal, and what concrete steps have been taken by the communities themselves to preserve their languages and make them suitable vehicles for educational and other purposes. Ethnic consciousness has indeed risen in west-central Nepal, and language issues have been politicized by activists. Nonetheless, it is safe to say that, at this stage, not very much has been accomplished, and the current political and economic climate of Nepal must necessarily divert attention from issues that no doubt seem to most people of secondary importance.

Globalization, broadly defined, has been responsible in large part for the growth of ethnic consciousness in west-central Nepal, but the forces of globalization have affected Nepal more recently than many parts of the world and their effect has been less profound to date. The extreme poverty of Nepal and its peculiar history – the fact that it had not been colonized by Europeans and had remained effectively isolated from global ideologies until relatively recently – explain this state of affairs and explain why the languages considered here have undergone so little development. Conventional print media and the newer electronic media (e.g. websites and chat rooms, so important in language preservation efforts elsewhere[20]) cannot do much to help preserve endangered languages without literate populations who can access them.

So, what can be done? If the minority languages of Nepal are to remain healthy, they will require some help from the state in the form of support for primary education in the minority languages, though as noted above there is not yet a consensus even among minority language speakers that this would be a good idea. Further, if the ethnic communities are to take a leading role in preserving and standardizing their languages, they need both time and trained personnel. For many of the minority languages of Nepal, time is running short: many languages will cease to be spoken by integral communities in the near future. Trained personnel are in still in short supply, though the faculty at Tribhuvan University are working valiantly to produce them.[21] For the time being, foreign linguists will necessarily play a leading role in documenting these languages, but in the final analysis, standardizing them and developing orthographies for them are tasks that must fall to the ethnic communities themselves.[22]

Notes

1. Until recently, the organization was known as the *Chhantyāl Parivār Saŋgha.*
2. The spelling <Chhantyal> is used when transliterating from Devanagari script and in the name of the ethnic organization, which has used this English spelling in its publications. In general references to the ethnic group and their language, I use here, as I have done in other published works, the spelling <Chantyal>. The work reported on in this paper has been supported by the following grants from the National Science Foundation of the United States: DBC-9121114, SBR-9600717, and SBR-9728369. See Noonan (1996) for a list of the villages in which Chantyal is spoken and a discussion of the current status of the language; Noonan (2003) is a grammatical sketch.
3. Chantyal, Gurung, and Tamang are members of the Tamangic branch of the Bodish languages, a section of Bodic. Magar is also Bodic, but is usually placed in the Kham-Magar branch of Himalayish, another section of Bodic.
4. See especially Anderson (1983), Brown (1989), Hobsbawm (1990) for discussion of these points.
5. For example, economic or religious oppression may lead to political action, as numerous peasant revolts and religious wars attest. Language issues, however, have seldom been a major source of political unrest until recent times: such unrest requires a modern type of ethnic self-consciousness, and this in turn requires, minimally, the leadership of a literate elite.
6. Ethnic identity in this part of Nepal is strong, but not unproblematic. As Gellner (1997: 15) notes: "the differences *within* [Nepalese] ethnic categories may be greater than those *between* them."
7. <http://www.nepalresearch.org/publications/resistance.htm>.
8. A close examination of the figures would likely reveal that they are inflated for many of the ethnic groups: the Chantyal figures, for example, report mother tongue identification that is approximately three times the actual number, as discussed below.
9. *Kantipur Online,* Dec. 7, 2002: "Govt announces 29-point education reform programme." The Maoists played a central role in pressuring the government to take this action.
10. In recent times there have been few genuine separatist movements in Nepal, the major exemplar being that of the Tarai-based Nepal Sadbhavana Party (Gurung 1997: 529).
11. One anecdote will suffice to illustrate how rapidly the status of a language may change. When I first began working with the Nar-Phu people in the mid-90s, their language was spoken by an integral community of young and old. By 2002, however, the young people had almost completely ceased to speak the language, though they could still understand it when older people addressed them in it. The young people used Nepali among themselves and explained to me that Nepali was the language they would need to make their way in the world, and that they wanted to feel comfortable speaking it. Small children are increasingly addressed in Nepali. At this rate, the language will effectively die with the death of the older generation.
12. See Grunow-Hårsta (in preparation) for documentation of the differences between the dialects.

13. Tofflin (1981) asserts that "classifications of the Tibeto-Burman hill tribes into Tamang, Gurung, Magar, Rai, Thakali, etc. correspond only very imperfectly to reality and can only be accepted as working hypotheses. In fact, none of these groups form a homogeneous ethnic group, either culturally or linguistically." It should be noted that questions about who is and who is not a Magar have not been settled: as De Sales (2000) shows, the Khams are still subject to claims on their Magar identity.
14. <http://www.magarstudiescenter.org/>.
15. See Noonan (2005) for some references.
16. For example, in the late 1980s, Chantyals who were serving in the Gurkha units of the British army began changing their surnames from the Magar name "Pun", which they had been using as long as they had been serving in these units, to "Chhantyal".
17. See Noonan (ms) for a discussion of these interesting case.
18. See also similar comments by Levine (1987: 73) and Holmberg (1989: 13).
19. The websites <http://www.tamangs.com/tmgorgs.htm> and <http://www.tamang-ghedung.org/> contain listings of publications in Tamang.
20. For an example of how such media can play an important role, see Mensching (2000).
21. See Yadava (2004) on the uses of digital technology in support of Nepal's minority languages. Linguists at Tribhuvan University have initiated a program to train Nepalese fieldworkers for the Linguistic Survey of Nepal.
22. I would like to thank John Manoochehri for helpful comments on this paper.

References

Anderson, Benedict
1983 *Imagined Communities*. London & New York: Verso.

Bhulanja, Ram Prasad, and Michael Noonan
1995 *Kyaratimaye Sastarma* [Children's Stories]. University of Wisconsin-Milwaukee.

Brown, David
1989 Ethnic revival: Perspectives on state and society. *Third World Quarterly* II (4): 1–17.

Chhantyal, Dil Bahadur Gharabja
2044 [1987]. *Chhantyal jati: Ek parichaya* [The Chantyal Ethnic Group: an Introduction]. Kathmandu: Sushi Jayanti Chhantyal.

Fuentes, Carlos
1997 The federalist way. In *At Century's End: Great Minds Reflect on Our Times*, Nathan P. Gardels (ed.), 104–108. La Jolla: Alti Pub.

Gellner, David
1997 Ethnicity and nationalism in the world's only Hindu state. In *Nationalism and Ethnicity in a Hindu Kingdom*, David Gellner, and Joanna Pfaff-Czarnecka (eds.), 3–31, Amsterdam: Harwood Academic.

Glover, Warren
2002 Choosing a Gurung orthography for a new dictionary. *Gipan: Texas University Papers in Linguistics* 2: 25–38.

Grunow-Hårsta, Karen
in preparation. A Grammar of Magar. University of Wisconsin-Milwaukee PhD dissertation.

Gurung, Harka
1997 State and society in Nepal. In *Nationalism and Ethnicity in a Hindu Kingdom*, David Gellner, and Joanna Pfaff-Czarnecka (eds.), 496–532, Amsterdam: Harwood Academic.

Gurung, Harka
2003 Trident and thunderbolt: Cultural dynamics in Nepalese politics. *The Mahesh Chandra Regmi Lecture*, 1–28. Lalitpur: Social Science Baha. <http://www.himalassociation.org/baha/Trident-and-Thunderbolt.pdf>.

Hobsbawm, Eric
1990 *Nations and Nationalism since 1780: Programme, Myth, Reality*. Cambridge: Cambridge University Press.

Holmberg, David H.
1989 *Order in Paradox: Myth, Ritual, and Exchange among Nepal's Tamang*. Delhi: Motilal Barnarsidass.

Levine, Nancy E.
1987 Caste, state and ethnic boundaries in Nepal. *Journal of Asian Studies* 46 (1): 71–88.

Macdonald, Alexander W.
1989 Note on the language, literature and cultural identity of the Tamang. *Kailash* 15 (3–4): 165–190.

Mensching, Guido
2000 The internet as a rescue tool of endangered languages: Sardinian. <http://www.gaia.es/multilinguae/pdf/Guido.PDF>.

Noonan, Michael
1996 The fall and rise and fall of the Chantyal language. *Southwest Journal of Linguistics* 15 (1–2): 121–136.
2003 Chantyal. In *The Sino-Tibetan Languages*, Randy LaPolla, and Graham Thurgood (eds.), 315–335. London: Routledge.
2005 Recent adaptions of the Devanagari Script for the Tibeto-Burman Languages of Nepal. In *Indic Scripts: Past and Future*, Peri Bhaskararao (ed.). Tokyo: Tokyo University of Foreign Studies.
ms. The great Chantyal number hoax. University of Wisconsin-Milwaukee.

de Sales, Anne
2000 The Kham Magar country: Between ethnic claims and Maoism. *European Bulletin of Himalayan Research* 19: 41–72.

Sonntag, Selma
1995 Ethnolinguistic identity and language policy in Nepal. *Nationalism and Ethnic Politics* 1 (4): 108–120.

Tofflin, Gérard
1981 L'organisation sociale des Pahari ou Pahi, population du centre Népal. *L'Homme* 21: 39–68.

Yadava, Yogendra
2004 Use of software support in Nepal's languages. *SCALLA 2004 Working Position Papers*. <http://www.elda.fr/ article 144.html>.

Why Ladakhi must not be written – Being part of the Great Tradition: Another kind of global thinking[1]

Bettina Zeisler

Globalization is generally understood as the increasing links across all parts of the world via modern means of transport and communication, and more particularly the increasing integration into the capitalist market economy. On the cultural plane, it is understood as an increase in the levelling of cultural and linguistic differences under the dominance of American language and (non-) culture, known also as Westernization or even "MacDonaldization".

Looking at India, and particularly at the Himalayan regions, one may observe that Ladakh, situated in the north of Jammu & Kashmir, instead suffers from "Bollywoodization", i.e. from the cultural impact of India's pop culture, which, although inspired by the American model, has developed its very own unmistakable style. Besides, western pop culture has for some time drawn quite a few inspirations from Indian spiritualism. Particularly Tibetan Mahāyāna Buddhism continues to attract people of the western world who are dissatisfied with a merely materialistic life style. Accordingly, Ladakh's tourist industry benefits from the Westerner's misconceptions and idealizations of the Buddhist world.

Globalization, thus, is not a one-way affair, nor is it solely a phenomenon of the past decades. Mahāyāna Buddhism, e.g., fostered by the rising Tibetan empire, was a global attractor in Central Asia in the 7th–9th centuries, leading to the Buddhization and Tibetanization of the Indo-European (Dardic) population of Ladakh and Baltistan (Pakistan). It is still an important global factor in Ladakh, competing with modern materialism on the one hand, and Shia Islam on the other. The Tibetan monastic tradition has a very strong impact on the self-conception of the Buddhist elites, and has until now hampered any development of literacy and literature in Ladakhi.

From a villager's perspective, one may further observe that there are quite different concentric as well as overlapping "globes" or spheres of economic or cultural integration, starting with the local centres (monasteries and masjids, villages with middle and higher level schools, etc.) leading to the local political centres, the two district towns, Leh and Kargil, and further beyond to the next economic and political centres, Jammu and Srinagar, and finally to Delhi.

On the religious plane, these centres include Mecca and the great Buddhist monastic institutions, formerly in Central Tibet, although now mostly located in South India. Depending on one's actual location and use of media, all these centres create their own sphere of influence. The lower ones may moderate or channel the influence of the higher ones and one might observe a greater sense of resistance towards the influences of the centre at the periphery.

Like any minor language, the Ladakhi language or *Ladakse skat* and its dialects, spoken by about 180 000 speakers in Ladakh,[2] is under strong pressure from the official state language (in this case, Urdu), the language of higher education (English), and the languages of mass media (Hindi and Urdu). The situation in the Leh district as described by Zeisler (1998) has not dramatically changed, but the trends have been re-enforced.

In the nineties, the strongest impact on the language came from education. Originally, teaching at governmental schools should have been through the medium of the local language up to the fifth class, followed by Urdu, and then by English after the eighth class. But since there were almost no Ladakhi or Tibetan textbooks, children merely learned the Tibetan alphabet, and otherwise had to rely on textbooks in Urdu or English without having any adequate training in these languages. I could witness students in their tenth year of school mechanically memorizing the content of their textbooks, obviously without much understanding. Most students would fail the exams, or would pass only by cheating.

The way out of this misery was demonstrated by the private schools in Leh, by starting with English medium from the first class (again, mostly based on mechanical reproduction). Two schools offered non-compulsory classes in *Budik*, the Tibetan script. There was virtually no further education in the Ladakhi *phalskat* (the spoken language), though there might have been classes in *choskat* (i.e. Classical Tibetan). Most students, however, preferred to have additional classes in Hindi, as this was more beneficial for their professional careers. This means that the younger generation did not get any formal training in its own mother tongue, and further did not learn Ladakhi words for modern concepts, be they political, social, or ecological. As a result, the use of Ladakhi is more and more restricted to the domain of traditional lifestyle and family affairs.

Even though the quality of the governmental schools has drastically improved in recent years, particularly through better teacher training and the involvement of the villagers in education committees, as well as through switching to English as the medium language from the first class, the tendency of sending children to private schools remains unbroken. The fees for the private schools[3] are usually beyond the means of an otherwise well-off family. The foreigner, who is rich by definition, sees herself confronted repeatedly with the moral obligation to sponsor the education of at least one child, and the demand comes not only from poor families but equally from families who just completed the building of a new house, or just bought a new car or perhaps even a bus. With an equal reliance on foreign generosity, several Buddhist organisations have meanwhile established private schools in the villages with classes for *choskat*.

Children at private and governmental schools in Ladakh get at least some explanations in Ladakhi, but more and more families tend to send even small children to schools in other parts of India where this is simply impossible, not to speak of learning the Tibetan alphabet. Not only do the parents believe that the schools in Chandigarh, Jammu, or Srinagar are better than those in Ladakh, but they also believe that if they invest so much money in their children, they would feel a stronger moral obligation to study well, than they might do in Leh or at local schools.[4] While it is generally true that many more students are getting a much better education than only some years back, and that some of them may develop a growing interest in Urdu or English written media, most of them are illiterate or merely alphabeticized in their own language.

For a long time, the Ladakhi program of All India Radio which was broadcast from the Leh radio station was the main modern and far-reaching medium of information throughout both the Buddhist dominated Leh district and the Muslim dominated Kargil district. The dialects spoken in Leh and its vicinity became influential everywhere in Ladakh, even in the Kargil district (and to some extent also in Baltistan). The Leh pronunciation was also promoted through the education system (as many teachers had either been to a higher secondary school in Leh or its vicinity, or at least had teachers that had been to school in Leh). Nowadays, the impact of the Leh dialect might be reduced by the upgrading of schools and civil services in remote villages, the installation of a second radio station in Kargil, and the switch to more fashionable media.

These days, TV, VCRs, and DVD players are widely spread, slowly reaching the remotest villages. The impact of these media is hardly counterbalanced by the one hour long TV magazine in Ladakhi, which has been scheduled three times a week since 2003. Hindi and Urdu, the languages of the media, as well as English, the language of higher education, are associated with high social prestige, so that educated townspeople may refuse to talk Ladakhi with the foreigner, and one may observe two Ladakhi families in Leh handling the matrimonial negotiations basically in Urdu.

The language, thus, is under strong pressure, and may soon reach the stage of endangerment. Unfortunately, the oral tradition of story telling has come more or less to an end. The children are too much occupied with their homework, and radio, TV, and videos help to bide the time in a more fashionable way. Furthermore, there is no literary tradition in Ladakhi *phalskat* that could slow down the trend, and all efforts to establish it are opposed by the dominant Buddhist scholars as being anti-Buddhist or as lacking in traditional scholarship.

One has to take into account that Buddhist scholars usually do not differentiate between language and script, and that the Tibetan script is in a way inseparable from its use for the Buddhist scriptures. The only "true" Ladakhi language (*asile skat*[5]), thus, is *choskat*, which should serve as a model for the literary language (*ikskat*[5]). This is also the official position of the members of the J&K Cultural Academy, Leh.[6] For many scholars, *phalskat* is but a deviation or even "rubbish", not worthy of being preserved, not to speak of being developed.

According to Tibetan historiography, the Tibetan script and the rules of grammar were introduced by a certain Thonmi Sambhoṭa in the first half of the 7th century, mainly for the codification of the sacred texts of Buddhism. From a Western academic perspective, this seems to be nothing but a pious legend, invented in the second half of the 11th century (cf. Miller 1963; Róna-Tas 1985: 183–303; Zeisler 2005), but for Buddhist scholars it has become an undeniable historical fact.[7] Accordingly, the script and the classical orthography have become sacrosanct, and should not be altered even when used for lay purposes.

The classical orthography reflects the pronunciation and grammar of some Tibetan dialects of the 9th century, but does not conform to the pronunciation and grammar of most modern Tibetan varieties. The nomadic Amdo dialects and the western dialects of Ladakhi, as well as Balti (spoken in Baltistan), come very close to the slightly different Old Tibetan spellings. But the grammar of these varieties, especially of Ladakhi, has considerably changed. To write the modern varieties according to the classical orthography and grammar would be the same as to write Italian or French according to the orthography and grammar of Latin, or to write Hindi or Bengali according to the orthography and grammar of Sanskrit, which simply means that one writes in a language different from that one speaks. Traditionally, only the monastic schools provided a good training in Classical Tibetan, though laymen might have learned the basics from their clerical relatives or friends. Nevertheless, even monks, whether Tibetan or Ladakhi, have great difficulties in writing the classical orthography correctly. The rules are no longer transparent, particularly since many prefixed consonants have become mute in the Central Tibetan dialects as well as in the monastic reading style. As the religious language had drawn from different dialectal sources, many *choskat* words do not have an equivalent in one or the other modern variety.

For this reason, new ways of writing have been adopted in Amdo and, most radically, in Bhutan. While the spelling reforms in Amdo have grown naturally, being backed or even initiated by the local scholars in order to spread the religion among the common people,[8] in Ladakh, unfortunately, the first attempts to reform the orthography seem to have come from outsiders such as Christian missionaries (cf. Bray 2001). This may be one of the many reasons why it has become a non-issue for most of the Buddhist scholars.

These days, the most enthusiastic adherents of a writing reform belong either to the younger generation such as the publishers of the bilingual magazine *Ladags Melong* (*Ladwags Meloŋ* 'Mirror of Ladakh'), SECMOL, who might easily be accused of not being firm in the classical language, or from some Muslim intellectuals, such as Molwi Muhammad Omar Gutu Nadvi, the Imam of the Leh Masjid, who might again be accused of as representing a non-Buddhist force.[9] The publishers of *Ladags Melong* promote a writing style for lay purposes close to the *phalskat* of Leh, but are by far not as radical as a linguist might want them to be. Although they are backed by at least one traditional scholar, Gelong Konchok Pande, they have been repeatedly accused (e.g. at IALS XI, the 11th Seminar of the International Association for Ladakh Studies, Choglamsar 2003) of intentionally spoiling the grammar of *choskat*. For the time being, it seems that the Molwi and his translation of the Quran into *phalskat* is better tolerated, possibly for political reasons (the Buddhist political leadership of the Leh district promotes a policy of reconciliation and unity) as well as because of his established reputation as a learned man, which forbids open criticism.

The fear that any writing reform would result in a disintegration of Buddhist identity might be explained by the increasing fear of being slowly but steadily outnumbered and overpowered by the fast growing Muslim population of Ladakh. Another factor might be the inherited trauma of the elder generation, who had its traditional education in Tibet, but lost its cultural ties with the Chinese occupation. There might be perhaps a certain influence from the Tibetan exile community or even from Tibet proper, where the need is felt to create a common language to maintain cultural and political identity. While some Chinese scholars have opted to use the language of the oral epic as a base for the common language (Jiangbian Jiacuo 1994; Wang 1994), certainly not without political afterthoughts, most Tibetan scholars have agreed to base the common language on the grammar and orthography of Classical Tibetan. The early attempts from the Chinese side to promote and develop the regional dialects were rather seen as an attack on the Tibetan national identity (for an overview on the ongoing discussion, cf. Prins 2002). One can also constantly hear that, due to the Chinese influence, the Tibetan language is in decline (Lhasa 1994; cf. also Ngawangthondup Narkyid 1992: 615). Developing the Ladakhi *phalskat* as a written language might thus be seen as treachery to the Tibetan cause. The Tibetan elites (including the Dalai Lama) do not appreciate moderate changes in the writing style for the Ladakhi *phalskat*, although Modern Literary Tibetan (the style developed in Tibet proper, as well as the style developed in the exile communities) has integrated quite a few grammatical features of the modern Central Tibetan varieties.

The strongest motivation for the conservative stance seems to be the pride or desire of being part of the Great Tradition. As an example one can take the Balti scholar Abbas Kazmi, who is involved in the revival of the Tibetan script for the Balti language.[10] As he explained to me at the 8th Himalayan Languages Symposium in Berne 2002, the Baltistan Cultural Foundation aims at establishing the classical orthography, not an orthography that would be suitable for the local variety, because the Balti people do not want to be just another negligible minority in the Northern Territories of Pakistan. They want to be recognized as one of the legal heirs of the Tibetan empire which once dominated Central Asia, and although they are Muslim, they nevertheless want to be associated with the fame of the cultural achievements of Buddhist Tibet.[11]

Although this motivation has not been expressed explicitly by the Ladakhi Buddhist scholars, it may be seen behind the claim that only the classical orthography allows inter-Tibetan communication from Ladakh to Bhutan (Chigmet Namgyal, IALS XI, Choglamsar 2003). This view is widely attested,[12] although there does not seem to be much need for communication between an average Ladakhi, Bhutanese, or Tibetan. The Bhutanese government, in particular, has developed a new orthography for Dzongka, the official language. From the point of view of Classical Tibetan, it must appear as *thestor* 'destruction', and many people hold that it would be better if the Bhutanese people did not use any script at all (Chigmet Namgyal, IALS XI). Even the most ridiculous statements concerning the Ladakhi language might become more understandable from the perspective of the Great Tradition. E.g., at IALS XI, one of the monks fiercely argued that the Ladakhi people should no longer use the traditional expression *jule*, universally used for greeting, pleading, apology, thanking, and goodbye, as this would be rather impolite and stupid, meaning merely 'good digestion' – in Tibetan, perhaps.[13] He was obviously identifying with a Tibetan perspective. According to many lay or clerical scholars, language change is a deviation from the true origins and should not be accepted.

The pride concerning Tibet's incomparable cultural achievements, particularly the invariant form of the Tibetan letters (in contrast to the modern Indian script that bears hardly any similarity with the script of the Gupta period), as well as the fear of cultural disintegration, was exemplarily formulated by the Amdo scholar Gedun Choephel (1978: 72–73[14]) in the late forties, well before the Chinese occupation or the challenges of modern globalization.

> 1,300 years have evolved since the time writing was introduced in Tibet. Yet, orthography and forms of writing have not witnessed much transformation through the years, and today, those with knowledge of Tibetan can decipher and comprehend inscriptions carved on stone pillars of old. In India, on the other hand, there is incomparable disparity between the Gupta scripts of a thousand years back with the script of the current era.
>
> Book printing systematically gained in popularity from the period of rJe Rinpoche [i.e. 11th century] and it is reputed that the volumes of block prints in Tibet find no parallel in the entire world.
>
> As mentioned above, orthography and literary forms [in the original: 'form of the script'] have retained their original structure. Therefore, as long as we adhere strictly to scriptural terminology, the unity of our diverse dialects will be preserved. If a written sample of our script travelled a regional cross-section from mNga-ris to A-mdo, every literate person would be capable of reading and understanding such a presentation. Conversely, if the colloquial languages of Ladakh and the Central provinces were encouraged to channel their growth into the compilation of dictionaries and religious works by a people possessing minimal aptitude in these languages, the unity of the common language would disintegrate, owing to the diversity of the colloquial languages in each province. An adjunct to this process would be the development of 'new ways of thinking' and distinct political characteristics as well as further debilitation of racial and political integration. Even if a new and common [in the original: 'such'] colloquial language were formed and developed all over Tibet, there will come a day when our regional dialects and literary language would be limited to surmise, and our voluminous literature, such as the Shastras and Tantras written in scriptural terminology, understood by none. This dangerous trend should be cautioned against and avoided.

Similarly, Ngawangthondup Narkyid from the exile community in Dharamsala writes (1992: 615):

> Historically speaking, Bonpos preserved the Bon religion and the Tibetan culture at first. Later, the Tibetan scholars and kings introduced Buddhism into Tibet, and Thonmi Sambota pioneered present Tibetan script. As a result of their great contribution, we Tibetans can proudly show our ancient civilized culture even in this very developed modern world.

Formal education and scholarship in Classical Tibetan is still considered more prestigious than a university degree in the modern sciences. For this reason, even lay people would not admit that they might have difficulties in understanding a classical text or might have experienced difficulties in learning the classical language. They obviously internalize the elitist point of view as expressed by Chigmet Namgyal (IALS XI, Choglamsar 2003) – even if that means admitting that learning *choskat* would be more difficult than learning *phalskat*: if the people do not understand the language of the religious books, it is their own fault, as they simply have not made the necessary efforts. Therefore, there would be no need to start teaching children with written *phalskat* or a simple version of Tibetan before teaching them *choskat*. Obviously, once one has mastered an average level of understanding Classical Tibetan, one tends to neglect the differences between the written and the spoken form. One can repeatedly hear that "*phalskat* and *choskat* are the same" (meme Tondup Tsering, Khalatse 2003).

One scholar, unintentionally, demonstrated the exact opposite to me by reading out a text in *choskat* format, explaining every second or third word by using its *phalskat* equivalent. Most illustrative might be his statement that the classical verb form *byed* 'does', which he gave in a pseudo Tibetan pronunciation *cet*, was nothing else than the Central Ladakhi spoken form *coat* (the written form of which could be *bcoḥad* or *byoḥad*). His argument proves to be heavily biased, since the Ladakhi pronunciation for the written form *byed* should be *bet*, as in the western dialects, and as in many other cases where written *py*, *phy*, or *by* + *i* or *e* becomes *pi*, *pe*, *phi*, *phe*, *bi*, and *be*, while the combination with *a*, *u*, and *o* yields *ca*, *cha*, *ja* etc. in the central dialects. Accordingly, the written past tense form *byas* would have to be pronounced *cas* (or even *bas* in the western dialects) in contrast to the Central Ladakhi spoken form *cos* (written *bcos* or *byos*). However, any linguistic argument about these features would be blocked as being based on Western concepts not applicable to the Ladakhi reality.

All of the conservative lay or clerical scholars I spoke to admitted that they do not have much or enough knowledge of the traditional grammar, and while their reading ability might be quite sufficient, many feel insecure about the correct spelling when writing. One person even made a fundamental mistake while explaining the difference between *choskat* and *phalskat* to me. Gelong Konchok Pande, on the other hand, who supports the language reform, seems to be one of the very few scholars who have studied the grammatical tradition in detail. He not only enjoyed the meta-theoretical discussion, but also told me frankly that the traditional grammar has certain shortcomings and that the commentary literature is not unanimous, particularly when it concerns the so-called "difficult points", such as case grammar.

In the context of the introductory remarks, it is quite interesting that scholars from Leh or the Upper Ladakh area generally opt for Classical Tibetan, while most if not all supporters of *phalskat* writing come from Lower Ladakh,[15] i.e. from the westernmost areas, where the dialects show a pronunciation closer to the Old Tibetan orthography than the Leh dialect. It seems that people from Lower Ladakh tend to have a greater awareness of linguistic issues. As Tsewang Tharchin (my informant for the Domkhar dialect in 1996, 2003, and 2004) told me repeatedly, he and other *Domkharpa*.s would often argue with the *Lepa*.s about the correct pronunciation of words like e.g. Classical Tibetan *rta* 'horse', *lta* 'look', and *starga* 'walnut', which are all pronounced as *sta (-rga)* in Leh and as *ta (-rga)* in the Upper Ladakhi dialects, while speakers from Lower Ladakh differentiate between *rhta*, *lhta*, and *starga*. When I experiment in writing Ladakhi e-mails, he often complains about expressions that are much too formal and *choskat*-like. Similarly, the lay historian, Sonam Phuntsog from Achinathang, does not get tired of opposing all claims of the conservative scholars that particular words or village names should be written according to Tibetan etymologies, arguing that they are of a non-Tibetan origin (IALS XI, Choglamsar 2003, cf. also Sonam Phuntsog 2004: 7). There might be historical reasons for this particular self-estimation and opposition to the "centre", as it was the king of Lower Ladakh, who reunited the kingdom of Ladakh and established the Namgyal dynasty (the split itself might have been related to an attitude of resistance against the influences from Tibet). People of the westernmost areas also tend to regard themselves as being of "Balti", i.e. Dardic or Turkic, rather than Tibetan origin (Skarma Namthak from Achinathang).

There are, of course, some linguistically more interesting arguments concerning the writing reform. One point is that once one made a reform, the dialects would further develop, so that one would soon need another reform, and so on. There has already been one language reform in the 9th century,[16] and this should be enough: "you cannot revise it again and again" (Chigmet Namgyal, IALS XI). Furthermore, many scholars believe that any change in spelling would lead to a change of meaning. If you write the name *Dbaŋmo* as *Aŋmo*, people will no longer know that it means 'mighty' (Nawang Tsering Shakspo).[17] Some scholars hold that by learning only the *phalskat* format, the students would be disabled to understand *choskat*, and, as they would never learn how to write "correctly", there would be "a lot of problems" (Geshe Konchok Namgyal, leaving it somewhat open as to what the problem was, exactly[18]).

Nawang Tsering Shakspo objected that you cannot just create a new (!) language. If you do so, you would need a new dictionary and a lot of text books, so who would do all the work? According to him, the Ladakhi people are not really interested in linguistic matters, so it would be a waste of time.

The main difficulty in establishing a standardized written *phalskat*, however, is the great diversity of the dialects, which differ in pronunciation as well as in grammar. As everywhere in the world, it seems to be difficult to decide which dialect should be given preference. Any debate on this issue might also enhance the particularistic tendencies and tensions between the Leh and Kargil districts. Many people thus argue that the dispute cannot be solved except by using the already established standard of Classical Tibetan. One response to this is that the most suitable dialects are either the dialect of Leh, because of its central position and, even more, its general prestige, or the dialects of Western Sham and Purik,[19] as it is generally admitted that their pronunciation comes closest to the classical orthography (cf. Thubbstan Dpalldan 2002: 237–238). Thus, they would need the least adaptation.

SECMOL's publications apparently use a compromise between these two options: while phrasing and grammatical markers mainly follow the Leh dialect, the classical orthography is retained as far as it is attested through the pronunciation of the western dialects. As a speaker of German, a language that shows great dialectal diversity, which is, to a certain degree, also reflected in the book language, I would like to add that a certain level of diversity in spelling and grammar may even add to the richness and beauty of the written language. Generally, changes in spelling in accordance to the phonology do not change the meaning any more than different pronunciations do.

Although the general fear that *phalskat* literacy would affect the understanding of *choskat* certainly has to be taken seriously, it is insubstantial, insofar as the two languages already differ to the extent, that ordinary people cannot understand *choskat*. Thus, for the sake of this argument, it would not matter whether they become literate in *phalskat* or not. However, if *phalskat* is not given the status of an official and literary language, and particularly as a written medium of instruction, people will switch to the more prestigious and practically more useful languages of Urdu/Hindi and English sooner rather than later. It goes without saying that education in the mother tongue proves to be more effective than education in any other language acquired at a much later stage. Literacy in *phalskat*, on the other hand, as well as an understanding of its grammar through adequate training in school, may well enhance the understanding of *choskat*, which after all is the younger cousin of Ladakhi and Balti *phalskat*. It could well turn out that at least a certain percentage of the students, once they are well versed in written *phalskat*, may develop a keen interest in the rich *choskat* literature.

This is also the opinion of Bakula Rangdol Nima Rinpoche who wrote a grammar for Ladakhi *phalskat* just because ultimately everybody should learn *choskat*. But he thinks, after all, that only by constructing a "bridge between the Ladakhi colloquial language and the classical literary grammar" one could heal the "weakness in Tibetan language among the younger generation" (Bakula Rangdol Nima 2005: 3).[20] This argument, however, is never accepted, perhaps not even understood by the conservative Buddhist scholars. Therefore, a formal congregation paid a visit to the Rinpoche after the release of his grammar in May 2005, humbly asking him to withdraw his grammar, while a more fundamentalist person went through all the bookshops telling the shopkeepers that they should not sell the booklet. At about the same time, the conservative scholars convoked a seminar in the main Buddhist temple where they passed the resolution to ban *phalskat* writing, imposing a fine on future publications. This was obviously directed against SECMOL, who also claim that they have been indirectly threatened with physical violence (cf. Sonam Wangchuk, open letter, *Ladags Melong* Summer 2005: 21).

While Chigmet Namgyal is one of the most conservative voices, and would probably not accept any compromise, most other scholars are ready for at least minor compromises. Most of the interviewed persons could accept a simplified spelling of grammatical morphemes as propagated by the Rinpoche as well as by SECMOL (e.g. *gi*, *ŋi*, *di*, *ni*, *bi*, *mi*, *ri*, *li*, and *si* in accordance to the last consonant of the lexeme instead of classical *kyi*, *gi*, and *gyi* for the Genitive) or the use of morphemes not attested in Classical Tibetan (such as the markers of evidentiality) while insisting on the "correct" orthography of the lexemes. Some of them would accept *phalskat* literacy if (and only if) *choskat* literacy is promoted at the same time. Geshe Konchok Namgyal, however, holds that it would not be necessary to train the children in *phalskat* writing; once they have learned *choskat* orthography, the ability to write in *phalskat* would come naturally. Most scholars would accept *phalskat* for the primary and pre-primary classes, but hold that *choskat* should be taught from the third class on. Only very few persons could be persuaded during the interviews that *phalskat* might be used in magazines and books on modern topics such as politics or technology, but even those vehemently opposed the use of *phalskat* in books on Ladakhi history.

The general view is that education in *choskat* is necessary to keep the standard of *phalskat* high. Obviously, the ideal of scholarship in the Great Tradition is deeply rooted. For the same reason, one may observe that until now, the issue of *phalskat* vs. *choskat* has been discussed in public only among male scholars, lay or clerical, and female voices are virtually absent (the foreigners Rebecca Norman from SECMOL and the author of this paper, of course, do not count). Women as well as "the ordinary man" might hide behind the back of the scholars, but if they are shown or read out a genuine piece of written Ladakhi *phalskat*, they might well appreciate it as *ŋati spera* 'our language', in contrast to *choskat*, which most of them do not really understand.

At present, it does not seem to be possible to convince the established scholars and elitists who simply do not want to be convinced. The only chance for the further development of Ladakhi (and Balti) *phalskat* is that the younger generation will develop the necessary interest and freedom of thought. Hopefully, the younger generation will understand that being part of the Great Tradition does not exclude being quite particular and, more specifically, that "you cannot sacrifice the welfare of your people for the ideal of some higher unity" (Gelong Konchok Pande).

Notes

1. By "Ladakhi" I refer to the spoken language of Ladakh, *phalskat* 'common language', and its various dialects. Ladakhi belongs to the Tibetan (or perhaps better: Tibetic) languages as part of the Tibeto-Burman language family. Tibetan as an linguistic or ethnic term is known in India as Bhotia, Bhutia, Bhotiya, Bhoti (with or without retroflex ṭ), Boti, or Bodhi. These designations as well as the Ladakhi word for the Tibetan script, *Budik* (from *bod* 'Tibetan' and *yig* 'letter, script') have led to the misinterpretation as 'Buddhist script' and thus to a certain reservation towards the use of the script in the Muslim community. On the request of various Muslim scholars "to change the name", as still formulated on the IALS XII, the 12th Seminar of the International Association for Ladakh Studies, Kargil, July 2005, it had been decided by an inofficial Ladakh Cultural Forum in March 2005 "that the spoken and written language of Ladakh region is to be called Ladakhi" (*Ladags Melong* Summer 2005: 8). The "written language of Ladakh", as far as it is propagated by the Buddhist scholars, is better known as Classical Tibetan, which will be termed here as *choskat* 'language of religion'.

   I am grateful to all those who, over the years, were willing to discuss the topic of language reform with me. Since the controversy between the traditionalist or elitist and the modernist, reformative stance tends to be very emotional, I will cite statements only when there is no risk that their author might be blamed or ridiculed for it, or if the statements were published or uttered in front of a greater public. Citations without indication of place and year are from interviews and conversations held in May 2004 in Leh, Ladakh in the context of the submission of this article.
2. No reliable data is available. The Indian Census of 2001 <http://www.censusindia. net> gives the population of the Leh district as 117 637 and that of the Kargil district as 115 227, totaling 232 864. No information is given on the linguistic and ethnic composition, as on the number of Tibetan refugees, army personal, seasonal or permanent migrant workers and employees. Apparently, many people

staying away from their native village have been counted twice, and there seems to be a sort of competition between Buddhists and Muslims and the two districts, respectively, to outnumber each other, leading to the unreasonable high figures.

Projecting the unbelievable growth rate of more than 30% per decade (21.34 % in India) backwards, the total number should have been 135 450 in 1986, but the Mini-Census of 1986–87 on behalf of the assignment of Scheduled Tribe status to the Ladakhis established eight "tribes" numbering altogether 152 035 persons (77 434 Buddhists and 74 601 Muslims). Even more, the Arghon, families of mixed Buddhist/Suni Muslim descent, were excluded from the tribal status and were obviously not counted, cf. VanBeek (1997: 35), so several thousand Ladakhis would have to be added. The home page of the state Jammu & Kashmir (India) <http://www.jammu-kashmir.com> gives the estimated population of Ladakh as 159 709 for 1998.

Generally, the data concerning the Muslim population seems to be problematic. While the 1981 Census (presented in Warikoo 1996: 189) counts 70 191 speakers of "Ladakhi" (i.e., mainly the Ladakhi Buddhist population), it counts only 46 890 speakers of "Balti" (i.e., the Ladakhi Shia Muslim population). An increase in population of 59% within 10 years (Mini-Census) is unlikely.

Even more confusing is the data given by SIL under <http://www.geocities.com> "Languages in Kashmir" and <http://www.ethnologue.com>, which count some 170 640 speakers in 1994 (97 000 for Ladakhi, 63 640 for "Balti", 8000–10 000 for Zanskari), but at least 311 000 in 1997 (102 000 for Ladakhi, 132 000 for Purik, 67 000 for "Balti", and 8000–10 000 for Zanskari, to which an unknown number for the speakers in the Changthang area would have to be added).

3. Up to 2000 Rs or 40 $ per month as against an average annual income from agriculture of about 10 000 to 12 000 Rs (in a good year) or a monthly salary between ca. 2000 and 6000 Rs for a teacher.
4. Nevertheless, the further away the children are sent, the less control the parents have over their success in school. There is a constant fear and gossip that the children might go astray. Moreover, some of the schools in Jammu are said to be even worse than in Ladakh (Rebecca Norman, Phey). Recently a case of child abuse shocked Ladakh. Ten Ladakhi children had been promised free education in Karnataka but ended up in a self-styled "orphanage". They were being treated badly and forced to work hard and beg for their living (*Ladags Melong* No. 21, April 2004: 20–22).
5. Geshe Konchok Namgyal. Note the use of the loan *asila* 'true', which is more and more replacing the word *denba* of Tibetan origin.
6. Cf. the statement in the *Daily Excelsior*, referring to the 8th Himalayan Languages Symposium in Berne, 2002 <http://www.dailyexcelsior.com/02oct29/state.htm>: "Ladakhi participants gave their thrust towards use of classical Tibetan for writing modern Ladakhi". All in all, there were three Ladakhi participants: apart from the two members of the Academy, Nawang Tsering Shakspo and Gelong Thupstan Paldan, cited by the newspaper, there was also Sonam Wangchuk from SECMOL, Phey, who certainly disagrees with that statement. SECMOL (Student's Educational and Cultural Movement of Ladakh) is a non-governmental organisation strongly concerned with the improvement of the education system and the promotion of *phalskat* literacy.

7. Bonpo scholars, however, usually claim that since the Bon religion existed prior to the introduction of Buddhism, there must have been also a script earlier than the time of Srongbrtsan Sgampo (cf. Gedun Choephel 1978: 71), and the more open minded scholars might be ready to accept this claim, even though it has even less historical evidence.
8. Already at the beginning of the 19th century, the Amdo scholar Guŋthaŋpa Dkonmchog Bstanpaḥi Sgronme (1762–1823) of Ndzorge (*Mdzoddge*) specified this need in the title of his religious text book: *Rdorjeḥchaŋdbaŋ miyi gar rolpaḥi* ‖ *yab rje Bstanpaḥi Sgronmeḥi žalsŋanas* ‖ *skyebo blodman kungyis go bdeḥi ched* ‖ *phalskad tshuldu gnaŋbaḥi zabchos bzhugsso* ‖ 'The Profound Dharma given in the vernacular so as to be well understood by all people of weak intellect. From the words of the honourable father Tenpay Dronme, who is a Vajradhara diverting himself in human form' (cf. Róna-Tas 1983 and Thubten J. Norbu 1983).
9. The Muslim standpoint seems to have changed over the years. Before the nineties, Muslims apparently experienced some pressure from the fundamentalist Sheiks to give up Ladakhi traditions on the pretext that they were non-Islamic. In this context, *Budik*, i.e. the Tibetan script, was interpreted and rejected as 'Buddhist script'. It seems, however, that more and more Muslims, due to the politics of reconciliation between the two communities (after the clashes of 1989), a growing sense of Indian nationalism (against the separatist movements in Kashmir), and a growing self-estimation of the Ladakhi culture, are at least willing to perceive *Budik* as what it is: a medium of writing, and the most suitable for the Ladakhi (and Balti) language. However, only a few persons have learnt the script.
10. For some basic information on this revival, cf. Siddharth Varadarajan (2002).
11. There are, of course, also different voices of people who are just "very concerned that Balti should not slowly be replaced by Urdu, that it should be read and written as an aid to preserving the language". The use of the Tibetan script could be "a nice ideal, but one which neither the government nor the great majority of Baltis are going to approve of". Particularly, if the original orthography "can't be of any use for present day spoken Balti [...] it simply won't be learned by the vast majority and will remain an interesting hobby for a tiny minority." A convenient romanization or a modified Persian/Arabic script would be even preferred as being "helpful to [the] children as they progress to higher education" (Eunice Jones, e-mail communication, December 2005).
12. As a variant of this common topic, Nawang Tsering Shakspo mentioned the need to send a letter to the Tibetan communities in South India. And Geshe Konchok Namgyal added that one could communicate even with the Mongolians, as they would use Classical Tibetan as the language of religion (which is not exactly the case for the common people).
13. Since this "etymology" has been promoted by authorities such as Tashi Rabgyas, it has become a sort of commonly accepted "fact" even for those who identify with the expression. E.g. Nawang Tsering Shakspo (s.a.: 3–4) writes:

    > *Julley* is the word which is used by every Ladakhi to greet both acquaintance and stranger alike. It sounds very sweet and literally means "good digestion" although it can have other similar meanings. This word can be used to say welcome, thank you and farewell.

    Besides the fact that an utterance has no other meaning than that intended by the speaker, the claimed origin of the expression is a case of folk etymology. As the word for 'to digest' *jucas* ~ *žucas* (Classical Tibetan *ḥjuba* ~ *žuba*) itself shows, *j* may vary with *ž* in Ladakhi (other examples for this variation are *kharži* for *kharji* 'food' and *žomšas* for *jomšes* 'to overcome, cure'). The people of Kargil as well as the Balti people again use *žu* as a greeting or as a polite interjection. There is another verb *žucas* (Classical Tibetan *žuba*) 'to request, talk (to a person of higher status)'. This verb is frequently used in Classical Tibetan as a performative verb, i.e. stating or accompanying a speech act, such as a request or any other kind of utterance. Thus, a formal phrase of welcome would have been *byonpa legspar žu* 'we wish you to have arrived well', and a pleading or apologizing phrase might have looked like *dgoŋspa stsalbar žu* 'I request you to give permission/leave', etc. The use of Ladakhi *ju* (Kargil and Balti *žu*) plus the honorific particle *le* is thus merely a conventionalized abbreviation (indepentently, Gelong Konchok Pande holds a similar view, Leh 2005). The phrase *in* + *ju* ~ *žu* 'pray, that's how it is/was' is commonly used as an affirmative interjection by the audience during the performance of a narration. The narrator him/herself might invite this interjection by terminating a sentence with *lo* + *ju* ~ *žu* 'pray, it has been said so'.

Sonam Phuntsog (2004: 7), on the other hand, suggests that *ju* could be an Indian loan, since one can find a honorific term of address *ju* 'Sir' in Rajasthan and Utarpradesh. This would be an elegant solution to a heated dispute, if only one could prove a closer relationship, e.g. through trade, between Ladakh and Rajasthan or Utarpradesh. There also seems to be a functional difference between a term of address and the multi-purpose expression in Ladakhi.

14. The Tibetan original is found in Dgeḥdun Chosḥphel, ed. 1979: 135–139.
15. E.g. Gelong Konchok Pande (Skyurbuchan-Achinathang) as well as Bakula Rangdol Nima Rinpoche (Lamayuru, originally likewise from Achinathang), to whom one could add the historian Sonam Phuntsog (from the same village, see also below). Achinathang is the last village on the right river bank before the Shina speaking enclave. Sonam Wangchuk's (SECMOL) native village Uley Trokpo is likewise located in Lower Ladakh, although in the eastern part.
16. Yet it was not a spelling reform, but a standardization of Buddhist technical terminology, see Zeisler (2005). Changes in spelling as attested between the Old and the Classical Tibetan texts must have occurred rather informally.
17. But, of course, they would, as they know the word *aŋ* (from *dbaŋ*) 'power, might' and the female derivative morphem *mo*.
18. He also referred to Sras Rinpoche of Ridzong monastery as having made a similar statement just some days earlier at the main temple of Leh.
19. The Purik dialects, however, appear to be somewhat less suitable, as they show a strong influx of Urdu loans as well as some differences in grammar and lexicon which are due to Balti influences.
20. Although he shows a strong interest in linguistics and in the particularities of Old Tibetan as well as of some modern Tibetan varieties, his modifications are rather moderate and concern only the grammatical morphemes. Lexemes should be written in the traditional orthography whether they corrospond to the Ladakhi pronunciation or not. At the most, one could use a spelling corresponding to the modern Amdo pronunciation or writing stile (discussion on behalf of the draft version, Leh, September 2004).

References

Bakula Rangdol Nima
2005 *Ladwagssi brdasprod bžugsso. A Ladakhi Grammar*. Achinathang, Leh: Cultural Preservation & Promotion Society.

Bray, John
2001 Christianity in Ladakh: The Moravian Church from 1920 to 1956. In *Ladakh Himalaya occidental. Ethnologie, écologie. Colloque organisé à Pau en janvier 1985 par Patrick Kaplanian et Claude Dendaletche. Recherches Récentes sur le Ladakh N° 2A / Recent Research N° 2A. Seconde édition entièrement revue, corrigée et augmentée par Patrick Kaplanian*. Patrick Kaplanian (ed.), S.l., s.a. [Paris 2001]: published by the editor.

Dgeḥdun Chosḥphel (Mkhaspaḥi dbaŋpo Mdosmadpa)
1979 *Bod chenpoḥi sridlugsdaŋ ḥbrelbaḥi rgyalrabs debther dkarpo žes bya* [The Genealogy Related to the Dominion of Great Tibet, Called the White Annals]. Dhasa: Bodgšuŋ Šesrig Parkhaŋ [Dharamsala: Tibetan Cultural Printing Press].

Gedun Choephel
1978 *The White Annals*. Translated by Samten Norboo. Dharamsala: Library of Tibetan Works & Archives.

Jiangbian Jiacuo
1994 <King Gesar> and Tibetan dialects. In *Current Issues in Sino-Tibetan Linguistics*, Hajime Kitamura, Tatsuo Nishida, and Yasuhiko Nagano (eds.), 1018–1026. Osaka: The Organising Committee.

Miller, Roy Andrew
1963 Thon-mi Sambhoṭa and his grammatical treatises. *Journal of the American Oriental Society* 83: 485–502.

Nawang Tsering Shakspo
s.a. *Notes on mask dance (cham) Ladakh*. Ayu Sabu [2004]: Center for Research on Ladakh Studies.

Ngawangthondup Narkyid
1992 A proposal: refinement of the Tibetan language and standardization of its writing system. In *Tibetan studies. Proceedings of the 5th seminar of the International Association for Tibetan studies, Narita 1989*, Shōren Ihara, Yūshō Miyasaka, Shigeaki Watanabe, and Shōkei Matsumoto (eds), 615–623. Narita.

Prins, Marielle
2002 Toward a Tibetan common language. Amdo perspectives on language standardisation. In *Amdo Tibetans in Transition: Society and Culture in the post-Mao Era. PIATS 2000: Tibetan Studies: Proceedings of the 9th seminar of the International Association for Tibetan Studies, Leiden 2000*, Toni Huber (ed.), 27–51. (Brill's Tibetan studies library, 2/5.) Leiden: Brill.

Róna-Tas, András
1983 Linguistic notes on an Amdowa text. In *Contributions on Tibetan Language, History and Culture*, vol. 1, Ernst Steinkellner, and Helmut Tauscher (eds.), 243–280. Wien: Arbeitskreis für Tibetische und Buddhistische Studien.

1985 *Wiener Vorlesungen zur Sprach- und Kulturgeschichte Tibets.* Wien: Arbeitskreis für Tibetische und Buddhistische Studien Universität Wien.

Siddharth Varadarajan
2002 Tibetan script makes a comeback in Pak. *The Times of India News Network* <http://timesofindia.indiatimes.com/articleshow/5135022.cms>. Thursday, March 28, 2002 12:26:40 AM (accessed).

Sonam Phuntsog
2004 *Ladwags debther. Ladakh annals.* Part two. *Šesbyaḥi baŋmdzod. General knowledge.* S.l: published by the author. 3rd edition.

Thubbstan Dpalldan
2002 *Ladwags.* Leh: Published by the author. 2nd edition.

Thubten J. Norbu
1983 Gungthangpa's text in colloquial Amdowa. In *Contributions on Tibetan Language, History and Culture*, Vol. 1, Ernst Steinkellner, and Helmut Tauscher (eds.), 221–242. Wien: Arbeitskreis für tibetische und buddhistische Studien.

VanBeek, Martijn
1997 The importance of being tribal. Or: The impossibility of being Ladakhis. In *Recent Research on Ladakh 7. Proceedings of the Seventh Colloquium of the International Association of Ladakh Studies, Bonn/Sankt Augustin, 12–15 June 1995*, Thierry Dodin, and Heinz Räther (eds.), 21–41. (Ulmer Kulturanthropologische Schriften, 9.) Ulm.

Wang, Xingxian
1994 Scientific value in the study of the language used in Tibetan oral version of Gesar. In *Current Issues in Sino-Tibetan Linguistics*, Hajime Kitamura, Tatsuo Nishida, and Yasuhiko Nagano (eds.), 1027–1033. Osaka: The Organising Committee.

Warikoo, K.
1996 Language and politics in Jammu and Kashmir: Issues and perspectives. In *Jammu, Kashmir & Ladakh: Linguistic Predicament*, P.N. Pushp, and K. Warikoo (eds.), 183–221. New-Delhi: Har-Anand.

Zeisler, Bettina
1998 Borrowed language: the impact of school education and mass media in Ladakh (Jammu & Kashmir India). In *Education in the North. Ways of Preserving and Enhancing Indigenous People's Languages and Traditional Knowledge*, Erich Kasten (ed.), 229–252. Münster: Waxmann.

2005 On the position of Ladakhi and Balti in the Tibetan language family. In *Ladakhi Histories: Local and Regional Perspectives*, John Bray (ed.), 41–64. (Brill's Tibetan Studies Library, 9.) Leiden: Brill.

Information and communication technologies and languages of South Asia

The impact of technology on language diversity and multilingualism

E. Annamalai

1. The ecological basis of language diversity

Diversity is intrinsic to the evolutionary process. This fact has been recognized since Darwin with regard to diversity in nature, which is the biological diversity of flora and fauna. That this is true of diversity in culture, which includes diversity in language, is a recent realization, which is reflected in the emergence of new fields of knowledge called Biocultural Diversity (Maffi 2001) and Ecolinguistics (Mühlhäusler 1996). The basis of this understanding is the thesis that, similar to how biological organisms select and develop unique features that strengthen their vitality by increasing their adaptability to their environment, humans also select and develop behaviors, beliefs and institutions which constitute their cultures, in order to increase their vitality within the environments in which they live. Language is a cultural product which enhances the ability to talk about the environment and strategize the best way of using it for the survival and perpetuation of the species that speaks it. The selection and development of unique cultural features, including language, leads to cultural diversity, just like speciation in the biological world.

Language becomes instrumental in representing the environment, communicating about it, and putting it to efficient use for furthering the interest of the cultural group as well as in disseminating this efficiency to the next generation. Language diversity is the natural outcome of cultural diversity, which in turn is the natural outcome of ecological diversity. It follows, as a corollary, that the protection of ecological diversity from environmental degradation is critical for the maintenance of language diversity. Differing languages codify and communicate efficiently different ecologies and the cultures based on them. Mühlhäusler (1995: 155) claims a dependency between speaking a language and living in an eco-system by arguing that "life in a particular human environment is dependent on people's ability to talk about it" and that the eco-system is coded and registered in the language as cultural "memory" to shape people's cognitive and behavioral ways of life.

2. Technological basis of language development

Tools, as extensions of the human body, redefine people's relation with the environment. They are a cultural product as well, whose creation and use involves designing for a purpose and learning from experience and coordination with others, all of which are facilitated by language. Visual-motor skills develop precision for tool making and using, and they refine the types of observation and articulation which support the development of language. Transmission of the skills in making and using tools is made efficient with language. Though the answer to the question "which came first?" in terms of the emergence of tools, language is speculative (Hewes 1993), it is a historical fact from the earliest times in the evolution of modern humans (Homo sapiens) that tools and languages are mutually reinforcing components of human cultural evolution. The tools made for improving the efficiency of hunting and gathering food and then cooking it for consumption are the beginnings of technology. Technology developed differently in different communities and ecological systems. It is the technology superior in creating surplus production and in intimidating others, be it arrows or atom bombs, that makes the difference in the power of the speakers of a language to control others. This power of the speakers possessing superior technology was transferred to their language by metonymic shift. The power acquired to use superior technology on another community, often against its will or consent, makes the language of the communities with inferior technology vulerable. This is a threat to language diversity. Indigenous languages are threatened more by cosmopolitan languages with impunity.

3. Language stability in India

This paper discusses the impact of technology on languages, and addresses particularly the question under what conditions technology destroys the multiplicity of languages in their form, that is, language diversity and in their function, that is, multilingualism. The discussion will particularly focus on the indigenous languages of India that are ecologically endemic and politically emaciated.

India is categorized as a developing country, which means, among other things, that agricultural production is the largest component of its economy, the consumption level of the majority of its population is low, and traditional ways of life have not been obliterated. These features correlate with the narrow spread of modern technology as means of production and consumption. This makes India a good place for a case study of the effect of the increasing use of technology to sustain the multiplicity of languages in real time. This paper does not produce empirical data to demonstrate this, nor does it make a control study of other variables such as ethnic or cultural identity or the economic interests of linguistic communities. It instead argues for the need for such a case study.

India is claimed to be a country of multiple linguistic communities whose natural adoption of the language of the community they come into contact with does not interrupt the transmission of their language to their offspring within their community (Pandit 1979). Nevertheless, language shift is reported, especially, within indigenous communities (Ishtiaque 1985; Khubchandani 1997). Though there are not many cases of total loss of speakers, and thus loss of languages, during the historical period (although some are close to disappearing), there is loss of speaker strength and functional spread. About half of the people of tribal communities do not speak their native languages. The difficulty in keeping count of lost languages in India, however, must be noted. Census reports, which give an account of languages, make languages "disappear" by subsuming some mother tongues under other, usually major, languages. For example, 1652 identifiable mother tongues are "reduced" to 193 languages in the 1961 census and 1576 to 114 in the 1991 census (Bhattacharya 2002). The difference, therefore, in the number of reported languages in 1961 (193) and in 1991 (114) is not an indicator of lost languages, but is a result of the difference in the officially subsumed mother tongues under languages. The arbitrariness of the language figures will be clear from the fluctuation in them between censuses: 105 in 1971 and 109 in 1981. Added to this difficulty of reduction in the number of languages by official classification is the political decision in 1971 not to bring to the census account unclassified mother tongues spoken by less than 10 000 speakers and to not even to raise the question of how many of them are distinct languages. They are deemed to lack language status because of their small speaker strength, and so are obliterated from official cognizance.

4. Non-local management of eco-systems

Real language loss occurs when the government's policy of development or the market's goal of fueling consumption brings ecological instability and destruction. The government relies on technology and on increased energy to use it for development; the market relies on the same to increase production and profit. The common interest of the government and the market can be best seen in the control of the forest. When timber emerged as an important commodity for use in building railways and ships to aid development of trade and industry (and military movement) the colonial government in India assumed monopoly of the forests in 1865. "When the colonial state asserted control over woodland earlier controlled by local communities, and proceeded to work these forests for commercial timber production, it represented an intervention in the day-to-day life of the Indian villager, which was unprecedented in scope" (Gadgil and Guha 1992: 147). Day-to-day life includes the ways the language is used. There is no systematic study in India as to how such political and commercial measures for the purpose of "development", which transforms ecology and the local communities' relation with it, impacted on the language use of those communities. But it can be extrapolated from documented cases that there is a real possibility of loss or mutilation of language and, consequently, loss or emaciation of multilingualism. The story of the language of indigenous people when forests became plantations is a case in point.

5. Ecological disruption and language

The documented case of language endangerment in India concerns Andamanese of the Andaman Islands (Annamalai and Gnanasundaram 2001). In addition to deaths of speakers, caused by diseases transmitted by settlers from the main land, or gunfire between British and Japanese troops in the Second World War, ecological and technological shifts in the islands are also important factors in the endangerment of this language. In the name of tribal protection and development, the hunting community of Andamanese is settled in tin-roofed concrete houses on one island with a government-ensured food supply. They are forced to abandon their traditional tools of hunting in the new environment and to adapt to modern scheduled means of moving from place to place. Making tools and moving freely is outside their control. This shift is sudden, and is external to the community. The language of the community does not have adequate time to change naturally, or to talk about the new environment and the way of life in it, thus becoming viable in the new environment. Without the above, the language becomes a relic of the past and shrinks in the present. Its speakers feel lost.

Andamanese is an extreme case of the negative impact on language from forced ecological change that includes a sudden introduction of new technology from outside. New technology could endanger a language in two ways: by denying time or opportunity for a language to adapt, as in the case of Andamanese, or by accepting another language for exclusive use of the new technology, as in the case of Indian languages in general with regard to high-level technology, for which English is favoured. In the former case, the language loses its vitality to survive; in the latter case, it loses its vitality to grow. Neither is conducive to stable language diversity or to viable multilingualism. The point of this paper, however, is not about language adaptation to new technology from an external source, but about the effect of technology on indigenous languages through disruption in ecology, i.e. disruption and its effects on the intrinsic relation between language and its environment.

6. Industrial development and language diversity

It is a fact that there is a difference in the nature of multilingualism in an industrialized environment and a primordial one, which is a result of different impacts of technology. The former environment involves urbanization, in which the migration of people of different linguistic communities to industrial centres takes place, giving rise to multilingualism. The urban ecology itself is created by technology, and therefore the question of technology being a disruptive agent is tautological. Technology may actually help in the maintenance of mulitingualism by facilitating speakers to keep in contact with their language through channels of communication with the home base and through entertainment delivered to their homes. The instability of urban multilingualism is caused by economic and socio-political factors which are only indirectly connected to technology. The technology of primeval ecology, on the other hand, is conditioned by the environment. Therefore, any sudden change in technology is potentially disruptive. The census figures concerning the mother tongues of indigenous communities in India suggest this potential of technology.

The density of endemic languages is high in areas with low technology. Of the 193 languages (of the 1961 census), 101 are tribal languages, i.e. languages of tribal communities with traditionally low technology. Of these, 63 are found in the hilly north-east region considered as least developed industrially. About 30 of these 101 languages are in the hills and adjacent areas in other parts of India, where modern industry is present in the periphery of their habitation. Many of these areas are rich in fossil energy and mineral resources. It will have to be seen how many of these languages will survive when high technology goes in to exploit these resources for trade and development. Even with industry in the periphery, language shift rate is higher in these regions (40–75% in the central region and 75% in the southern region) compared to the north-eastern region (11–17%) (Khubchandani 1997). Sudden introduction of new technology, which creates a new, unequal economic relation with the players in the primeval ecology, will be a threat to the multiplicity of languages.

7. Twin edges of technology

The following is a paradox. We indicated the mutual reinforcement of development of tools and language earlier in the text. We now point to the disruptive potential of technology to language. We already noted the emergence of some languages as dominant and threatening to other languages, due to differential development of technology and the imposition of technology from their ecology onto another. The latter is an ecological misapplication of technology resulting in the maladjustment of the ecological system to the imposed technology. The apparent paradox can be explained if we change the focus of the question from whether technology is detrimental to language diversity and multilingualism to how technology is used, by whom, and for what determintal purpose. Technology aids the maintenance of languages for migrated speakers, as mentioned above, and it makes communication possible between speakers and readers of different languages without recourse to one dominant language through swift, automated translation. It also aids in the teaching of second languages to a large number of learners, taking down the cost of learning. It helps to record endangered languages and to preserve them in a permanent form, if the language could not be saved. All this is possible using the same information technology with whose predominance English has strengthened its position as a global language for communication and information. If other languages are not lost totally to English, they in any case lose their speakers or their public functions to it.

8. Manner of using technology

The destructive power of technology comes from its users. To prevent it from destroying language diversity and multilingualism, the prevailing paradigm of economic development and political hegemony must change, both of which are based on the self-serving use of technology. It must change in such a way that it does not disrupt the ecological balance of a particular place, including the role of the local language in it. Uncritical use of technology, unmindful of whether it results in deforestation or submersion under water, will displace languages as it makes ecosystems disappear. It must also change in such a way that it will not place a premium on quick returns and will allow, a gestation period for languages to absorb the new technology and adapt it to the local conditions. Not global technological advances, but technological appropriation and application in sync with the local systems, is what will support diversity in languages and their use.

References

Annamalai, E., and V. Gnanasundaram
2001 Andamanese: Biological challenge for language reversal. In *Can Threatened Language be Saved? Reversing Language Shift, Revisited*: A 21st Century Perspective, Joshua A. Fishman (ed.), 309–322. Clevedon: Multilingual Matters.

Bhattacharya, S. S.
2002 Languages in India: Their status and function. In *Linguistic Landscaping in India with Particular Reference to the New States*, N.H. Itagi, and Shailendra Kumar Singh (eds.), 54–97. Mysore: Central Institute of Indian Languages and Mahatma Gandhi International Hindi University.

Gadgil, Madhav, and Ramachandra Guha
1992 *This Fissured Land: An Ecological History of India*. Berkeley: University of California Press.

Hewes, Gordon
1993 A history of speculation on the relationship between tools and language. In *Tools, Language and Cognition in Human Evolution*, Kathlenn R. Gibson, and Tim Ingold (eds.), 20–31. Cambridge: Cambridge University Press.

Ishtiaque, M.
1985 Language shifts among the tribal communities: A case study of Santals, Mundas and Oraons in Bihar. *The Geographer* 32 (2): 69–78.

Khubchandani, L.M.
1997 Demographic indicators of language persistence and shift among tribals: A sociolinguistic perspective. In *Languages of Tribal and Indigenous Peoples of India: The Ethnic Scene*, Anvita Abbi (ed.), 71–90. Delhi: Motilal Banarsidass.

Maffi, Luisa
2001 Introduction: On the interdependence of biological and cultural diversity. In *On Biocultural Diversity: Linking Language, Knowledge and the Environment*, Luisa Maffi (ed.), 1–50. Washington D.C.: Smithsonian Institution Press.

Mühlhäusler, Peter
1995 The interdependence of linguistic and biological diversity. In *The Politics of Multiculturalism in the Asia/Pacific*, David Myers (ed.), 154–161. Darwin: Northern Territory University Press.
1996 *Linguistic Ecology: Language Change and Linguistic Imperialism in the Pacific Region*. London: Routledge.

Pandit, P. B.
1979 Perspectives on sociolinguistics in India. In *Language and Society: Anthropological Issues*, W. C. McCormick, and S. A. Wurm (eds.). The Hague: Mouton.

The impact of technological advances on Tamil language use and planning

Vasu Renganathan and *Harold F. Schiffman*

1. Introduction

The influence of technology on language use and maintenance in the Tamil situation has never involved any imposition upon the speakers of a language, but has rather been one where changes evolved involuntarily and naturally along the way as the technology developed. From the beginning, with palm-leaf manuscripts, to the introduction of printing,[1] and up to recent electronic developments, languages have adapted themselves to the development of technologies in various stages. For example, the use of the Tamil script in inscriptions on stone and writing on palm-leaves made it necessary to eliminate the dots in the script,[2] but later this hurdle was overcome in the development of printing technology for Tamil, and the dot could be reintroduced. Printing technology helped standardize the shapes of Tamil letters by making new typefaces, first in wood and later in metal. The developments in electronic media impacted the standardization of a digital code for every letter in the language. The graphic rendering of Tamil scripts and typesetting technology further changed the way the Tamil letters looked as far as their aesthetical appeal.

One of the main forms that the electronic development of computerized printing took was that of competitive entrepreneurship in the newspaper and magazine industry, which made use of internet technology in a rapid and exceptionally comprehensive manner. This aided in the evolution of many online newspapers, with the result that the use of the Tamil language, as well as the dissemination of knowledge via "electronic" Tamil, was able to construct a new dimension of the globalizing media. However, the electronic method of presentation of the language required sufficient skills in keyboarding in the language using Tamil fonts, which required a specialized literacy from the point of view of the users. Many users were handicapped at first as far as the inputting aspect of the language use in electronic medium, and thus were left the possibility of being involved only in the reading but not in the writing medium, an identical circumstance to that which arose when the use of Tamil typewriters developed. Only those who could acquire skill in typing

Tamil were able to use the typewriters, and others had to depend upon them. There have been instances where there is an implicit need for the submission of language materials in the print media as opposed to handwritten documents. In such circumstances, technological development has in fact produced more illiteracy with respect to the use of the language on a par with the emerging technological developments.[3] Unlike high school graduates in the US, who are expected to learn to type in order to succeed in college, in India and other countries people may *never* learn to type, and then, with the introduction of computers and the need for keyboarding, are force to "hunt and peck" or else not be able to produce any electronic documents.

2. Natural and artificial language

When the British came to India, language became a central issue in their ability to govern the country, and they imposed their preference of language and culture onto the various peoples of the region. Sanskrit, Persian, and English were the three languages they cared most about, and they mostly ignored the other vernacular languages. Thus, Tamils evolved a practical alternative of using English to facilitate better communication with the rulers of that time. This trend continues to date; the English language has influenced the way Tamil is used in a number of different ways, and there are some bottlenecks which have developed in the use of this language at various levels. The two dominant languages, Sanskrit and English, influenced the Tamil language both at linguistic and cultural levels. Publication of many Tamil language magazines and books in the *manipravala* style from the middle of the nineteenth century onwards resulted in the development of the Tamil purism movement,[4] whose main goal was to safeguard the Tamil language from the excessive use of Sanskrit loan words. Supplementing the efforts of the purist movement, the Tamil Nadu government, ruled by the Dravidian movement (especially the DMK), created a separate department for Tamil development, whose primary goal was to promote the use of "pure" Tamil in government offices, notice boards, and name boards in busses, trains etc. This, however, made much less impact on the common people, who often tend to assume that "pure" Tamil is always artificial, and that the *manipravala* style language that they use quite consistently is the natural one. Have the recent advancements in electronic media impacted this particular attitude of the people in any manner? In other words, can anyone claim that the recent developments in technology or the achievements of the Tamil purist movement have helped people in any way to adapt to the "purest" form of Tamil, as opposed to Sanskritized and/or Anglicized Tamil language? The answer seems to be that no developments in technological advances seem to have affected this basic attitude of the common people about the way spoken Tamil is used, but rather these advances have only triggered the implementation of policies that affect the more formal and technical variety of the language.

Since the late eighteenth and early nineteenth centuries, the Tamil language (as written) has become relatively standardized and is now used for most levels of administration, business, and social media. British rule was an impetus for the official codification of the regional tongues, and the colonial administrators and missionaries learned regional languages and often studied Tamil literature. Their translations of English-language materials and the Bible encouraged the development of written, standard language. Teaching materials, prose compositions, grammars, dictionaries, and textbooks were often commissioned in the standard literary Tamil, but much less frequently in the spoken variety of the language. Industrialization, modernization, and printing gave a major boost to the vocabulary and standardization of the Tamil language, especially by making the wide dissemination of dictionaries possible. These developments, however, helped create a variety of sophisticated and educated Tamil that is opposed to the day-to-day use of the less sophisticated variety of Tamil, which is influenced heavily by Sanskrit and Hindi at the lexical level, and English at both lexical and grammatical levels.

3. Localization efforts

A need to develop local resources was long felt in order to develop online news publishing, online education, and electronic commerce. Though the driving force for the internet market comes from the US, according to a report by one of the online magazines, the percentage of US internet users as a percentage of worldwide users is dropping: from 65% in 1994 to 55% in 1997, and further down to 40% in 2000. What this implies is that the proportion of non-US, non-English-speaking internet users is growing rapidly. Hence many US companies, including the Microsoft Corporation, have realized the need to publish web content in languages other than English, and to also provide fonts in these languages. Various attempts in this line of thinking have ventured into multilingual web publishing, with varying degrees of success.

Besides these special efforts, the development of internet and electronic technology have in general implied furthering the use of English among speakers of other languages. The vernacular languages always had to compete with the already well developed technology using English. The 8-bit font technology and the most recent Unicode technology, for instance, are still in the developmental stages, and have not yet completely created an environment wherein one can use the Tamil language effectively to promote local language content. However, in order to cope with the technological developments, many language policy and planning decisions were taken by the Tamil Nadu government in order to try to meet the compelling needs of the emerging technology.

4. Efforts of governmental and non-governmental agencies

One such attempt is the announcement by the Tamil Nadu government in 1999 of a Tamil software development fund to facilitate the use of the Tamil language in all aspects of computing, such as the development of Optical Character Recognition (OCR), Tamil accounting software, educational software to teach science and math in Tamil, and so on. At the same time, the government also standardized the coding schemes for Tamil fonts and keyboards, hoping to better coordinate the efforts of software developers and to share resources without major impediments. The global initiative of the International Forum for Information Technology in Tamil[5] (INFITT) founded by Tamil supporters throughout the world helped coordinate efforts with the Tamil Nadu government to boost local language content and to access the infrastructure of the technological developments in an efficient manner. Efforts to digitize the already extant Tamil literature that are now only in printed form have been among the major planning strategies of adapting the technological developments. This has changed the accessibility of various genres of Tamil literature both to Tamils and to the Tamil researchers around the world in a number of different ways.

The Tamil Nadu government also sponsored the establishment of a Tamil Virtual University,[6] which has as its goal the teaching of the Tamil language online, on a global scale, as well as to develop a comprehensive online archive of Tamil literature, lexicon, and other related materials. These developments are in fact nurturing, in many respects, the already evolving globalization efforts in the areas of language, literature, business, entrepreneurship, economy, and so on. In this sense, the government's support for online initiatives and keyboard and font standardization efforts in the local language is expected to be instrumental in tapping into Tamil-speaking rural and home markets in India, and in the Tamil Diaspora.

Efforts of the Unicode consortium,[7] along with Microsoft's initiatives in promoting localization supports in computing have, had a tremendous impact on the way the Tamil language is being used by the Tamils. The Unicode font technology helped solve the problem with the previously implemented 8-bit font technology, and as a result the creation and sharing of online resources have benefited from them in a tremendous manner. The Microsoft Community Glossary on Tamil (an INFITT sponsored Microsoft initiative)[8] has developed a new way to create technical glossaries for Tamil through their online input method, so Tamil language experts throughout the world can help build and use the database in order to ease the use of Tamil language in computers. It must be noted, however, that technical vocabulary (known in linguistics as "register development") tends to be developed by *users* first, i.e. the people who actually develop the technology and the knowledge also create the vocabulary, and this vocabulary then tends to spread into other languages, because users from those languages learn it from them. Since computing vocabulary and knowledge has developed primarily through English, English terminology spread, and no amount of "translation" into other languages seems to help. Often the English terminology is short and snappy, using abbreviations (e.g. DOS) or whatever other term came to mind (the "floppy" disk) whereas translated terminology tends to be long and complicated. Translation also takes time, and often the technology and the knowledge are light years ahead of the translations, so that by the time the terminology is provided, the technology has outstripped it. Users then try to keep abreast of developments by using the original language, i.e. English.

These independent efforts have been supplemented by the Tamil Nadu government's regular ongoing efforts to publish new technical vocabularies for Tamil through their expert groups, led by the Department of Tamil development. However, these technical vocabularies, containing both the genuine Tamil words as well as loan translations, tend to create an artificial style of the language which is in no way compatible to the intuitive knowledge of the speakers of the language. This, then, tends to create a new technical variety of Tamil, which is often quite incomprehensible and not very usable by native speakers of the language, unless special training is undertaken to teach them to understand and use them. In other words, the competence of a speaker of Tamil to handle spoken or written varieties of Tamil is not sufficient to handle this new variety of technical Tamil. It is also not clear who the audiences of the newly created technical vocabularies are, but presumably they are intended for everyone, in the same way that both literates as well as illiterates are expected to have an understanding of literary Tamil, as Schiffman (1978: 105) notes in the context of diglossia in Tamil.

5. Evolution of standard varieties and internet communication opportunities

Popular use of radio, television, the movies, and the print media throughout the twentieth century have fostered a standardization of the Tamil language at both spoken and written levels. The existence of these standardized varieties contributed to the strategy of code-switching to a common (standardized) language among the speakers of various dialects. Yet the language of instruction and administration maintains the very formal classical variety of Tamil, and as a result there is an even wider divide between the formal and the informal variety of Tamil. This divide seems to be getting even wider and deeper. What is questionable in the context of corpus planning in Tamil, thus, is whether these efforts of "elaboration" in terms of expansion of vocabulary actually allow the language to function in a greater range of circumstances, including that of using computers. Growing interests in computer usage and internet technology among the Tamils does not in fact seem to be promoting Tamil language use, nor does the knowledge of Tamil help them to better understand the technology. Instead, the knowledge of English language has always been a prerequisite in adapting to the new trends in technology, and computer use may actually tend to make more people switch to English rather than less.[9]

Email and listserv communications are the two other media that extensively use the Tamil language. Many Tamil word processors and Unicode fonts help render Tamil in word processing documents and emails more extensively than ever before. Marathadi,[10] Thendral magazine,[11] Tamil Araychchi[12] (Tamil Research) etc., are some of the electronic groups that engage quite extensively in a Tamil mediated communication using Unicode and other standardized font encoding methods such as TSCII. The style and the variety of language the members use in these groups is obviously uncontrolled, and the informal spoken variety is gaining currency in these medium of communication. Further, this new trend of cyber communication is quite different from the earlier methods of handling print and postal medium, in that exchange here is very interactive and almost equivalent to day-to-day oral communications. Also, the threading facilities in these methods of communication make the exchanges more lively on a wide variety of topics. Seemingly, this method of using informal varieties in a formal setup can have the potential of narrowing the gap between the formal and informal varieties of the Tamil language, and can thus help break the notion of the *de facto* standard Tamil as projected by some of the official and unofficial Tamil language special interest groups.

6. Agencies affecting the language use

One of the press releases from the Information Technology department of the Tamil Nadu government indicates that Tamil Nadu has the largest number of English medium schools in any Indian state, and that there are more than 245 Engineering colleges, 12 Medical colleges and 400 Arts and Science colleges. As English is the most commonly used business and social language, the majority of these educated people are fluent in English. Premier educational institutes like the Indian Institute of Technology, Madras University, Anna University, Regional Engineering College, and others produce some of the best talent in the country. It is further claimed that through innovative and progressive schemes students are graduating from these schools and colleges computer literate, and of course are taught exclusively in English medium, and that Tamil Nadu was the first State to introduce computer education in schools and colleges with Public-Private participation (PPP). Under this program, students in classes 9 to 12 are given computer literacy classes during school hours, and the public in the neighborhood are trained on computer skills after school hours and during holidays. According to the government, this program is being successfully implemented in 1200 High schools (Higher Secondary schools) spread all over the State, and that it has provided, through the PPP route, computer labs in 65 Arts and Science Colleges which provide computer education as an optional subject. Furthermore, about 28 000 students are undergoing training under this scheme. Beyond this, 11 Government Medical Colleges, one Dental College and 5 Law colleges covered under this scheme train about 2000 professionals annually.

It must be noted that all of these governmental schemes pay much attention to the use of the English language as opposed to the Tamil language, because of popular demand and the future opportunities that are provided. Unless these money-making prestigious institutions transform themselves in a way that involves adaptation of Tamil to its the fullest extent, we can not expect them to have any promising impact using the relatively tentative policies such as those which involve the government software development fund, the creation of databases of technical vocabularies, or online Tamil education.

Notes

1. Tamil was the first Indian language to have movable-type printing technology, introduced by the Portuguese at an early stage, in order to Christianize the Tamil population. (See Renganathan and Schiffman 2005).
2. Tamil script uses a dot (puLLi) placed over a letter, to indicate that no vowel (usually /a/) is present.

3. In fact, writers in other languages often had to provide "fair" copies of their writing in order to get them printed; in the current centennial year of James Joyce's *Ulysses*, one reads that Joyce had to produce this himself, since he could not type, and could not afford to pay anyone else to do it.
4. Known by various names, but the *tani tamiR iyakkam* covers all bases.
5. <http://www.infitt.org>.
6. <http://www.tvu.org>.
7. <http://www.unicode.org>.
8. <http://members.microsoft.com/wincg/ home.aspx?langid=1097>.
9. There is perhaps a parallel here with the development of literacy in previously illiterate societies. At first literacy rises, as the population learns to read in their own language. But then this literacy allows people to learn a colonial or more widely used language, and thus leads to language shift. (Mühlhäusler 2000)
10. <http://groups.yahoo.com/group/Maraththadi/?yguid=586001>.
11. <http://groups.yahoo.com/group/thendralmag/selprof?yguid=586001>.
12. <http://groups.yahoo.com/group/tamil_araichchi/?yguid=586001>.

References

Mühlhäusler, Peter
2000 Language planning and language ecology. *CILP* 1 (3): 306–367.

Tamil Nadu Government
2002 Press release. Draft Information Technology Enabled Services (ITES) policy of Tamil Nadu. IT Department, Tamil Nadu Governement.

Renganathan, Vasu, and Harold F. Schiffman
2005 Tamil script reform and glyph rendering approach in Unicode: Past and present attempts to simplify Tamil writing system. In *Indic Scripts: Past and Present*, Peri Bhaskavarao (ed.). Tokyo: Tokyo University of Foreign Studies.

Schiffman, Harold
1978 Diglossia and purity/pollution in Tamil. *Contributions in Asian Studies* XI: 98–110.

Corpus-building for South Asian languages

Andrew Hardie, Paul Baker, Tony McEnery and B. D. Jayaram

1. Introduction

Many language technologies are based on or utilize language corpora. However, until recently, corpora of South Asian languages were not available to the international research community. This chapter discusses some issues related to the construction of such corpora, in the context of efforts undertaken by a team at the University of Lancaster in constructing a 96-million word corpus covering fourteen major languages of South Asia. In particular, we will address some of the difficulties involved in building corpora for these languages, and the solutions which we found to these difficulties.

This corpus building was undertaken in the context of the EMILLE[1] (*Enabling Minority Language Engineering*[2]) project. In sections 2 and 3, the background and basic aims of this project are briefly explained. We then describe the most important features of our corpus (section 4). In section 5, we discuss a range of problems that we encountered in developing corpora of parallel texts, of spoken data, and of monolingual written data (5.1 to 5.3 respectively), elaborating on how we solved or circumvented these difficulties. Section 6 discusses another set of difficulties, those which accompanied our work to develop automated part-of-speech tagging for the Urdu part of the corpus.

The EMILLE project is now complete, and the full corpus is available free of cost for non-profit research use.[3]

2. The EMILLE project: Background

Corpus construction and language technology development for the languages of South Asia have, for some years now, constituted an area of growing interest within the field of language engineering (LE). This has been the case both in South Asia and in areas such as the US, the UK and continental Europe, where many South Asian languages such as Hindi-Urdu, Punjabi and Gujarati are important minority languages.[4]

However, until relatively recently there was a dearth of corpora for both the Indo-Aryan and the Dravidian languages.[5] This became clear to us in the course of an earlier project, MILLE[6] (*Minority Language Engineering* – see Baker and McEnery 1999). This project undertook a major review of the resources available to language engineers wishing to work on South Asian languages, as well as looking at the resources that the LE community would like to have available. EMILLE was undertaken in order to provide the desired resources.

3. EMILLE: An overview

The goals of EMILLE were threefold: to develop an LE architecture suitable for working on the South Asian languages; to develop corpora for the languages shown to be "most wanted" in the LE community by the Baker and McEnery (1999) survey; and to undertake some basic automated linguistic analysis of this corpus data.

To develop a suitable LE architecture for the languages in question, a team at the University of Sheffield extended the GATE[7] architecture to make it fully Unicode-compliant. This research is discussed in depth by Tablan et al. (2002) and will not be discussed further here.

The corpus that we originally envisioned building consisted of 9 million words of written data for each of Bengali, Gujarati, Hindi, Punjabi, Sinhala, Tamil and Urdu. We also planned 500 000 word corpora of spoken data for those languages with a large enough UK speaker community to support data collection – namely Bengali, Gujarati, Hindi, Punjabi and Urdu. Finally, we planned a parallel corpus consisting of 200 000 words in each of these languages plus English.

The final make-up of the EMILLE spoken and parallel corpora is generally in line with what was planned. A major change, however, was that the planned make-up of the monolingual written corpora changed substantially as the project progressed. Thanks to a series of grants from the UK EPSRC,[8] EMILLE was able to establish a dialogue with a number of centres of corpus building and language engineering research in South Asia. The most important outcome of this was a very fruitful collaboration with the Central Institute of Indian Languages (CIIL) in Mysore. By combining our corpus-building efforts, we have been able to produce a wider range of monolingual written corpora than was originally envisaged on the EMILLE project. Although the equal word-counts originally planned for each language could not be achieved in the final version, we were able to extend our coverage to cover seven further languages (Assamese, Kannada, Kashmiri, Malayalam, Marathi, Oriya, and Telugu). We were also able to increase the overall extent of the monolingual written corpus by an additional 30 million words.

Table 1. Word counts for the EMILLE- CIIL monolingual written corpora

Language	***Words***
Assamese	2 620 000
Bengali	5 520 000
Gujarati	12 150 000
Hindi	12 390 000
Kannada	2 240 000
Kashmiri	2 270 000
Malayalam	2 350 000
Marathi	2 210 000
Oriya	2 730 000
Punjabi	15 600 000
Sinhala	6 860 000
Tamil	19 980 000
Telugu	3 970 000
Urdu	1 640 000
Total	***92 530 000***

The final word-counts for the EMILLE-CIIL Monolingual Written Corpora are given in Table 1.

The final goal of EMILLE was to build basic tools for linguistic analytic annotation of our various corpora. In fulfilment of this goal, three types of analysis were applied to the data: part-of-speech annotation of the Urdu corpora, alignment of the parallel corpus, and annotation of demonstrative anaphora in a subsection of the Hindi corpus.[9] The next section discusses the makeup and format of the corpora in some greater detail.

4. An overview of the EMILLE corpora

4.1. Text encoding

As will be discussed in greater detail in 5.3.3 below, there are a large number of incompatible text encodings of the various South Asian writing systems in use today. The texts in the EMILLE corpora are drawn from sources that make use of over twenty incompatible encoding systems. To harmonize the data and thus make the corpus more coherent and more easily useable, the texts in all of the EMILLE corpora have been transferred to a common encoding, namely 16-bit Unicode.[10] Using Unicode allows the same encoding system to be used for all the files, no matter what the source of the text, and no matter what writing system is used. All South Asian alphabets, such as Perso-Arabic, Devanagari and Sinhala, are available in Unicode.

Another advantage of Unicode is its status as a major international standard. Since its inception it has been incorporated into a number of widely-used software systems. For instance, Microsoft Windows and Microsoft Word have become, in recent versions, Unicode compliant. The Java programming language[11] also allows the full use of Unicode.

4.2. Textual mark-up

The various EMILLE corpora are marked up using SGML/XML,[12] which has, over the past twenty years, emerged as a mark-up standard for text corpora. To enhance the standard nature of the mark-up, we made it compliant with the level-one recommendations of the Corpus Encoding Standard (CES[13]). To implement the higher-level CES recommendations would have involved additional processing and / or post-editing; for this reason, we have not gone beyond the level one recommendations.

The main SGML elements in use within the body of the corpus texts are as follows:

- <p>, <s>, <head>, <gap> *in written texts*
- <u>, <unclear>, <omit>, <event>, <vocal> *in spoken texts*[14]
- <foreign> *in both*

In written texts (including the texts in the parallel corpus), the most commonly used tags are <p>…</p>, which delineates the paragraphs of the original text, and <s>…</s> which delineates sentences.[15] Also used fairly frequently is <head>…<head> for headings of various kinds (for instance, headlines in news text). The <gap> element is used to indicate an element of the original text that is not represented in the corpus text (for instance, illustrations.)

The spoken text is marked up differently. The main division of the text is into *utterances* <u>…</u>; the <u> tag also indicates which of the speakers specified in the header was responsible for the utterance. Non-linguistic sounds on the recording are indicated by the <vocal> element (for vocalisations such as coughs) and the <event> element (for other noises, e.g. musical breaks in a radio programme). Where speech has had to be omitted, if it was insufficiently clear on the recording to be transcribed, this is indicated with an <omit> element; speech that was unclear and has been uncertainly transcribed is surrounded by an <unclear>…</unclear> element.

Table 2. Language codes used in the EMILLE corpora, drawn from ISO-639

Language	***Code***
Hindi	Hin
Bengali	Ben
Punjabi	Pun
Gujarati	Guj
Urdu	Urd
Tamil	Tam
Sinhala	Sin
Marathi	Mar
Oriya	Ori
Assamese	Asm
Kashmiri	Kas
Malayalam	Mal
Kannada	Kan
Telugu	Tel
English	Eng

The <foreign>…</foreign> element, which indicates words that are in a language other than the primary language of the text, is used throughout, but most particularly in the parallel and spoken corpora. Since all of our parallel and spoken data was drawn from a UK context, the incidence of code switching was high throughout these two parts of the corpus. Code switching most typically occurred between the main language of the text and English, but also between the main language of the text and another South Asian language (for instance, a short section of Hindi in a text which is mostly in Gujarati). In cases such as this, the words in the secondary language are enclosed in <foreign> tags, and the language is indicated using codes from ISO-639, as listed in Table 2.

Figure 1 shows an example written text from the corpus, and Figure 2 shows an example of spoken text.[16] In both cases, a number of typical SGML tags are shown (including the <foreign> tag).

The full SGML/XML mark-up was post-edited and validated at Lancaster prior to the release of the corpus.

<p><head><s>

سوتے وقت جسم کی پوزیشن

</s></head></p>

<p><s>

حالیہ تحقیق سے معلوم ہوا ہے کہ جو بچے پیٹ کے بل سوتے ہیں ان میں اس قسم کی "کاٹ ڈیتھ" زیادہ ہوتی ہے۔

</s> <s>

اگر آپ اس امر کا خیال رکھیں کہ آپ کا بچہ درست پوزیشن میں سوئے تو "کاٹ ڈیتھ" کا خطرہ کم ہو سکتا ہے۔

</s></p>

<p><s>

بچوں کو اس حالت میں سلانا چاہئے:

Figure 1. Example of written text (from the Urdu parallel corpus)

<**u** id="32" who="BU420">ঠিক এখানে এমনি এই ধারাটাই মানে চলছে</**u**>

<**u** id="33" who="BM400">আপনার মতে তার কি পরিমান সমর্থন আছে !</u>

<**u** id="34" who="BU420">না সমর্থন তেমন নাই সমর্থন খুবই কম। সমর্থন <foreign lang="eng">very few</foreign> । আপনার আমি যদি বলি <foreign lang="eng">ten percent</foreign> এরও বেশী হবে। মানে হ্যাঁ কিন্তু <foreign lang="eng">ten percent</foreign> ও হবে না। মানে সমর্থন নগন্য এবং এই জন্য এই নগন্যটা মানে নগন্যটা আপনার আমার ভাই জাতির কাছে ওরাই।</u>

<**u** id="35" who="BM400">জনাব চৌধুরী আপনাকে আপনার মুল্যবান সময় দেওয়ার জন্য অনেক অনেক ধন্যবাদ ।</u>

<**u** id="36" who="BU420">আচ্ছা ঠিক আছে। <foreign lang="eng">thank you</foreign> ভাই। আমি সব শ্রোতা বন্ধুদের আমার সালাম বলবেন এবং সবাইকে আমাদের সবাইকে সাদলম মাসের জন্য <omit extent="1 syllable" cause=" unclear dialect"></omit> করতে বলবেন আগামী সোমবার যেটি কাল বাদ পরশু দিন। আমি সাধারনবাসীদের <foreign lang="eng">complex</foreign>এর কাজ আরম্ভ করেছি। বিড়লা সাইটে আসতেছে আগামী সোমবার। আপনাদের সবাইয়ের দাওয়াৎ রইল। এই উপলক্ষে একটা <omit extent="1

Figure 2. Example of spoken text (from the Bengali spoken corpus)

4.3. Textual information

Each file in the EMILLE corpora has a full CES header with bibliographical and other information about the source text. In the case of spoken texts, information about each of the speakers is also given in the header. An example of an EMILLE corpus header is given in the Appendix.

However, a novel feature of the EMILLE Corpus is that information about the text-type is also encoded in the filename of each text. This allows users to see at-a-glance the most crucial details of genre, source, and date of the text.

Each filename consists of a series of codes chained together with hyphen characters. These codes specify the main language of the file, the source of the text, its subcategory in terms of subject matter if such information is available, and an identifying number.[17] In the case of sources from which data was gathered on a periodical basis (e.g. news text or radio programmes) the identifying number is a date. For other files it is simply an arbitrary distinguishing number. For example, the file *hin-w-ranchi-news-01-03-22.txt* is a written file in Hindi, containing news stories published on the *Ranchi Express* website on the 22nd of March 2001. The file *ben-s-cg-asiannet-02-07-23.txt* is a context-governed spoken text in Bengali, consisting of a transcript of a radio programme broadcast on the BBC Asian Network on the 23rd of July 2002. A full key to the codes used in the filenames is given in the corpus user manual.[18]

4.4. Corpus composition

In this section, we give a brief description of the types of text that make up the EMILLE corpora. In many cases, the final composition of the corpora was dictated by the solutions we found to some of the problems discussed in section 5 below, and therefore additional details of the make-up of the corpus will be found in section 5. However, the types of text in each corpus will be summarised here.

4.4.1. The parallel corpus

The parallel corpus consists of seventy-two information leaflets published by various UK governmental or quasi-governmental[19] bodies, including the Department of Social Security, the Department of Health, the Home Office, the Department of Education, the Office of Fair Trading, the ministry responsible for housing law,[20] and Manchester City Council. These leaflets were published in a range of UK minority languages as well as English. The research value of this data is very high in our view, as the data represents well the type of material which is frequently translated into South Asian languages in the UK, and it is in a genre which is term-rich.

4.4.2. The spoken corpora

The texts in the spoken corpus consist largely of transcribed radio broadcasts. However, a small proportion of the Hindi and Bengali spoken corpora consist of speech that has been tape-recorded by volunteers and then transcribed.

The original target size for the spoken corpus was 500 000 words per language, i.e. 2.5 million words overall. The final size of the spoken corpus was actually 2.6 million words; the Bengali spoken corpus is slightly smaller than the others, at only 442 000 words, while the Hindi corpus is the largest, at 588 000 words. The original recordings of the texts remain in our possession, and we are currently digitizing and editing them to remove songs and other such material. The resulting audio files will be made available in conjunction with a future release version of the corpora.

4.4.3. The written corpora

For reasons that will be discussed below (see 5.3.1), the sizes (see Table 1 above) and composition of the different monolingual written corpora vary greatly. However, some general comments can be made. The majority of the texts in these corpora comes from news websites; however, a significant minority of texts (those integrated from the CIIL Corpora) are from books or non-news periodicals.

5. Problems and solutions in South Asian language corpus building

The major end-product of EMILLE was the corpora described in the previous section. We will now move on to describe some aspects of the process by which these corpora were built. Our experiences in this process have been illustrative of the difficulties that must be faced in building corpora for the languages of South Asia – difficulties that one might not anticipate based solely on the experience of corpus building for languages such as English or Spanish.

In discussing the problems we have faced and the solutions we found to them, we have two aims. Firstly, we wish to explain how these issues have impacted on the final form of the corpus. Secondly, it is our hope that by sharing our experiences and solutions, researchers making future efforts in the field of corpus building for South Asian languages may be aware of the problems that must be faced and will have the option of availing themselves of the strategies we devised to get around these problems. In the following subsections we discuss the problems associated with each part of the EMILLE corpus in turn.

5.1. Problems in constructing the parallel corpus

The first problem we faced with the parallel corpus was a problem of permissions. Although the UK government gave us permission to use the texts, the company that produced the electronic versions of the texts refused to give us the electronic originals. Therefore, the texts were only available to us in PDF format or as paper documents, not as word-processed text that could be mapped to Unicode and added to the corpus. However, since the parallel corpus was to be relatively modest in size (1.2 million words), it was economically viable to pay transcribers to keyboard in electronic versions of these printed documents.

The second major problem was a UK-specific issue: there is no pan-agency agreement on what languages government information leaflets are required to be translated into. Most leaflets for which translations exist were available in all of the languages we needed for the parallel corpus (English, Bengali, Gujarati, Hindi, Punjabi, and Urdu). Indeed, some were available in many other UK minority languages as well – for instance Arabic, Chinese, Persian, Polish, Russian, Somali, Vietnamese, and Welsh. However, in a substantial minority of cases, a text was not available in the full set of languages that we required. We could not identify a sufficient number of texts available in all of the six languages required to make up the necessary 200 000 words of the parallel corpus. Therefore we had to use some texts for which some languages were "missing".

In these cases, rather than leave a "hole" in the parallel corpus, we commissioned a translation into the missing language from one of the EMILLE project's transcribers (all of whom were fluent in English as well as one or more of the South Asian languages in question). While far from ideal, this is not unprecedented, as the English Norwegian Parallel Corpus project also commissioned translations (see Oksefjell 1999). Even so, we recognize that for some purposes it would be preferable to exclude these "non-official" translations from the dataset. Therefore, it is always indicated in the header if a text is a non-official translation, so that these texts may be excluded if a user so desires.

One final issue arising from the parallel corpus is that, as construction of the corpus proceeded, we gathered a certain quantity of anecdotal evidence from our transcribers that some of the translations in the corpus were of poor quality.[21]

Since our primary purpose in developing the parallel corpus was to create a resource for translators and researchers in the field of translation, the presence of poor quality translations in the parallel corpus may be seen as an advantage, as it allows the issue of translation quality to be investigated empirically with greater ease than was previously the case. There are clear benefits for a researcher if the corpus reflects accurately the range of quality of materials that a speaker of a South Asian language in the UK must deal with.

5.2. Problems in constructing the spoken corpora

The spoken corpus, as has already been described, consists mostly of transcribed radio broadcasts. This was not the original intention of the project. We had initially explored the possibility of following the BNC (British National Corpus) model of spoken corpus collection by demographic sampling (see Crowdy 1995). We piloted this approach by inviting members of South Asian minority communities in the UK to record their everyday conversations. In spite of the generous assistance of radio stations broadcasting to the South Asian community in the UK, notably BBC Radio Lancashire and the BBC Asian Network, the uptake on our offer was dismal. Two of the Hindi- and Bengali-speaking transcribers working on the project agreed to record their own everyday conversations with family and friends; this data has been included in the release version of the corpus.[22] However, not nearly enough participants volunteered for us to gather the full 2.5 million words in this way. Furthermore, the feedback from this trial was decisive – members of the South Asian minority communities in Britain were uneasy with having their everyday conversations included in a corpus, even when the data was fully anonymized.

Again, this is a significant difficulty inherent in South Asian language corpus-building, especially when studying the languages in diaspora, as it means that an important source of data is cut off to the researcher; and again, this is a difficulty that would not be encountered in building a corpus of, say, English. Even if the reluctance to be recorded that we encountered does not occur among speakers living in South Asia,[23] and it proves possible to gather spoken data from communities where the target language is not a minority language, it still remains the case that the UK-specific forms of Hindi, Urdu, Gujarati, Bengali, Punjabi, and so on are inaccessible to this methodology.

Our solution to this difficulty was to gather data from Asian radio programmes broadcast in the UK. The BBC Asian Network[24] was our main source of spoken data.[25] The BBC readily agreed to allow us to record their programmes and use them in our corpus. The five languages of the EMILLE spoken corpora are all covered by programmes on the BBC Asian Network. At least four and a half hours in each language (and more in the case of Hindi-Urdu) are broadcast weekly. The programmes play Indian music – the lyrics of which have not been transcribed – as well as featuring news, reviews, interviews, and phone-ins. As such the data allowed a range of speakers to be represented in the corpus, including listeners and interviewees from the UK and from South Asia, as well as professional broadcasters. In consequence a significant proportion of the data is made up of spontaneous, unscripted speech. Some minimal encoding of demographic features for speakers has often been possible, as at least the sex of the speaker on the programmes is usually apparent.

In summary, it has been possible to work around the problem of speakers' reluctance to let their conversations be recorded by resorting to radio recordings; however, this is not a complete solution, as the corpus user does not have access to the variety of spoken language that would be found in a non-broadcast setting.

The process of the orthographic transcription of the radio programmes has brought out two interesting issues, both, arguably, related to dialect.

The first issue arose from the variety of Bengali spoken in the UK. Our main Bengali transcriber lived in India for most of her life. She had no problems transcribing the conversations of other Bengali-speaking Indians, but when faced with tapes of the radio programme which featured Bengali speakers who lived in the UK, it became apparent that British-born Bengali speakers speak a variety of Bengali rarely heard in India. UK Bengali speakers are overwhelmingly from the Sylhet region of Bangladesh and speak Sylheti, which one may either view as a separate language or a dialect of Bengali (Baker et al. 2000). As some of these words were unfamiliar to our non-Sylheti speaking transcriber, they were not transcribed. Instead the CES code <omit> has been used on such occasions; for instance, *<omit extent="1 syllable" cause="unclear dialect">*. Our intention is that, at a later date, we will return to these points in the data with a Sylheti speaker and correct the transcription.

The second problem related to prescriptive attitudes. As noted, the phone-in radio data is of particular use as it means that a number of speakers are represented in the corpus, not all of whom are speakers of a nominal standard form of the language in question. This observation is not restricted to

Bengali/Sylheti. It is apparent in all of the languages that we gathered data for. This caused some transcribers who happily worked on typing parallel corpus data to refuse to work with the spoken material at all. They objected to the representation of South Asian languages in the corpus. For example, one Hindi-speaking transcriber from India refused to transcribe recordings of the BBC Asian Network's Hindi Programme, saying that linguists should only study "classical Hindi texts and not the bastardized slang" that was used by South Asians living in the UK. Some of the differences that the transcribers objected to related to the code-switching practices of the UK South Asian community. However, there were also objections to non-standard and non-prestige forms such as Sylheti being studied by linguists. While this was a manageable problem in the context of the EMILLE project, this experience served as a useful reminder that, while linguists may be happy studying all forms of a language, the willingness of speakers of that language to help corpus builders may be very much influenced by their attitude to the forms of the language that a corpus linguist is seeking to represent and study.

5.3. Problems in constructing the written corpus

5.3.1. Poor availability of electronic texts

The first major challenge facing any corpus builder is the identification of suitable sources of corpus data. In most corpus-building exercises (as in ours) it will be neither economical nor practical to rely on mass keyboarding of written texts to convert texts to electronic form; therefore, the corpus must be built from texts which are available in a suitable electronic form already.[26] This causes problems in corpus building for the languages of South Asia, as the availability of electronic texts for these languages is limited. This availability does vary by language, but even at its best it cannot compare with the availability of electronic texts in English or other major European languages.

In theory, when designing a large-scale written corpus, one would like to choose among sources of electronic text to create a corpus that is balanced across a range of media and genres. However, in a situation where the possible sources of electronic text are restricted, such design criteria may simply not be practical. This was the case on EMILLE. This is not a problem which can be "solved" as such, although we may hope that the increasing global spread of information technology will eventually ameliorate this difficulty.

What sources of text, then, were actually available to us? Several publishers were prepared to give us permission to take samples from books they published for inclusion in the corpus, but the prevalence of hot-metal printing methods in South Asia meant they could rarely supply us with electronic versions of these documents. We therefore had to rely on documents published in an expressly electronic medium, i.e. on the internet. There are many websites produced in South Asian languages. To focus our efforts, and to reduce the number of script encoding systems that would need to be decoded (see below) we decided that we would only gather data from websites that could yield significant quantities of data. We therefore excluded small and/or infrequently updated websites from our collection effort.

In practice this meant that we were collecting data from news websites,[27] since these are typically updated daily or at least weekly with several tens of thousands of words of data. This was an acceptable decision to a degree, since some corpora for languages such as English that were not balanced have consisted solely of news text;[28] so heavy use of news text is to some extent in compliance with established practice. That is to say that a news corpus is in some ways the "next best thing" to a balanced, representative corpus: news periodicals are typically written by a range of individuals, on a range of topics and in a range of styles (for example, news reports, entertainment news, sports news, feature articles and even some fiction[29]). Of course, whenever we were able to acquire data from a source other than the news websites, we took advantage of that opportunity.[30]

The other approach that we took to lessen the impact of the narrow range of text-types in our collection was to seek collaborative links with researchers in South Asia who could share with us, or assist us in accessing, text collections of more diverse genres. For instance, Mr Vincent Halahakone of the University of Moratua, Sri Lanka, undertook to collect the majority of the written Sinhala corpus directly from the text providers in Sri Lanka; this assistance was absolutely crucial to the completion of our goals for this language. As a result, the Sinhala corpus is rather more diverse in text types than the corpora that are entirely reliant on news data.

However, as has been discussed above, the largest contribution to our text collection activities by one of our collaborators is that made by the Central Institute of Indian Languages, who granted us permission to integrate their written corpora into the overall text collection. As discussed in section 3, the primary benefit of merging the EMILLE and CIIL corpora has been to vastly increase the size of the final joint collection and the number of languages that it covers – in these respects the joint EMILLE-CIIL Monolingual Written Corpora are superior to either the EMILLE data or the CIIL Corpus considered separately. However, a secondary benefit[31] is that the integration has greatly broadened the genre reach of the collection.

By a process of serendipity, the corpus data being provided by CIIL covers a wide range of genres,[32] but not news material. The CIIL and Lancaster data are thus complementary in terms of text-type. In short, by making our corpus-building a collaborative effort, we were able, to some extent, to circumvent the practical limits imposed on the spread of genres in our collection by the poor availability of electronic texts in South Asian languages.

5.3.2. Problems of mark-up

The data which we gathered from the web was initially marked up using HTML. During the process of mapping the files to Unicode (discussed in depth in the following section), this mark-up was converted to the CES-compliant SGML mark-up that is used throughout the EMILLE corpora. However, this conversion was not unproblematic. Firstly, the HTML of the original webpages contained large quantities of non-textual material – menu bars, advertisements, and so on. We obviously did not wish to include this material in the corpus. While we did look at the feasibility of using web robot programs to extract the text only from a news webpage, this turned out to be impractical due to the large number of different webpages that we were working with and the often considerable complexity of the HTML surrounding the actual story. The solution that we employed was to manually copy and paste the text we wanted from a web browser window to an MS Word document, which was then saved as HTML to create a cleaner HTML source text.

However, doing it in this way[33] meant that although most of the unwanted HTML was filtered out, certain features of the HTML text in the final version were dependent on how the original web page was coded. An example of this is the encoding of headings and headlines. On some news sites, they were marked up with the HTML tags <h1>, <h2>, <h3> and so on. On other sites, they were marked up as normal paragraphs (<p>) but with font formatting features that made them more prominent (for example, large size or a contrasting colour). This encoding was carried through to the HTML version saved using MS Office, which preserves all formatting information. When it comes to mapping the HTML files to Unicode SGML, it is easy enough to automatically detect <h*N*> tags and replace them with the <p><head>… </head></p> tags which are appropriate in a CES document. However, because of the virtually unlimited variety of font styles used for headings and headlines across our range of data sources, it was not possible to use formatting information to determine which <p> elements represent headings and which represent actual paragraphs. The upshot of this is that the <head> element is only used in these files if the original webpage used an <h*N*> element for headings, and not if the original webpage used a <p> element plus formatting for headings.

Similarly, the <s> element is not used throughout the written corpus, since there was no element in the original file that corresponded to a sentence marker. It would in theory have been possible to insert the <s>…</s> tags automatically. However, this could only have been done by reference to the punctuation, and we did not have access to sufficient native speaker input for all the languages we were collecting to have confidence in our judgements of what punctuation does and does not indicate a new sentence. For example, we might have inserted the tags for a new sentence after every full stop (or equivalent punctuation mark). However, we know that for English at least, this would not always produce the correct results. It would fail, for example, when three full stops are used as an ellipsis, or when the full stop indicates an abbreviation rather than a period. We therefore could not assume that a similarly simplistic rule applied to another language would not have similar chaotic results.

In consequence, <s> tags have only been applied in the written corpus where we could be absolutely confident of applying them correctly; assigning

```
<p><head>અભિષેક ચાલે કે ન ચાલે પણ કીર્તિ રેડ્ડી બોલીવુડ પર છવાઈ જશે</head></p>

<p><head>હમણા તો કીર્તિ મુંબઇ અને બેંગ્લોર વચ્ચે
આંટાફેરા મારી રહી છે. પરંતુ નજીકના ભવિષ્યમાં મુંબઇમાં ઘર લઇ સ્થાયી થવાની ઇચ્છા છે.
હિન્દી ફિલ્મો મળવાને કારણે દક્ષિણની ફિલ્મો હાલમાં સ્વીકારવાની ઇચ્છા નથી પરંતુ કોઇ
સારી ફિલ્મ મળે તો સ્વીકારી પણ લે.</head></p>

<p>રિફ્યુજી ફિલ્મની રિલિઝ પછી બોલીવુડના
માંધાતાઓની નજર આજે રિલિઝ થયેલી ફિલ્મ તેરા જાદુ ચલ ગયા પર હતી. ટીવી પર રજૂ
થતા

<gap desc="illustration"></gap>

પ્રોમોએ આ ફિલ્મ અંગે લોકોનું કુતૂહલ ઘણું વધાર્યું હતું. રિફ્યુજી
ફિલ્મે ભલે ઉજળો દેખાવ કર્યો નહીં. પરંતુ અભિષેક બચ્ચનની ડિમાન્ડ છે. આ ઉપરાંત
દક્ષિણની કીર્તિ રેડ્ડી પણ લોકોની નજરમાં વસી ગઇ છે. મોટા મોટા બેનરના નિર્માતાઓ
તેની પાછળ ભાગે છે. હજુ તેની એક પણ ફિલ્મ રિલિઝ થઇ ન હોવા છતાં પણ તેને રાજીવ
રાયની ફિલ્મ ઇશ્ક, પ્યાર, મહોબ્બત મળી છે જેમાં તે આફતાબ સાથે કામ કરી રહી છે. આ
```

Figure 3. Example of written text (from the Gujarati monolingual written corpus)

<s> tags to the remainder of the written corpus is a matter to which more detailed effort will be addressed in the course of future research into and development of the EMILLE data. This is admittedly a minor issue, but it nicely illustrates the inherently challenging nature of building corpora of languages where there is little previous corpus-based work to go on.

Figure 3 shows a sample of a text that lacks sentence tags as described here – note the comparison to the mark-up in the parallel corpus (see Figure 1 above).

5.3.3. Difficulties standardizing the text encodings

Once we had realized that it would be necessary to gather the greater part of the data for the monolingual written corpora from news websites, it soon became clear that the issue of text encoding would be critical. Since the corpus was to be in Unicode, we would ideally have liked to include texts that already existed in Unicode format in our corpus. However, when we first started to collect data, we were unable to locate documents in the relevant languages in Unicode format on the web.[34] Rather, we found that when a document in a South Asian language is released online, the publisher typically relies on one of the following five methods of representing the text:

- They use online images, usually in GIF or JPEG format. Such texts would need to be keyed in again, making the data of no more use to us than a paper version;
- They publish the text as a PDF file. Again, this made it almost impossible to acquire the original text in electronic format. We were sometimes able to acquire ASCII text from these documents, but were not able to access the fonts that had been used to render the South Asian scripts. Additionally, the formatting meant that words in texts would often appear in a jumbled order, especially when acquired from PDF documents that contained tables, graphics or two or more columns;
- They use a specific piece of software in conjunction with a web browser. This was most common with Urdu texts, where a separate program, such as *Urdu 98*, is sometimes used to handle the display of right-to-left text and the complex rendering of the *nasta'liq* style of the Perso-Arabic script;
- They use a single downloadable True Type (TTF) 8-bit font. While the text would still need to be converted into Unicode, this form of text was easily collected;
- They use an embedded font. For reasons of security and user-convenience, some site-developers have started to use OpenType (eot) or TrueDoc (pfr)

font technology with their web pages. As with PDF documents, these fonts no longer require users to download a font and save it to his or her PC. However, gaining access to the font is still necessary for conversion to Unicode. Yet gathering such fonts is difficult as they are often protected. We found that owners of websites that used embedded fonts were typically unwilling to give those fonts up. Consequently, using data from such sites proved to be virtually impossible.

There are a number of possible reasons for the bewildering variety of formats and fonts needed to view South Asian scripts on the web. For example, many news companies who publish web pages in these scripts use in-house fonts or other unique rendering systems, possibly to protect their data from being used elsewhere, or sometimes to provide additional characters or logos that are not part of standard South Asian character sets. However, the obvious explanation for the lack of Unicode data is that, until relatively recently, there have been few Unicode-compliant word-processors available. Similarly, until the advent of Windows 2000, operating systems capable of successfully rendering Unicode text in the relevant scripts were not in widespread use. Even if a producer of data had had access to a Unicode word-processing/web-authoring system, they would have been unwise to use it until recently, as the readers on the web were unlikely to be using a web browser which could successfully read Unicode and render the scripts.

Given the complexities of collecting this data, we chose to collect text from South Asian language websites that offered a single downloadable 8-bit TTF font. This meant that the issue of encoding had an impact on the choice of data sources, which as we outlined above was limited to start with. For example, some websites who had given us permission to use their texts in the corpus, and from which we had collected data, switched from the use of an 8-bit font to the use of PDF files halfway through the project, meaning we could gather no more data from that source, even though the texts were still available to download.[35]

As well as dictating what sources we could and could not use, the encoding systems instantiated by the fonts used on the web presented the practical difficulty that each of them was an isolated, incompatible encoding of a script. Unlike fonts that encode the Latin alphabet, such as Times New Roman as opposed to Courier, South Asian fonts are not merely repositories of a particular style of character rendering. They represent a range of incompatible glyph encodings. In different English fonts, the hexadecimal code 0x42 is always used to represent the character "B". However, in various fonts which allow one to write in Devanagari script (used for Hindi among other languages), the hexadecimal code 0x42 could represent a number of possible characters and/or glyphs.

While the ISCII standard (Bureau of Indian Standards 1991) has tried to impose a level of standardization on 8 bit electronic encodings of South Asian writing systems, ISCII is ignored by South Asian TTF font developers and is hence largely absent from the web. Thus, almost all of the TTF 8-bit fonts have incompatible South Asian glyph encodings (McEnery and Ostler 2000). To complicate matters further, the various 8-bit encodings of South Asian writing systems have different ways of rendering diacritics, conjunct and half-form characters. For example, the Hindi font used for the online newspaper *Ranchi Express* tends to only encode half-forms of Devanagari, and a full character is created by combining two of these forms together. For example, to produce प (Unicode 0x092A – Devanagari letter PA) in this font, two keystrokes need to be entered (ट+Γ, hexadecimal codes 0x68 and 0x65, corresponding to ASCII characters "h" and "e"). However, other Devanagari fonts use a single keystroke to produce प.

This meant that for every additional source of data using a new encoding that we wished to include in the corpus, an additional conversion function had to be written in order to map that data to the Unicode standard. Thus, the difficulties of mapping between character encodings for South Asian scripts further constrained our choice of data sources. Not only did we have to restrict data collection to websites using a single 8-bit font, we had to ensure that the overall number of 8-bit fonts we had to deal with remained manageable.

The task of mapping this data to Unicode was, despite our efforts to minimise it, a fairly difficult one. Whilst it is fairly simple to write a program that will map every character in a given font to one or more given Unicode characters, this basic algorithm will not handle any other than the simplest of the systems used to encode South Asian alphabets. The full set of formats we had to deal with (considering now all texts, not just those gathered from the web) fell into three broad groups.

- Texts in Urdu or western Punjabi required one-to-one or one-to-many character mapping. This was due to the nature of the alphabet[36] in which they were written, which does not contain conjunct consonants as the Brahmi-based alphabets such as Devanagari and Bengali do.
- Texts in ISCII[37] required one-to-one character mapping. These texts, primarily those from the data provided by CIIL, could be mapped very simply because the Unicode standard for Indian alphabets is actually based on an early version of the ISCII layout.

– Texts in the specially-designed TTF fonts discussed above required the most complex mapping. They typically contain four types of characters.
  The *first* type need to be mapped to a string of one or more Unicode characters, as with ISCII and the Urdu texts.
  The *second* type have two or more potential mappings, conditional on the surrounding characters. Some of these conditional mappings could be handled by generalised rules; others operated according to character-specific rules.
  The *third* type of characters required the insertion of one or more characters into the text stream prior to the point at which the character occurred.[38]
  The *fourth* type, conversely, required characters to be inserted into the text stream *after* the current point (in effect, into a Unicode stream which does not yet exist).[39] In neither this case nor the case of the third character type was it simply a case of going "one character forwards" or "one character back"; the insertion point is context-sensitive.

The third type of text in particular could not be dealt with using simple mapping tables – each font required a unique conversion algorithm. No software existed that was capable of performing such a complicated mapping between encoding systems prior to our work on EMILLE. It was therefore necessary for us to devise one. The *Unicodify* software suite[40] developed at Lancaster is currently capable of mapping HTML files in fifteen different fonts to Unicode, as well as converting to Unicode plain text encoded as ISCII, PASCII or the text output of the popular *Inpage* Urdu word-processing software. All the data in the monolingual written corpora has been mapped using Unicodify, which also performs the mapping from HTML elements to CES-compliant SGML mark-up discussed earlier, and generates appropriate file headers from the filenames of the texts it processes.

6. Corpus annotation and analysis tools: Part-of-speech tagging for Urdu

On the EMILLE project we wished to develop a part-of-speech (POS) tagger for at least one of the languages covered by the project, and use it to annotate the relevant sections of the corpus. We selected Urdu because there are a number of factors that we anticipated would make tagging Urdu more complicated than tagging any other EMILLE language. For example, the right-to-left directionality of the Indo-Perso-Arabic script, and the presence of grammatical forms borrowed from Arabic and Persian, which are structurally quite distinct from Indo-Aryan forms, mean that Urdu represents a unique challenge in our data. It seemed that the best course of action was to confront these problems by choosing Urdu as the language for which to develop POS tagging.

The main difficulty involved in implementing POS tagging for a language such as Urdu is simply that it has not been done before. Therefore, one cannot rely on resources for tagging such as a tagset, pre-tagged training data, tagging guidelines, electronic lexicons, or modules of software for automated tagging – as one could if one were working on English. Indeed, it has proven necessary to develop these resources for Urdu from scratch.

The first resource that was needed was a categorization scheme for words in Urdu texts and corpora.[41] To create the linguistic categories of a tagset, it is necessary to have a model of the language to categorize. We relied on the current standard grammar of Urdu by Schmidt (1999) to furnish a model of the language. Using this model, the U0 tagset for Urdu[42] was devised in accordance with the major international standard on POS tagsets, the EAGLES guidelines on morphosyntactic annotation (Leech and Wilson 1999). These guidelines were designed to help standardize tagsets for the official languages of the European Union. However, the categories in the attribute-value system outlined in the EAGLES guidelines were suitable for application in the design of the U0 tagset. There was no major group of Urdu words for which there was no equivalent category in EAGLES. Furthermore, the EAGLES guidelines proved able to easily describe the gender, case, and number system of Urdu.[43] The verbal system was slightly more problematic, in the sense that the mood, tense, and finiteness features outlined in the EAGLES attribute-value system do not map easily onto those found in Urdu.[44]

However, the greatest difficulty arose in dealing with the minor, idiosyncratic features of Urdu – whilst the idiosyncratic features of the EU languages are covered by the EAGLES guidelines, this is not the case for Urdu. These features include: the appearance of case on some verbal elements;[45] the distinction between "marked" and "unmarked" nouns; the Urdu honorific pronoun *āp*, which does not fit easily into any of the EAGLES categories for pronouns; the borrowed Persian enclitic called *izāfat*; and the problem of bound derivational suffixes which appear in some contexts as independent tokens, but not in others.[46] None of these problems were insurmountable. EAGLES proved a robust and useful framework within which to approach Urdu tagset construction. Table 3 shows some example tag definitions from the U0 tagset.[47]

The next resource that is required to create a successful tagger is some tagged text. This is needed for training purposes by many types of tagging software (for example, taggers that use the frequencies of pairs of tags in the training data to construct a probabilistic model to choose the correct tag when

Table 3. Some example tags from the U0 tagset

Tag	Example	Description
AL	ال	Arabic definite article
FF		Foreign word
II	پر، میں	Unmarked postposition
IIM1N	کا	Marked masculine singular nominative postposition *kā*
JJF1N	بڑی	Marked feminine singular nominative adjective
JDNM2O	تسرے	Masculine plural oblique ordinal number
JDYF2N	اِتنی ، ایسی	Feminine plural nominative proximal demonstrative adjective (*itnī*, *aisī*)
NNMM1N	لڑکا	Common marked masculine singular nominative noun
NPUF2V		Proper unmarked feminine plural vocative noun
PPT1O	تجھ	Second person singular oblique personal pronoun (*tujh*)
PJ2N	جو	Plural nominative relative pronoun (jō)
PA	آپ	Honorific pronoun (*āp*)
QQ	کیا	Question marker *kyā*
RR	ہمیشہ	General adverb
VVYF2O	سنی	Feminine plural oblique perfective participle lexical verb
VVSM1	سنوں	First person singular subjunctive lexical verb
VXNF1	پڑنی	Infinitive general auxiliary verb, feminine singular nominative
VHHV1	ہے	Third person singular indicative present hai

analysing a new text[48]). Even if a type of software is used which does not require training data (for example, one based that employs disambiguation rules written by the user to choose the correct tag[49]), pre-tagged data is needed to test the output of the system. In the case of Urdu, one major difficulty was that no pre-tagged data existed; it was therefore necessary for us to undertake *manual tagging* of a small set of texts drawn from the EMILLE Urdu corpus. In the process of manually tagging these texts, a set of comprehensive tagging guidelines were developed, to ensure that the tagset categories were being applied consistently. Due to practical and economic limitations, we were only able to have about 45 000 words hand-tagged in this way. This is a relatively small amount of data. By contrast, if one wished to develop a tagger for English, millions of words of data are available.[50] This had implications for the success of the tagger, as noted below. Figure 4 shows an example of some of our hand-tagged text.[51]

The next resource is an electronic lexicon which lists all of the possible tags that a word-form may have. The use of such a lexicon is the most efficient way to assign tags to tokens when analyzing a text (the ambiguities in the resulting analysis must then, of course, be dealt with). When we began working on this stage of the project, no such lexicons were available. Nor were there any electronic dictionaries available that could readily be converted into such a lexicon. Some poor-quality lists of Urdu words in the Latin alphabet were available on the Internet, but there was nothing suitable for POS tagging purposes. There are two possible ways around this problem.

Firstly, an alternative way for a tagger to deduce the potential tags of a token is by morphological analysis of the form of the word (i.e. looking for affixes that are indicative of a particular word category). However, this is not as reliable as a lexicon of broad coverage – particularly in the case of Urdu, where the many loanwords display morphology characteristic of Persian or Arabic, which is significantly different from the morphology characteristic of "native" (i.e. Indo-Aryan) words. Furthermore, all morphological rules have exceptions, and these exceptions still need to be stored in some form of lexicon.

The second way around this problem is the fairly obvious step of creating a lexicon. This is in fact what we did, although we also built a morphological analysis module into the tagging system to handle cases where the lexicon fell short. However, we did not have any native speaker input for this part of the project. Therefore, the lexicon had to be derived automatically from the hand-tagged data, with some manual additions for closed-category words such as pronouns; but as mentioned above, we only had a relatively small amount of data, which yielded a lexicon of around 8000 word-types.

The final and most critical resource for the development of a tagger is, of course, the tagging software itself. Many language-independent taggers have been developed and are available. However, there were a number of drawbacks to using any of these. Firstly, many require more training data than we had. Secondly, most operate on 8-bit ASCII text, and the EMILLE Urdu cor-

s00007 w009 ویسٹ	RAh NNUM1O
s00007 w010 کے	RAh IIM2N
s00007 w011 سب	RAh PNN
s00007 w012 ھی	RAh XHC
s00007 w013 سننے	RAh VVNM1O
s00007 w014 والوں	RAh JXVM2O
s00007 w015 کو	RAh II
s00007 w016 ممتاز	RAh NNUM1O
s00007 w017 حسین	RAh NNUM1O

Figure 4. Example of manually-tagged Urdu text

pus is in Unicode. While mapping from Unicode to an 8-bit format is possible, it seemed a little counter-productive, given that providing corpora and resources *in Unicode* was one of the major guiding principles of the EMILLE project. For this reason, it was decided to develop new, Unicode-compliant tagging software, using well-established tagging methodologies, and a standardised input-output format. This software, the Unitag system, consists of a number of separate modules to perform tokenization, word-form analysis and tag disambiguation. Although built with the demands of South Asian language data in mind, it is designed to be fully language-independent, and could be applied to tag English or Chinese as well as Urdu or Gujarati.

The version of Unitag which we developed for Urdu uses a custom-made analyser to supply possible analyses to tokens, working from the lexicon described above and also performing morphological analysis. It then removes contextually inappropriate analyses using a rule-based disambiguator, which applies hand-written rules to reduce ambiguity. We created 270 of these rules.

The accuracy of the tagger is circa 90% with a very high ambiguity level. This is rather poor by the standards of many contemporary taggers for languages such as English; however, an analysis of the output shows that this relatively poor performance is due largely to the inadequate size of the lexicon. When most word-forms are not found in the lexicon, a greater weight is thrown onto morphological analysis. However, there is a high degree of syncretism in many Urdu affixes, and therefore the process of morphological analysis typically yields a large set of candidate tags for each token. This increases the final ambiguity, and makes accurate disambiguation difficult. However, despite the limitations of the lexicon, this result represents a good start in this area and a useful basis for future work.[52]

7. Conclusion: Current status and future directions

In this paper, we have described our work in creating the 96-million word EMILLE Corpus of South Asian languages. In the process, we have described a number of difficulties which we encountered and which, we believe, are likely to impact on any project to construct or apply analytic annotation to corpora for these languages. Some of these problems, such as the issue of the wide range of 8-bit encodings for South Asian scripts, we have been able to solve satisfactorily. Others we have not been able to resolve completely and have had to find ways of working around, such as the hesitancy of UK speakers of South Asian languages to contribute to the spoken corpora.

We envisage that our future work in this area will be focussed on finding more complete solutions to some of these problems. Furthermore, we intend to extend the type of corpus-building work we did on EMILLE to other languages of South Asia for which there is currently minimal corpus coverage, for example Nepali. Our experience on EMILLE of the difficulties one is likely to experience in such an undertaking (for instance, in identifying sources of text) should maximize the efficiency of future undertakings of this type. We will also be looking at ways in which we can improve the corpus described here by extending and enriching the analyses annotated on it. For example, as discussed above, audio files of the spoken corpus are currently in preparation for a release alongside the corpus. However, we have not to date looked at the issue of time-aligning transcriptions and recordings. Techniques for performing such alignment are already well-established for English. To work on such techniques would be one obvious avenue of future activity that would enhance the utility of the corpus annotations.

We also wish to explore ways of exploiting the corpora we have developed in linguistic research; there are several other research areas which are opened up by the existence of the corpora and the analytic annotation applied to them. We anticipate that exploring these areas will be increasingly productive over the next few years, both for us and for all researchers using corpus-based approaches to the languages of South Asia.

Appendix

Example of the CES header for a monolingual corpus text in the EMILLE Corpus. Note that in all parts of all headers in the corpus, the date format *yy-mm-dd* is employed.

```
<cesDoc id="guj-w-samachar-news-01-05-23" lang="guj">
    <cesHeader type="text">
        <fileDesc>
            <titleStmt>
                <h.title>guj-w-samachar-news-01-05-23.txt</h.title>
                <respStmt>
                    <respType>Electronic file created by</respType>
                    <respName>Department of Linguistics, Lancaster University</respName>
                    <respType>text collected by</respType>
                    <respName>Andrew Hardie</respName>
                    <respType>transferred into Unicode by</respType>
                    <respName>"Unicodify" software by Andrew Hardie</respName>
                </respStmt>
            </titleStmt>
            <publicationStmt>
                <distributor>UCREL</distributor>
                <pubAddress>Department of Linguistics, Lancaster University, Lancaster, LA1
                    4YT, UK</pubAddress>
                <availability region="WORLD"></availability>
                <pubDate>02-12-18</pubDate>
            </publicationStmt>
            <sourceDesc>
                <biblStruct>
                    <monogr>
                        <h.title>"Gujarat Samachar" internet news
                            (www.gujaratsamachar.com), news stories collected on
                            01-05-23</h.title>
                        <h.author>Gujarat Samachar</h.author>
                        <imprint>
                            <pubPlace>Gujarat, India</pubPlace>
                            <publisher>Gujarat Samachar</publisher>
                            <pubDate>01-05-23</pubDate>
                        </imprint>
                    </monogr>
                </biblStruct>
            </sourceDesc>
        </fileDesc>
        <encodingDesc>
            <projectDesc>Text collected for use in the EMILLE project.</projectDesc>
```

```
        <samplingDesc>Simple written text only has been transcribed. Diagrams, pictures
            and tables have been omitted and their place marked with a gap element.
        </samplingDesc>
        <editorialDecl>
            <conformance level="1"></conformance>
        </editorialDecl>
    </encodingDesc>
    <profileDesc>
        <creation>
            <date>02-12-18</date>
        </creation>
        <langUsage>Gujarati</langUsage>
        <wsdUsage>
            <writingSystem id="ISO/IEC 10646">Universal Multiple-Octet Coded
                Character Set (UCS).</writingSystem>
        </wsdUsage>
        <textClass>
            <channel mode="w">print</channel>
            <constitution type="composite"></constitution>
            <domain type="public"></domain>
            <factuality type="fact"></factuality>
        </textClass>
        <translations></translations>
    </profileDesc>
    <revisionDesc></revisionDesc>
</cesHeader>
```


Notes

1. Funded by the UK EPSRC, project reference GR/N19106. The project commenced in July 2000 and ended in September 2003.
2. For earlier progress reports on EMILLE, see Baker et al. (2002, 2003).
3. To obtain a copy of the corpus, see <http://www.emille.lancs.ac.uk/>.
4. For instance, although official figures are not available to confirm this, it has been reported to us by language activists working in this field that Punjabi may now be the second most commonly spoken language in England.
5. Other research institutes across the world have also set themselves to addressing this dearth: for example, the Central Institute of Indian Languages (CIIL; see <http://www.ciil.org/>), who also collaborated with us in the EMILLE Project, are engaged in corpus building for a variety of Indian languages. Likewise, the Centre for Research in Urdu Language Processing in Pakistan (CRULP; see <http://www.crulp.nu.edu.pk/>) is working on Urdu corpora, and the University of Columbo, Sri Lanka, has begun project to create a Sinhala corpus. A number of smaller-scale projects have also been set in motion to create various corpus resources, for example the 50 000 word corpus of written Urdu described by Becker and Riaz (2002).
6. MILLE was, like the later EMILLE Project, funded by the UK EPSRC (grant number GR/L96400).

7. General Architecture for Text Engineering. See also Cunningham et al. (1999) and Gaizauskas et al. (1996).
8. Grants GR/M70735, GR/N28542 and GR/R42429/01.
9. The part-of-speech tagging of the Urdu corpora is discussed in section 6 below; for information on the alignment project, see Roy (2003) and Singh et al. (2000), and for information on the work on demonstrative anaphora in Hindi, see Sinha (2003).
10. See Unicode Consortium (2000), and also the website <http://www.unicode.org>.
11. Indeed, the GATE architecture (see Tablan et al. 2002) is now coded in Java.
12. While the EMILLE mark-up was originally planned as SGML, in view of the growing popularity of XML, we have not used any features of SGML that are not features of XML. Therefore the texts may be treated as XML for practical purposes.
13. See <http://www.cs.vassar.edu/CES/>.
14. The CES recommendations for spoken texts also include codes for pauses, overlapping speech, and shifts in intonation. However, although the training material for our transcribers documented these tags, in practice transcribers appear to have mostly ignored them, so they are not very prevalent in the texts.
15. Since the parallel corpus consists of government advice leaflets, there are many bulleted lists in these texts (this appears to be a common feature of this text type). These have also been represented with <s>...</s> elements.
16. Note that in Urdu texts such as the one shown here, tags are given on separate lines to the surrounding text, since some Unicode word processing software does not handle text directionality correctly if left-to-right and right-to-left text (such as Urdu) are mixed on a single line.
17. In the parallel corpus, no identifying numbers are used, since there are a relatively small number of documents per language; single-word names are used instead.
18. This is distributed with the corpus and can also be viewed on the Internet at <http://www.emille.lancs.ac.uk/manual.pdf>.
19. An example of a quasi-governmental body whose publications are included in the EMILLE parallel corpus is the Low Pay Commission.
20. At various points during the EMILLE Project, this ministry was the Department for Environment, Transport, and the Regions; the Department for Transport, Local Government, and the Regions; and the Office of the Deputy Prime Minister.
21. Our transcribers were all (bar one) natives of India rather than members of UK South Asian language communities. Therefore, some of the perceived infelicities they report in the texts they transcribed may be dialect differences rather than translation errors. However, it seems unlikely that this can be the case for all the reported cases of poor-quality translation.
22. 50 000 words of spoken Bengali and 40 000 words of spoken Hindi have been gathered in this way.
23. At the time of writing we are aware of projects in the South Asian nations to develop spoken corpora; for instance, the CIIL is active in this area, as is CRULP.
24. See <http://www.bbc.co.uk/asiannetwork/>.
25. Programmes broadcast in Bengali and Urdu on BBC Radio Lancashire make up the remainder of the spoken corpus.
26. Another option is converting print documents to electronic text using optical char-

acter recognition (OCR). However, at the time at which we began the EMILLE Project, OCR systems for South Asian scripts were still in their infancy, and were not considered stable and robust enough for this project to use gainfully. Over the past five years, progress has been made in the field of OCR for South Asian scripts, so this might be a viable alternative when approaching corpus-building in the future. However, it should be kept in mind that a scanned text may still require post-editing, at least some of which may have to be manual, to remove errors made by the OCR program or to insert mark-up.

27. In some cases, though not all, these websites were associated with each other.
28. An example of such a corpus would be the Wall Street Journal Corpus, distributed by the Linguistic Data Consortium at <http://www.ldc.upenn.edu>.
29. All the sub-genres mentioned here are attested in the EMILLE collection of Gujarati written texts, for instance.
30. The Punjabi written corpus, for instance, as well as containing data from news websites, also contains the full text of the Guru Granth Sahib. It also contains a set of articles from a UK magazine written in the Indo-Perso-Arabic alphabet.
31. There are of course also purely practical advantages to the integration of our corpus-building efforts, such as giving the research community access to both text collections in a common format.
32. The data in the CIIL corpora covers a very large number of genres, including – to give three randomly chosen examples – ayurvedic medicine, novels, and the physical sciences.
33. This simple approach was selected to minimize the quantity of training necessary for the analysts copying the news data from the web.
34. At that time, the only website we found that uses Unicode for South Asian languages was the BBC's; see for example <http://www.bbc.co.uk/urdu> or <http://www.bbc.co.uk/hindi>.
35. In fact, we perceived something of a minor trend away from the use of 8-bit fonts and towards the use of PDF, particularly on Bengali news websites, over the lifespan of the EMILLE project. This caused us considerable difficulty, since there were no alternative sources of Bengali text on the web.
36. We have come to refer to this alphabet as "Indo-Perso-Arabic", although it is more widely known simply as the "Urdu alphabet" or "Urdu script" (e.g. by Nakanishi 1980: 36), or in the case of the various forms of western Punjabi that use it, "Shahmukhi". The term "Indo-Perso-Arabic" is used because many Indo-Aryan languages that use the Arabic alphabet share certain features not found in Arabic, Persian, etc. – for instance, characters for retroflex consonants, the use of many Arabic consonant symbols with altered phonetic values, or the use of the *nasta'liq* style of calligraphy.
37. The same is true of text in PASCII, the (Indo-)Perso-Arabic equivalent of ISCII.
38. This is primarily the case for those Indian alphabets which allow conjunct consonants whose first component is the letter "ra". When this letter is the first half of a conjunct, it takes the form of a diacritic which appears *after* the second half of the conjunct. In Unicode, the text stream contains the logical order of the characters, but in the TTF fonts, the graphical order is almost always the order that is held in the computer's memory.
39. This is primarily the case for certain vowel diacritics which indicate vowels that follow the consonant but which appear before the consonant. Again, Unicode fol-

lows the logical order, whereas TTF fonts almost always follow the graphical order of the glyphs.

40. A generalised (i.e. lacking features tailored specifically to the EMILLE Corpus) version of this software is freely available on the internet (from <http://www.emille.lancs.ac.uk/unicodify.php>), and the source code (in C) is available on request.
41. A more comprehensive description of the creation of the tagset is given by Hardie (2003).
42. Two slightly smaller subsets of U0, U1 and U2, were also defined for use in particular applications.
43. Urdu has masculine and feminine gender, singular and plural number, and nominative and oblique case, all expressed in a single fusional suffix on each noun / adjective.
44. Urdu verbs have one simple finite verb form (the subjunctive), two simple forms that may be finite or non-finite (the perfective and imperfective participles), and two further non-finite simple forms (the root and the infinitive). There are, however, a large number of complex verb forms using irregular auxiliary elements.
45. The participles and the infinitive can all display case.
46. For example, the morpheme *dār* in *zimmah dār*, 'responsible', and *samajhdār*, 'sensible'. Since every orthographic space was treated as a token break, a special tag (LL) was created for tokens like *zimmah* in *zimmah dār*, which are really bases rather than morphological words. This is only partially analogous to the well-known problem of multi-word idioms in English and similar languages that leads, for example, to phrases such as *given that* being tagged as the two parts of a single subordinating conjunction. In these cases, there is also an analyzable internal syntactic structure (in this case, verbal past participle followed by conjunction). In the Urdu case, the internal structure of *zimmah dār* would of necessity be morphological rather than morphosyntactic, and therefore lies beyond the scope of POS tagging.
47. A listing of the entire tagset (U1 version) is available on the internet at <http://www.emille.lancs.ac.uk/u1tagset.pdf>.
48. An example of a tagger that takes this approach is the CLAWS tagger (see Garside, Leech and Sampson 1987).
49. An example of a tagger which takes this approach is the Constraint Grammar tagger described by Karlsson et al. (1995).
50. For instance, the Brown Corpus (see Francis and Kučera 1982) is often used by researchers developing new tagging technologies.
51. The columnar format shown here is used for manual tagging and for communication between different modules of the *Unitag* system discussed below; the system outputs SGML/XML word tags.
52. The Urdu POS tagger is described in greater depth in Hardie (2005).

References

Baker, J. P., M. Lie, A. M. McEnery, and M. Sebba
2000 Building a corpus of spoken Sylheti. *Literary and Linguistic Computing* 15: 419–431.

Baker, P., and A. M. McEnery
1998 Needs of language-engineering communities: Corpus building and translation resources. MILLE working paper 7, Lancaster University.
Baker, P., A. Hardie, A. M. McEnery, H. Cunningham, and R. Gaizauskas
2002 EMILLE, a 67-million word corpus of Indic languages: Data collection, markup and harmonisation. *Proceedings of LREC 2002: Third International Conference on Language Resources and Evaluation*, 819–825. Las Palmas: ELRA.
Baker, J. P., A. Hardie, A. M. McEnery, and B. D. Jayaram
2003 Corpus data for South Asian language processing. *Proceedings of the EACL Workshop on Computational Linguistics for the Languages of South Asia: Expanding Synergies with Europe*, 1–8. Budapest: ACL. <http://www.aclweb.org/anthology/>.
Becker, Dara, and Kashif Riaz
2002 A study in Urdu corpus construction. *Proceedings of the 3rd Workshop on Asian Language Resources and International Standardization at the 19th International Conference on Computational Linguistics.* Taipei: ACL. <http://www.aclweb.org/anthology/>.
Botley, S. P., and A. M. McEnery
2001 Demonstratives in English: A corpus-based study. *Journal of English Linguistics* 29: 7–33.
Bureau of Indian Standards
1991 *Indian Standard Code for Information Interchange* IS13194.
Crowdy, S.
1995 The BNC spoken corpus. In *Spoken English on Computer: Transcription, Mark-up and Application*, G. Leech, G. Myers, and J. Thomas (eds.), 224–235. London: Longman.
Cunningham, H., R. G. Gaizauskas, K. Humphreys, and Y. Wilks
1999 Experience with a language engineering architecture: Three years of GATE. *Proceedings of the AISB'99 Workshop on Reference Architectures and Data Standards for NLP*. Edinburgh. <http://www.dcs.shef.ac.uk/~hamish/GateAisb99.html>
Francis, W. N., and H. Kucera
1982 *Frequency Analysis of English Usage*. Boston: Houghton Mifflin.
Gaizauskas, R, H. Cunningham, Y. Wilks, P. Rodgers, and K. Humphreys
1996 GATE – An environment to support research and development in Natural Language Engineering. *Proceedings of the 8th IEEE International Conference on Tools with Artificial Intelligence (ICTAI-96)*, 58–66. Toulouse.
Garside, R., G. Leech, and G. Sampson (eds.)
1987 *The Computational Analysis of English*. London: Longman.
Hardie, A.
2003 Developing a tagset for automated part-of-speech tagging in Urdu. *Proceedings of the Corpus Linguistics 2003 Conference.* UCREL technical papers, vol. 16, 298–307. Department of Linguistics, Lancaster University.
2005 Automated part-of-speech analysis of Urdu: Conceptual and technical issues. In *Contemporary Issues in Nepalese Linguistics*, Y. Yadava, G.

Bhattarai, R. Lohari, B. Prasain, and K. Parajuli (eds.), 49–72. Kathmandu: Linguistic Society of Nepal.

Karlsson, F., A. Voutilainen, J., Heikkilä, and A. Anttila (eds.)
1995 *Constraint Grammar: A Language-Independent System for Parsing Unrestricted Text.* Berlin: Mouton de Gruyter.

Leech, G., and A. Wilson
1996 *EAGLES Recommendations for the Morphosyntactic Annotation of Corpora.* <http://www.ilc.pi.cnr.it/EAGLES96/ annotate/annotate.html> .

McEnery, A. M., and Nicholas Ostler
2000 A new agenda for corpus linguistics – Working with all of the world's languages. *Literary and Linguistic Computing* 15: 401–418.

Nakanishi, A.
1980 *Writing Systems of the World.* Tokyo: Charles E. Tuttle Company.

Oksefjell, S.
1999 A description of the English–Norwegian parallel corpus: Compilation and further developments. *International Journal of Corpus Linguistics* 4: 197–219.

Roy, S.
2003 The alignment of English, Bengali and Hindi. Unpublished MA thesis, Lancaster University.

Schmidt, Ruth Laila
1999 *Urdu: An Essential Grammar.* London: Routledge.

Singh, S., A. M. McEnery, and J. P. Baker
2000 Building a parallel corpus of English/Punjabi. In *Parallel Text Processing*, J. Veronis (ed.), 335–347. Dordrecht: Kluwer.

Sinha, S.
2003 Demonstrative anaphors in Hindi newspaper reportage: A corpus-based study. Unpublished MA thesis, Lancaster University.

Tablan, V., C. Ursu, K. Bontcheva, H. Cunningham, D. Maynard, O. Hamza, and M. Leisher
2002 A Unicode-based environment for creation and use of language resources. *Proceedings of LREC 2002: Third International Conference on Language Resources and Evaluation*, 66–71. Las Palmas: ELRA.

The Unicode Consortium
2000 *The Unicode Standard 3.0.* Harlow: Addison Wesley.

Digitized resources for languages of Nepal

Boyd Michailovsky

1. Introduction

The object of the present paper is to describe some currently available resources reflecting the application of information technology (IT) to languages of Nepal, with particular emphasis on linguistic documentation and research. Three categories of resources will be considered:[1]

– tools for the coding and rendering of Nepal languages and scripts
– spoken and written corpora, in particular annotated speech corpora
– dictionaries and wordlists

The first category comprises general-purpose software tools. The second and third categories, which will be my main focus, cover properly linguistic resources. Since one of my aims is to show the possibilities of information technology, the focus of my paper will be on resources that take advantage of this technology beyond text processing. To illustrate this point, I have given a rather detailed description of the Lacito Archive, for which I am responsible together with my colleague Michel Jacobson. Given the evident conflict of interest, I do not pretend to evaluate the resources covered.

Resources are considered here from the point of view of linguistic research and language study. Readers who are interested in language engineering in the South Asian context can refer to the proceedings of the SCALLA ("Sharing Capability in Localisation and Human Language Technologies") conference, held in Kathmandu in January 2004. Another source of information is the newsletter *VishwaBharat@tdil* of the Indian Ministry of Communications and Information Technology's Technology Development for Indian Languages (TDIL) project.

2. Tools for Nepal languages and scripts

The IT industry was slow to establish standards for coding character sets beyond ASCII (96 printable characters) and some European language extensions. Hence, users of phonetic characters and of Devanagari and many other scripts adopted a variety of unstandardized codings and associated fonts. Non-standard Devanagari fonts like Preeti, Kantipur, Himal, etc., are still very widely used in Nepal and in India, but new development is generally based on the Unicode standard (Unicode Consortium 2003), which has been adopted by the World Wide Web Consortium (W3C). Some developments relevant to the Devanagari and the Limbu (Sirijanga) scripts, to transliteration, and to the International Phonetic Alphabet in the context of Unicode will be mentioned below.

2.1. Nepali Unicode

The Madan Puraskar Library (MPP) in Nepal has developed and made freely available a software package facilitating the use of standardized Unicode coding for Devanagari according to Nepali typographic usage. This includes:

Installation instructions
TrueType fonts covering the relevant portions of Unicode
two Windows keyboard layouts, based roughly on:
- the Nepali Remington layout familiar to Nepali typists (but inevitably requiring adaptation on the part of typists)
- romanization

a utility for converting to Unicode from existing, non-standard fonts

These developments are part of an ambitious program of software localization in Nepali which is outside the scope of the present article. See Chalmers and Gurung 2004 for a progress report on Nepali Unicode and further activities designed to promote its adoption.

2.2. Using Devanagari for lesser-known languages

Other Nepalese languages which use Devanagari, like Newari, and, more recently, Tamang and Wambule, can take advantage of Devanagari Unicode. However, certain combinations which do not occur in the languages on which the standard is based may not be handled correctly by rendering software.

2.3. Limbu script

The Limbu, or Sirijanga script has been included in Unicode version 4.0 (Unicode consortium 2003: 260–262, based on Michailovsky and Everson 2002), although most current fonts are based on earlier Unicode versions. Limbu

Unicode defines codepoints for all current Limbu characters, and for some that are obsolete.

2.4. Romanisation; Phonetic fonts

The Unicode standard provides for the coding of the International Phonetic Alphabet and other characters commonly used by linguists, including a wide variety of spacing and non-spacing diacritics. Roman transliteration of Devanagari orthography can be coded using Unicode diacritics and combinations of diacritics, but certain combinations, like macron and tilde on the same letter (used for Nepali nasalized long vowels), may not be handled correctly by rendering software.

3. Corpora

The design of annotated speech corpora has been the object of considerable interest in recent years, having spread to language research and documentation from the language engineering world, where digitized speech corpora (also known as speech databases) are used to build, test, and evaluate automatic speech processing applications. Two large current programmes for the study of endangered languages, the Volkswagen Foundation Dobes project and the Hans Rausing Endangered Languages Project, require grantees to prepare and make available digitized speech corpora; both support research projects in Nepal. At the same time, fifty years after the use of sound recording became widespread in field research, a number of research institutions have taken steps to conserve and make available existing speech recordings and transcriptions made in the course of field research. This activity is often referred to as "archiving", with the understanding that digitized archives should be more accessible and less dusty and expensive to maintain than traditional archives. The Lacito Archive, which John B Lowe, Martine Mazaudon and I started at the French CNRS in the early 1990s, is an example. The architecture of this site is described in some detail below.

Written language corpora are very useful for many kinds of linguistic research, and have become indispensable for lexicography. Large corpora, important for the study of relatively infrequent phenomena, are relatively easily attained since transcription is not required. Unfortunately, there do not appear to be any proper written language corpora of any language of Nepal available at present. However, there are a few computerized sets of journalistic and literary material in Nepali, which will be mentioned below.

The Bhasha Sanchar project, inaugurated in 2005 by the Madan Puraskar Library and Tribhuvan University, has as one of its objectives the constitution of a freely accessible, web-based Nepali National Corpus of both written and spoken Nepali (see website).

3.1. The Lacito Archive

The purpose of the Lacito Archive is to (1) conserve and to make available speech recordings in little-known languages, with synchronized transcriptions, translations, and other annotation and (2) to develop an architecture for such documents and tools for their exploitation, using standard information technology. 4 of the 18 languages currently covered by the Archive are spoken in Nepal. 2 corpora, in Limbu (10 texts) and in Hayu (26 texts), 2 texts in Tamang, and 1 text (less scientifically annotated than the others) in a western dialect of Nepali are currently available on the Internet.

The Lacito Archive is a fairly representative example of current thinking on the design of speech archives and the annotation of recorded speech. It is structured in a client-server architecture and accessed using a standard browser. The underlying data is coded in Unicode and marked-up in XML (eXtensible Markup Language), the W3C's metalanguage for structured text. In response to client requests, the data is processed on the server and the response is furnished to the client, along with associated digitized sound data.

The user interface on the Lacito Archive website proposes a number of "views", which do not exhaust the possibilities of the archived data. In this interface, the user chooses the language, the document to browse, and a "view" on the data. If he chooses the "text" view, he can choose among different transcriptions, and, for the more thoroughly annotated documents, among translations at different levels (e.g. utterance-level "free" translations, morpheme-level glosses, etc.) or in different languages. When the document is displayed, morpheme-level transcriptions and translations are displayed in aligned interlinear format. The user can choose to hear the recording corresponding to a single sentence, or to hear the whole remainder of the text while scrolling through the annotation. Figure 1 shows an interlinear view of a Limbu document on the Lacito Archive site. The user can request a "search list" of all morphemes or glosses occurring in the text. Selecting an item from this list (which spares the user having to enter Unicode characters) brings up all the utterances in which it appears. If he chooses the "concordance" view, a concordance of the entire text is displayed. Any time that an utterance is displayed, the recorded sound is immediately accessible.

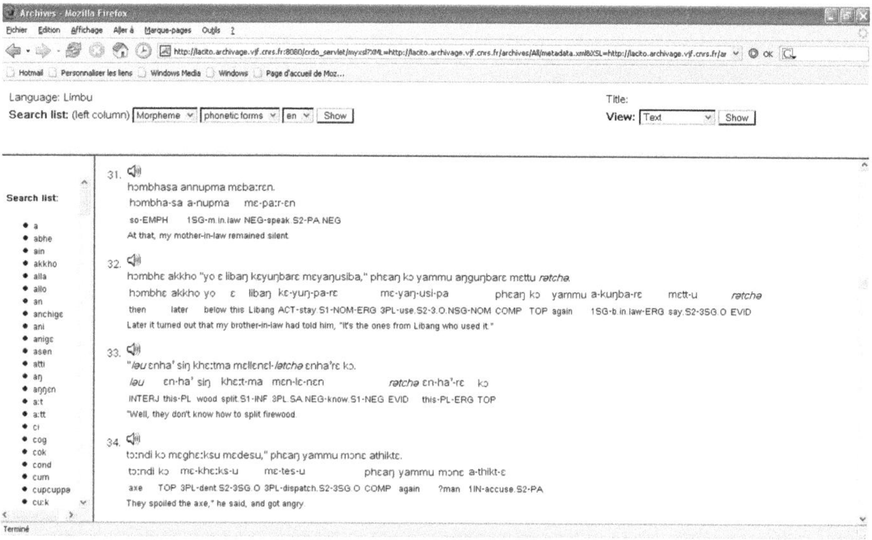

Figure 1: Interlinear view of a text in the Lacito Archive

Figure 2 shows a fragment of a concordance of the document shown in Figure 1, accessed by selecting the "concordance" view. Five concordance entries, including the three occurrences of the verb stem *pa:r* (including one in utterance 31, shown in fig. 1) are shown. The preceding and following context are shown to the left and right of the concorded items (underlined), with a reference identifying the text and utterance number. Clicking on the concorded item causes the sound recording of the utterance in which it occurs to be played. "Talking concordances" of this kind, particularly of whole corpora, have proved a useful tool for verifying transcriptions.

NUPPAs98	ləu alla kɔ peg-innε lɔ'r-ε yammu khunε	paŋs	-i yus-igε
NUPPAs31	hɔmbha-sa a-nupma mε	-pa:r	-εn
NUPPAs82	na rɔt kε-tha-pa lɔ'r-ε-aŋ kɔ anigε sor cu:k-pa	pa:r	-i-aŋ yuŋ-igε
NUPPAs93	alla kɔ hara ni phi:ŋ-ma-siŋ-aŋ pek-ma him-mu pek-ma phεaŋ anigε kɔ yammu anigε s'wa?	pa:r	-igε mεn-khεm-ε khunchi
NUPPAs21	na th'y-ε-aŋ kɔ mεn	-pa:t	-ε s'wa? yammu kεŋs-usigε-rɔ cupcuppə yo sɔŋmaŋ yo la-si pe-sigε

Figure 2: Concordance view showing the stem *pa:r* and adjoining lines.

```xml
<?xml version="1.0" encoding="UTF-16"?>
<!-- copyright 1977-2000 B.Michailovsky -->
<!DOCTYPE TEXT SYSTEM "Archive.dtd">
<?xml-stylesheet href="http://lacito.archivage.vjf.cnrs.fr/archives/styles/default.xsl" type="text/xsl"?>
<TEXT id="LIF:NUPPA" xml:lang="x-sil-LIF">
	<HEADER>
		<TITLE xml:lang="en">Father-in-law</TITLE>
		<TITLE xml:lang="fr">Beau-père</TITLE>
		<SOUNDFILE href="Nepal/Limbu/NUPPA.wav"/>
	</HEADER>
	<S xml:lang="x-sil-LIF" id="NUPPAs31">
		<AUDIO start="148.5400" end="150.6601"/>
		<FORM kindOf="phono">hɔmbhasa annupma mɛba:rɛn.</FORM>
		<TRANSL xml:lang="en">At that, my mother-in-law remained silent.</TRANSL>
		<TRANSL xml:lang="fr">Du coup, ma belle-mère n'a plus rien dit.</TRANSL>
			<W>
				<M class="misc">
					<FORM kindOf="phono">hɔmbha</FORM>
					<TRANSL xml:lang="en">so</TRANSL>
				</M>
				<M class="misc">
					<FORM kindOf="phono">sa</FORM>
					<TRANSL xml:lang="en">EMPH</TRANSL>
				</M>
			</W>
			<W>
				<M class="misc">
					<FORM kindOf="phono">a</FORM>
					<TRANSL xml:lang="en">1SG</TRANSL>
				</M>
				<M class="misc">
					<FORM kindOf="phono">nupma</FORM>
					<TRANSL xml:lang="en">m.in.law</TRANSL>
				</M>
			</W>
			<W>
				<M class="vprefix">
					<FORM kindOf="phono">mɛ</FORM>
					<TRANSL xml:lang="en">NEG</TRANSL>
				</M>
				<M class="v" sclass="pastem">
					<FORM kindOf="phono">pa:r</FORM>
					<TRANSL xml:lang="en">speak.S2</TRANSL>
				</M>
				<M class="vsuffix">
					<FORM kindOf="phono">ɛn</FORM>
					<TRANSL xml:lang="en">PA.NEG</TRANSL>
				</M>
			</W>
	</S>
	</TEXT>
```

Figure 3: XML source for “Father-in-Law” utterance 31 (see Figure 1).

All of the data displays described – straight transcription, interlinear glossed text, morpheme lists, lists of utterances containing a particular morpheme, concordance – are simply different "views" on a single set of structured data – the "annotation" – associated with a recording. Figure 3 shows an annotation document, in the form of structured text marked up in XML, whose contents have been reduced to contain only the first sentence (no. 31) seen in fig. 1. The annotation is structured hierarchically into logical units, text, utterance ("S"), word ("W") and morpheme ("M"), and elements at each level marked up as transcriptions ("FORM"), translations, etc. Views are produced by selection and rearrangement of these logical data elements, a classic IT paradigm.

Although the immediate print-medium ancestors of such documents (without the sound recordings) are collections of texts in interlinear format, going back to early examples like Boas 1911, it is misleading to characterize the computerized documents as "interlinear". The interlinear display may be the view that shows the largest proportion of the data at one time, but the data structure is logical, not typographical. If the underlying data were in fact marked up typographically, in lines and aligned tabulations or table cells, as in a word processor or in the HTML which is actually furnished by the Lacito server in answer to requests, it could not be searched intelligently or transformed automatically into concordances, wordlists, etc. for linguistic research.

Metadata, or cataloguing information, has been mentioned above, in the context of locating linguistic resources on the web. The Lacito Archive is one of 29 "data providers" which are members of the Open Language Archives Community (OLAC, part of the more general Open Archives Initiative) and provide metadata in the OLAC format. This metadata is "harvested" by two OLAC "service providers", the Linguist List and the Linguistic Data Consortium, which provide search interfaces and serve as portals to the participating archives. Thus linguists do not need to know where a resource is archived in order to find it.

The term "open" in the context of "open archives" refers to metadata, not to the data itself. In fact, data providers are only obliged to provide metadata. However, in the case of the Lacito Archive, the metadata for each catalogued resource includes a web address (URL) where the data resource itself is accessible. The Lacito Archive website provides an interface with a limited number of "views" on this data, as described above. But since the data itself is accessible, any user can write a script providing a view on it to satisfy a particular research need. A script accessing the Lacito annotation – like those on the Lacito server – would normally be written in XSLT, the W3C's stylesheet and transformation language for XML, the main means for presenting XML documents in various ways depending on the presentation medium and requirements. The data can also be downloaded, but downloaded copies risk becoming obsolete if the data is updated in the archive.

3.2 *Himalayan Linguistics* online journal and archive

Himalayan Linguistics is a free, refereed online journal and archive in PDF format. The format allows for the inclusion of sound resources. Genetti and Slater (2004) take advantage of this possibility in an article on Dolakha Newar intonation: linguistic examples cited in the body of the article are linked to sound recordings, and a glossed and translated text in Dolakha Newar with recorded sound is provided as an appendix. One item in the *Himalayan Linguistics* archive is an interlinear, glossed corpus of 21 Chantyal texts in PDF format, without sound recordings (Noonan 2005).

3.3. Written language corpora

The online archives of back issues of Nepali periodicals constitute corpora which could be used for linguistic research, although they are not designed for this purpose. Collections of periodicals in non-standard codings for easily available fonts can be found on the NepalNews, KantipurOnline, and Gorkhapatra sites, among others.

An interesting collection of early twentieth century Nepali texts, unfortunately practically impervious to automatic treatment because it is in the form of scanned images, is the complete run of the periodical *Gorkha Sansar* (Dehra Dun 1926–1929) on the DSAL site.

A corpus of Nepali literary works is available on a Nepali site at the Indian Institute of Technology (IIT), Kanpur, a regional outpost of the TDIL project mentioned above. Data on the site is stored in a MySQL database, coded in the Indian ISCII standard coding (see *Vishwa Bharat* 10: 22, July 2003).

4. Lexical resources

A number of dictionaries and wordlists of languages of Nepal are available in digital form. Some are coded as databases, some as structured text, and some in display formats like PDF or HTML, or as word-processing (e.g. Word) files. Most use non-standard character codings.

4.1. The Digital South Asia Library (DSAL)

The Digital Dictionaries of South Asia (DDSA) project, which is part of the Digital South Asia Library at the University of Chicago, has dictionaries of a large number of South Asian languages available or in preparation. Most are computerized versions of large, authoritative print dictionaries. Nepali is well-represented by digitized versions of Turner's classic *Nepali Dictionary* (1931) and of Schmidt (1993), a recent Nepali-English dictionary specifically addressed to non-speakers. Turner's comparative Indo-Aryan dictionary (1962–1966) is also available.

Users of the digital dictionaries can choose between Unicode and non-Unicode data display. The non-Unicode display omits diacritics in roman transliteration, so that serious users will want to use Unicode. Headwords in Devanagari, coded according to the Indian ISCII standard, are present in the underlying digitized data for the Turner dictionary (of which James Nye, director of the DDSA project, kindly provided a sample), but Devanagari is not displayed by the current interface. Headwords in Devanagari script are displayed by the Unicode interface to the Schmidt dictionary. I am not sure whether their underlying encoding is ISCII or Unicode; the two codings are in principle interconvertible. (On ISCII, see "standards" on the TDIL site.)

The search interface proposes to match strings of characters (with the options "starts with", "ends with" and "exact word match") either (1) in the entry headword or (2) anywhere in entries, without regard to the type of information searched for. The underlying data markup is essentially typographical (its notation resembles HTML) rather than logical, which would seem to rule out more precise search options.

4.2. The Newari Lexicon

As a non-Sanskritist I cannot fully appreciate the *Newari Lexicon*, but I cannot leave it out of a review of computerized lexical resources. It is marked as "under construction". Here is how it is described by its authors, the Nepal Bhasa Dictionary Committee:

> The *Newari Lexicon* is compiled from a group of Nepalese manuscripts of a single text, the *Amarakosa*. The manuscripts date from the end of the fourteenth to the early nineteenth centuries. Each manuscript contains the text of the *Amarakosa*, a popular Sanskrit thesaurus written in verse, with notes in Newari containing brief Newari glosses of each group of Sanskrit synonyms. The *Newari Lexicon* is a compilation of words from the Newari glosses, with reference to the original Sanskrit and to English glosses.

The lexicon is in the form of a database in which every Newari word cited in the manuscripts is indexed as to its manuscript source and to a standard printed edition of the *Amarakosa*. The sources can be browsed online. All the manuscripts are indexed so that they can be compared and referred to the standard edition. The underlying data has been entered into a MySQL database in two roman transliterations, one idiosyncratic using only ASCII characters, the other using diacritics in the Sanskritist tradition, in a non-standard coding for display with a supplied font.

4.3. The Tower of Babel

The late Sergei Starostin's Tower of Babel site holds a vast and growing array of lexical resources including bilingual and comparative dictionaries covering a significant proportion of the world's languages in a database format. The Nepal material is the following:

Limbu dictionary (van Driem, 1987)
Dumi dictionary (van Driem 1993)
Yamphu dictionary (Rutgers 2001)
Kulung dictionary (Tolsma n.d.)
Kiranti comparative dictionary (an original set of reconstructions by Starostin)

The bilingual dictionaries are computerized versions of glossaries included in published descriptions of Kiranti languages written by members of the Himalayan Languages Project at the University of Leiden. The data fields are the following (with some variation): headword, part of speech, English definition, Nepali gloss, comments (which may include example sentences), and cross-references to Starostin's comparative Kiranti dictionary.

The comparative Kiranti dictionary draws on the bilingual dictionaries mentioned above and on published sources for Khaling (Toba and Toba 1975), Sunwar (Bieri and Schulze 1971) and Thulung (Allen 1975). It is not a dictionary of synonyms but of etymologies, relating words which continue hypothesized protoforms. The data fields are: protoform, English gloss, forms by language (drawn from the 7 languages), comments. No table of phonological correspondences or justification of the reconstructions is available.

The lexical data for both the bilingual and comparative dictionaries are in Starostin's StarLing database format, in non standard character codings. The search interface is quite complete: string matches can be specified for any field or independently for any combination of fields. This is particularly useful for comparative databases. Data is displayed either in Unicode or in a special font. The complete databases and the database software itself (for DOS or Windows platforms) can be downloaded.

4.4. Limbu dictionary of the Mewa Khola dialect

The Limbu dictionary of the Mewa Khola dialect is the online version of Michailovsky (2002), coded in Unicode and marked up in XML. The markup, loosely inspired by the Text Encoding Initiative (Sperberg-McQueen and Burnard 2001), distinguishes a large number of data fields, including headword, part of speech, morphological and derived forms, grammatical information, definitions, English glosses for inverse indexing, Nepali glosses, botanical binomials, example sentences with references to a text corpus, etymological and derivational relations (coded as hyperlinks to other items in the dictionary), etc. Roughly half of the example sentences are drawn from the Lacito Archive Limbu corpus (see above). References to the corpus are displayed as links giving access to the original recorded sound and glossed transcriptions. The dictionary is also accessible from the corpus: clicking on a morpheme in the Limbu corpus brings up the corresponding entry in the computerized dictionary.

XML seems well suited for capturing a rich structure in which one entry does not necessarily resemble the next. The current Internet interface provides for sequential browsing, or for selection of articles via Limbu and English indexes.

4.5. Wambule and Jero dictionaries

Dictionaries of the closely related Kiranti languages Wambule and Jero have been put online as MySQL databases. These are comptuerized versions of the dictionaries accompanying linguistic analyses by Jean-Robert Opgenort (2004, 2005) of the Himalayan Languages Project. Language forms are coded in Unicode. The interface provides for literal word or string search (limited to English text) in the dictionary database; in response, the dictionary articles in which the specified item occurs are displayed. The articles are quite complete, with phonetic and Devanagari transcriptions of headwords, example sentences, and Nepali glosses (all coded in Unicode), and with a reference to the published source where appropriate.

4.6. Thangmi

Mark Turin's (2004) 1700-word *Thangmi-Nepali-English Dictionary* can be consulted online or downloaded in PDF format. The online interface provides for string-matching in any or all of the three language fields, Thangmi (Devanagari Unicode), Nepali (Devanagari Unicode), English.

4.7. Other lexical resources

An 800-word monolingual Nepali dictionary and a synonym dictionary (thesaurus) are available for download on the MPP site. These are coded in Nepali Unicode.

A computer version of Oja et al. (2004), a 1400 entry Nepali-English and English-Nepali glossary keyed to Oja and Oja (1992), a language primer, is available on the Tibetan and Himalayan Digital Library site. Fields in the Nepali-English alphabetical browsing view are: headword (Nepali Unicode), part of speech, English gloss, subject area, primer lesson reference. The glossary can be viewed in English, Nepali, or primer reference order. String search on the English gloss field is available.

The Nepal Research website offers a number of dictionaries in PDF format. These are bilingual dictionaries (1400 entries) of Sherpa in English, French, and German, compiled by Lhakpa Doma Salaka-Binasa Sherpa and Chhiri Tendi Salaka Sherpa in cooperation with Karl-Heinz Kraemer, and Nepali-English and Nepali-German dictionaries (4300 entries) compiled by Karl-Heinz Kraemer.

Appendix: Sites and abbreviations/acronyms of institutions and projects:

Bhasha sanchar: <http://www.bhashasanchar.org>
DSAL (Digital South Asia Library): <http://dsal.uchicago.edu>
Gorkhapatra: <http://www.gorkhapatra.org.np>
Himalayan Linguistics: <http://www.uwm.edu/Dept/CIE/HimalayanLinguistics>
IIT (Indian Institute of Technology) Kanpur Nepali site: <http://www.nepali.iitk.ac.in>
ISCII (Indian Script Code for Information Interchange):
<http://tdil.mit.gov.in/standards.htm>
Kantipur Online: <http://www.kantipuronline.com>
Lacito Archivage: <http://lacito.vjf.cnrs.fr/archivage>
LDC (Linguistic Data Consortium): <http://www.ldc.upenn.edu>
Linguist List: <http://linguistlist.org>
MPP (Madan Puraskar Library): <http://www.mpp.org.np>
Nepal Bhasa Dictionary Committee: <http://www2.pair.com/webart/mysqllex>

Nepalnews: <http://www.nepalnews.com>
Nepal Research: <http://nepalresearch.org/index.htm>
OAI (Open Archive Initiative): <http://www.openarchives.org>
OLAC (Open Language Archives Consortium): <http://www.language-archives.org>
SCALLA (Sharing Capability in Localisation and Human Language Technologies): <http://www.elda.org/rubrique70.html>
TDIL Technology Development for Indian Languages: <http://tdil.mit.gov.in> Nepali site: <http://www.nepali.iitk.ac.in>
THDL (Tibetan and Himalayan Digital Library): <http://iris.lib.virginia.edu/tibet/>
Tower of Babel: <http://starling.rinet.ru>
Unicode Consortium: <http://www.unicode.org>
W3C (World Wide Web Consortium): <http://www.w3c.org>

Note

1. I apologize in advance for any omissions. The resources described in this paper are not generally easily located, since very few of them are provided with metadata, the cyberspace equivalent of library catalogue information, available on the web (see 3.1 below).

References

Allen, Nicholas K.
1975 *Sketch of Thulung Grammar*. Ithaca, N.Y.: Cornell University.

Bieri, Dora, and Marlene Schulze
1971 A vocabulary of the Sunwar language. Kirtipur: SIL [mimeo].

Boas, Franz
1911 *Handbook of American Indian Languages*. Washington: Government Printing Office.

van Driem, George
1987 *A Grammar of Limbu*. Berlin: Mouton de Gruyter. [Glossary: <http://www.starling.rinet.ru>].
1993 *A Grammar of Dumi*. Berlin: Mouton de Gruyter. [Glossary: <http://www.starling.rinet.ru>].

Genetti, Carol, and Keith Slater
2004 An analysis of syntax and prosody interactions in a Dolakha Newar rendition of the Mahabharata (with appendices and sound files). *Himalayan Linguistics* 1: 1–91.

Michailovsky, Boyd
2002 *Limbu-English Dictionary with English-Limbu Index and Notes on the Maiwa-Mewa Dialect*. Kathmandu: Mandala Book Point. <http://lacito.vjf.cnrs.fr/archivage/dico>.

Michailovsky, Boyd, and Michael Everson
2002 Revised proposal to encode the Limbu script in the UCS. ISO Working Group Document/Expert Contribution. <http://std.dkuug.dk/jtc1/sc2/wg2/docs/n2410.pdf>.

Noonan, Michael, with Ram Prasad Bulanja
2005 Chantyal discourses. *Himalayan Linguistics Archives* 2: 1–254.
Oja, Banu, and Shambhu Oja
1992 *Nepali: A Beginner's Primer, Conversation and Grammar.*
Oja, Banu, Shambhu Oja, Mark Turin, and Elisabeth Uphoff
2004 *Nepali–English, English–Nepali Glossary.* 2nd edition. Ithaca, N.Y.: Cornell University South Asia Program. <http://iris.lib.virginia.edu/tibet/reference/dictionary/nepali/index.php>.
Opgenort, Jean Robert
2004 *A Grammar of Wambule.* Leiden: Brill. <http://www.opgenort.nl>.
2005 *A Grammar of Jero: With a Historical Comparative Study of the Kiranti Languages.* Leiden. Brill. <http://www.opgenort.nl>
Rutgers, Roland
1998 *Yamphu.* Leiden: CNWS. [Glossary: <http://www.starling.rinet.ru>.]
Sperberg-McQueen, C. Michael, and Lou Burnard
2001 *Guidelines for Electronic Text Encoding and Interchange.* 4th edition. TEI Consortium. <http://etext.virginia.edu/teip4/>.
Toba, Sueyoshi, and Ingrid Toba
1975 *A Khaling-English English-Khaling Glossary.* Kirtipur, Nepal: SIL. Tribhuvan University.
Tolsma, Gerard
n.d. *Kulung Dictionary.* <http://www.starling.rinet.ru>.
Turin, Mark, with Bir Bahadur Thami
2004 *Nepali-Thami-English Dictionary.* Kathmandu: Martin Chautari. <http://iris.lib.virginia.edu/tibet/reference/dictionary/thangmi/index.php>.
Turner, Ralph Lilley
1931 *A Comparative and Etymological Dictionary of the Nepali Language.* London: K. Paul, Trench, Trubner. <http://dsal.uchicago.edu/dictionaries>.
1962–1966. *A Comparative Dictionary of Indo-Aryan Languages.* London: Oxford University Press. Includes three supplements, published 1969–1985. <http://dsal.uchicago.edu/dictionaries>.
Schmidt, Ruth Laila
1993 *A Practical Dictionary of Modern Nepali.* Delhi. Ratna Sagar. <http://www.dsal.uchicago.edu/dictionaries/schmidt>.
Unicode Consortium
2003 *The Unicode Standard,* version 4.0. Boston: Addison Wesley. <http://www.unicode.org>.
VishwaBharat@tdil
Newsletter published by Technology Development for Indian Languages, (TDIL). Ministry of Communications and Information Technology. Government of India. ISSN 0972-6454.

Multimedia: A community-oriented information and communication technology

David Nathan and Éva Á. Csató

1. Introduction

The articles in this volume discuss the linguistic situation of lesser-known languages in South Asia with special emphasis on the employment of information and communication technology (ICT). Other articles in this section include discussions on how language technology can be used in the documentation of lesser-known languages (Borin, Trosterud) and on the necessity of providing linguistic support for the maintenance of linguistic diversity (Allwood). The present article contributes to this discussion by giving examples of how ICT has been applied in communities speaking lesser-known languages for preserving their intellectual heritage and for training in community culture and language.

Experience gained through work on lesser-known languages has convinced linguists of the importance of long-term relationships, co-operation, and interaction with the speech communities (Grinevald 2003: 56 and in this volume). Such sustained relationships have implications beyond the ethical principles that govern fieldwork. Ethical issues define relationships between linguists and speech communities when making agreements about the type of work to be done and the utilisation of its results. Long term relationships provide the understandings of the community's history, social fabric and aims that enable projects to deliver outcomes that *can* support language and cultural maintenance.

Grinevald (2003: 58) presents a set of approaches to fieldwork, or "frameworks" (adapted from Deborah Cameron, cited in Grinevald 2003), that have successively evolved over time:

- fieldwork *on* a language
- fieldwork *for* the language community
- fieldwork *with* speakers of the language community
- fieldwork *by* speakers of the language community

Each framework sums up responses to factors including changing political trends and fieldwork practice. Grinevald characterises fieldwork *on* a language as an activity "carried out by individual linguists for purely academic purposes, with individual speakers". In fieldwork *for* a language, linguists make themselves "useful" to communities; this has typically focused on community advocacy. The application of ethics and new social science methodologies in the 1980s saw community members embraced as participants in research – hence a shift to carrying out fieldwork *with* the collaboration of speakers. More recently, with increased recognition of community control and the provision of training, fieldwork is being carried out *by* community members. In this paper, we focus on the pathway from community partnership in language work to the tangible products that meet their aims for language maintenance or strengthening, in other words, the delivery of usable resources. We therefore add a new framework to the set:

– fieldwork delivered *to* a language community

This is a distinct framework from fieldwork *for* a language, which Grinevald identifies with political activism from the 1960s, and can also be seen in more recent emphasis on community control of the initiation and conduct of research (e.g. AIATSIS 2000). The "deliver *to*" framework is less concerned than the other frameworks in describing the input side of the process or distinguishing between community and linguists' contributions: it is more concerned with the form and effectiveness of the outputs. It complements the other four frameworks and together they describe the full set of perspectives within which endangered languages fieldwork takes place.

We believe that this framework is particularly important for working with endangered languages. It directly addresses language endangerment through a commitment to participating in countermeasures. It has the potential to ameliorate some of the polarity in current debates about who initiates and controls projects because it adds an alternative priority for actual outcomes and their effectiveness. We think that it can add a much-needed dimension to the evolving understanding and practice of the discipline of language documentation (Nathan 2006). On-going interchanges between linguists and communities in the context of developing and testing linguistic resources can steer the documentation process towards the most important aspects of the language while ensuring that the products designed in this framework are attuned to the actual needs of the community.

A "deliver *to*" framework, then, focuses our minds on the urgency of turning research and fieldwork into usable products. Anticipating the needs of "philologists 500 years from now" (Woodbury 2003: 45) is not its task. Within this framework we might plan for what the speech community needs in the next 50 years; but we are even more interested in responding to the fragile state of languages and their speakers by supporting language maintenance and strengthening within the learning time span of youngsters – in other words, a year or two, or three.

2. Language documentation and community linguistics

During the last two decades there has been intense debate about the theoretical, practical and ethical aspects of relationships between linguists and speech communities. Different positions have been taken by various individuals and groupings within the wider field. Two main groupings have regarded the task of the linguist to be primarily concerned with the elicitation, recording, and preservation of linguistic data: "traditional descriptivists", and, more recently, those involved in defining data cataloguing standards and building data banks of various types, e.g. for typological studies.

On the other hand, some linguists and field researchers, such as Ken Hale (1992) have advocated linguists' engagement in the communities' efforts to maintain their linguistic heritage. Community linguistics is becoming more influential among linguists and its principles have played a part in initiating new approaches such as documentary linguistics (Himmelmann 1998). The concept of documentation has been widened to include a range of data much greater than that handled in traditional grammar-dictionary-text models. Today, a state-of-the-art documentation is expected to include multimedia representation of a range of aspects of community life and to present the speakers' attitudes towards and reflections on the language (Himmelmann 1998:166). Such work is, of course, difficult to do if linguists have not developed trusted relationships with community members or participated in meaningful ways in community life. Thus, documentation in this sense presupposes a practice of linguistics *for*, *with*, *by* and *to* language communities.

3. Documentation and ICT

We focus here on local, customised multimedia-based ICT projects rather than generalised or large-scale IT development. Two large IT-centred projects, OLAC and DoBeS, have developed encoding systems and data banks with access infrastructure using metadata to enable "resource discovery", and have developed software for supporting researchers.[1] Both see the production of resources suitable for community use as a secondary task for local research teams (Wittenburg 2003: 123; Simons and Bird 2000). However, the bottleneck for endangered languages support is not a lack of encoding or metadata; text in standard interlinear format, or even a simple orthographic text file "marked up" by punctuation, has far more standardised and encoded structure than the simplest digital images or sounds. The real problem is a lack of bandwidth for interchange between knowledge holders and end-users of that knowledge. In a "deliver *to*" framework, the primary and urgent task is to seek methodologies that can mobilise the partnerships between speech communities and linguists. Communities will not need a system for resource discovery when resources are *delivered to* them.

Therefore, the challenge facing language documentation in the "deliver *to*" framework is less one of cataloguing, archiving and dissemination of materials for research and data processing, but rather the discovery and evolution of software and interfaces to assist in the collection, construction, and flexible usage of resources by a wide range of users, especially language community members. This does not mean simplifying or trivialising data or the way we work with it; it means recognising that language resources are not merely data but are embedded in a variety of social processes, and working harder to create sophisticated but friendly, usable, and pedagogically effective software to a level comparable to that already found in mainstream computer games and office applications.

While documentary linguistics has been widely embraced as a new approach to language endangerment, it is remarkable that in an era of global networking, seamless multimedia and communication technologies, and improving understandings of the value of multimedia and hypertext in learning, we continue to assume that traditional printed, linear text is adequate for most purposes. The documentation agenda must include exploration in new genres of products that can deliver the diversity and richness of materials, and support a range of users (Csató and Nathan 2003: 74).

Later in this paper we describe some of our experiences of cooperation with communities to design the form, content and usage of ICT language resources. Although these communities are not located in South Asia, they represent typical cases also relevant for the Indian context.

4. The multimedia pathway

Multimedia is a powerful vehicle for working in a "deliver *to*" framework. The process of developing multimedia naturally directs attention to the nature of linguistic events and performances and to the quality of recording. Multimedia products tangibly present the community's relationships to the language and language performances that appear in the product. As a result they implement pathways between documentations, community members as actors, and end users of products.

Bird (1999) noted that providing original recordings alongside an analysis can provide a more scientific linguistic account, because any user can examine the analysis in the light of the actual "data". For language community members, the advantages of providing ready access to rich and contextualised representations of actually occurring language events are even greater. In Csató and Nathan 2003: 74 we described some of these advantages for those using the *Spoken Karaim* CD,[2] which places its recorded language events at the very *centre* of the product architecture. There are an almost unlimited number of ways that multimedia can connect an individual or community to recorded linguistic events through social, emotional, intellectual and learning pathways.

Two key design principles of *Spoken Karaim* implement such pathways between community input and the user of the CD:

- the users' perception and navigation of the interface depends on their place in, and relationships to, the community
- sounds are never merely data but are performances of fully identified speakers, accompanied by a variety of links to biographical, social, cultural, geographical, and historical information.

5. Multimedia linguistic and cultural resources

5.1. The Karaim communities

Spoken Karaim is a multimedia CD-ROM developed by the two of us in co-operation with Karina Firkaviciute, a Karaim musicologist and representative of the speech community. The CD gives an introduction to the linguistic and cultural heritage of the Karaim community of Lithuania. The community consists of less than three hundred people, of whom the majority still follow the cultural and religious traditions of their ancestors, but only about forty still have a good command of their Karaim language. The current generation of Lithuanian Karaims are, thus, in a crucial position in regard to transmitting the language and culture to the next generation.

Currently, there are four main Karaim communities, and together with Karaims living in diasporas in other parts of the world, they total about three thousand people. The communities are scattered far apart, but they have a tradition of keeping in touch with each other. The Lithuanian Karaims, based in the village of Trakai near Vilnius, are in close contact with the other communities in Russia, Ukraine and Poland.

There were previously three Karaim varieties, spoken in Lithuania, Crimea and Ukraine respectively. Only the Lithuanian Karaim variety is still spoken. In the 19th century, the Crimean community shifted its language to Crimean Tatar, a Turkic language spoken by a larger Turkic community on the Crimea. In the Ukraine, most of the west Ukrainian community left the traditional Karaim settlements in the political turmoil of World War 2 and only a handful of Karaims remained. The last fluent speaker of this variety died recently.

5.2. *Spoken Karaim* and the Karaim community

The linguistic, cultural and community-based material included in the *Spoken Karaim* CD was recorded and collected by Csató during fieldwork in the Karaim community of Lithuania financed by the Deutsche Forschungsgemeinschaft. Members of the Karaim community participated in the documentation work. Subsequently, Nathan and Csató started to design the CD while they were researchers at the Institute of the Languages and Cultures of Asia and Africa (ILCAA) at the Tokyo University of Foreign Studies. Initial development of the CD was funded by ILCAA (Nathan 2000a), and it was further developed by Nathan while at the Australian Institute of Aboriginal and Torres Strait Islander Studies, and by Csató with the support of the University of Cologne and the University of Uppsala. For more information about the architecture of the CD see Csató and Nathan 2003.

The Karaim CD provides an integrated presentation of many different types of information: recordings of language events, linguistic descriptions and analyses; descriptions of the community's history, religion, literature, social structure, and food; secular and religious music; and images and videos of people and local features in Trakai.

Karaim users, therefore, find themselves immediately at home when encountering the CD. The CD opens with a photo of a large community gathering outside their most important building, the *kenesa* (temple). In the middle of the photo is the late Mykolas Firkovicius, the *hazzan* or religious and administrative leader, who was well known to every Karaim. This opening image invokes the interest of community members; here, or elsewhere on the CD, most Karaims find either themselves, family members, relatives or close friends.

The four Karaim communities are in continuous contact with each other, and marriages across the communities are common, so the social significance

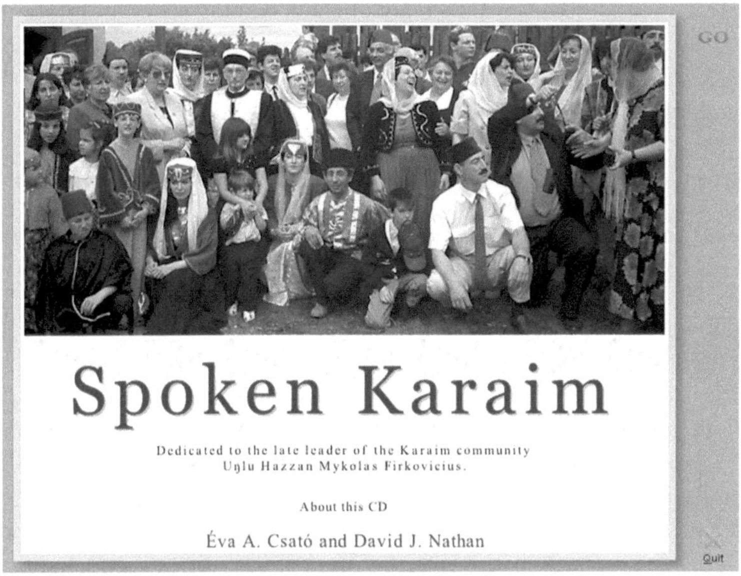

Figure 1. The community opening to *Spoken Karaim*

of the CD transcends the local Trakai community. In addition, family lineage is a common topic of discussion whenever and wherever Karaims get together. It seems, therefore, that *Spoken Karaim* will continue to be significant in all the communities and potentially for future generations.

5.3. Reflecting language semantics

The CD is focused on the village of Trakai, which is located between two beautiful lakes. A stylised map of Trakai, its lakes, and the surrounding landscape serves as the navigational frame in which all the information is presented. The user first navigates along the "Karaim street" running along a peninsula to the castle. This street is where for 600 years most of the local Karaims have traditionally lived. The user can visit various sites: the entrance to the Karaim street, the *kenesa*, the house of a Karaim speaker, the Karaim restaurant, the Karaim cemetery, the house of a famous writer and teacher, and finally the castle of Vytautas, the Grand Duke of Lithuania who invited the Karaims to Trakai at the end of the 14th century in order to defend the newly built castle.

Knowledge of this landscape is not only an indispensable part of Karaim identity; it is also deeply embedded in the semantics and pragmatics of the language. Small, dominated languages such as Karaim carry an important semantic load. While national languages used for communication between heterogeneous communities must comprehensively "mirror the world", small, local languages tend to elaborate on

> … semantic features relating to aspects of their cultural or geophysical environment. Since they are not in need of more general semantic resources, they may develop greater lexical complexity in certain semantic fields… (Johanson 2003: 28)

Understanding spatial deixis in Karaim presupposes knowledge of the local landscape. Like many other smaller languages, Karaim has environmentally defined deictic reference points. Spatial orientation in Trakai is described in relation to one of the two lakes, e.g. *göl artxarï* (lake behind: 3POSS) 'behind the lake', *göl katnï* (lake at:3POSS) 'at the lake'. The Karaim community in Halich, Ukraine, on the other hand, is settled on the shore of the Dniester River. Speakers refer to the river to describe locations in their area: *ezen katïn* (river at) 'near the river', *ezen ašarï* (river beyond) 'beyond the river'.

In both communities, *göl* 'lake' is used in a restricted sense to refer to the "Karaim lake" in Trakai (to refer to a lake in general, Halich speakers use the Slavic *ozero* 'lake'). Similarly, *ezen* 'river' is always the Dniester, the "Ka-

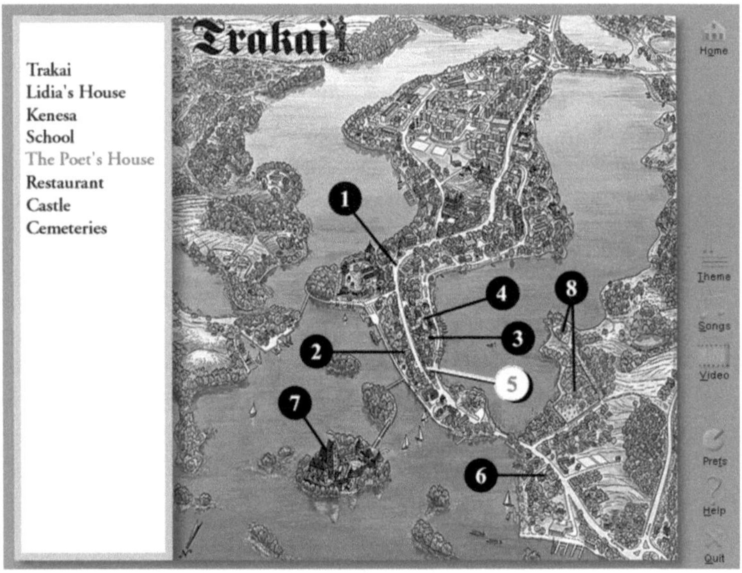

Figure 2. Spoken Karaim configures the user's experience in terms of locations

raim river". This example of complementary lexical restriction found in two dialects of Karaim illustrates the important role of the geographic features in community life.

There could be no better way to document this aspect of Karaim language and community than the representational and navigational system used in the *Spoken Karaim* CD that configures the user's experience in terms of locations around the lake.

5.4. Reflecting authentic lexical copying and morphological strategies

The language content of *Spoken Karaim* is based around recordings of natural colloquial speech. Such speech contains frequent occurrences of words that are copies (borrowings) of words from Slavic and Baltic languages. Lithuanian Karaims have long been intensely multilingual, speaking Lithuanian, Russian, Polish, and today, among the younger generation, English. Lexical and grammatical features from these languages have been copied into Karaim over many centuries (Csató 1999b, 2000a, 2000b, 2002a), and have contributed to the communicative potential of the language. The strategies involved in copying words into Karaim are part of the native speakers' knowledge. Here is an example: a verb can be copied into Karaim with the help of the Karaim auxiliary verb *et-* 'do', which in turn can carry verbal suffixes. From the Slavic verb *peresest'* 'to change (buses)' comes Karaim *p'er'es'es't' et'–* and 'I changed (buses)' is *p'er'es'es't' et' t'im* (change do:PAST:1SG), 'you changed (buses) *p'er'es'es't' et' t'iy* (change do:PAST:2SG), etc. Within the CD, users can easily hear, read, and search for many such examples.

In contrast, older materials, such as textbooks, typically present purified, prescriptive forms of the language. This has several drawbacks. "Purist" materials foster pressure against fluent speakers' creative copying strategies by giving the impression that "good Karaim" does not employ "foreign" elements; this, in turn, both inhibits speaking and discourages older speakers from passing on their language (Csató 1999a). Mastery of copying strategies is essential not only to Karaim fluency, but also to its survival, and the presentation in *Spoken Karaim* of authentic contemporary speech containing many examples reflects another community-oriented emphasis.

5.5. Reflections of performances

The *Spoken Karaim* CD documents several significant community events. These include video of a drama written and performed by community members, and of the 600th anniversary of Karaim settlement in Lithuania, which was celebrated by costume theatre in the Karaim street. Other videos show young Karaims reciting poems, counting, and learning songs from their relatives. Such examples may inspire young Karaims to involve themselves in similar types of cultural activities.

Songs play an important role in the maintenance of Karaim identity. When community members gather together they sing in Karaim. For young members, who are often ashamed because they do not know the words, the CD functions as a "jukebox" that gives them easy access to listen to and learn the songs.

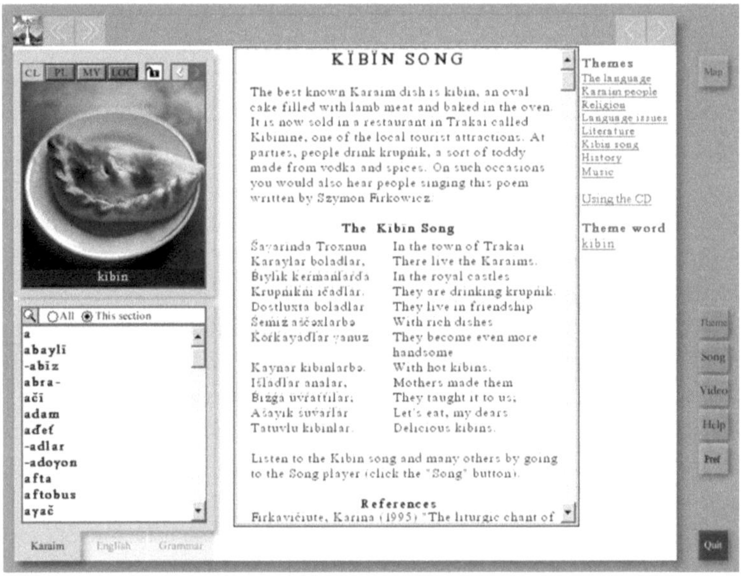

Figure 3. Cultural expression in multimedia: Karaim's favourite food, the *kïbïn*, in sound (the song is playing), text, and image.

5.6. In-community development: The Trakai Summer School CD

The *Spoken Karaim* CD, and associated multimedia resources, have been used as teaching resources in community training at summer schools. These events were organized by the Karaim community of Trakai. We have participated in three of them; the 2001, 2004 and 2005 summer camps sponsored by the Visby Programme of the Swedish Institute, Stockholm.

The new ICTs provide many opportunities to serve the needs of speech communities, scholarly communities, and the wider public. The Open Language Archives Community states:

> Today, language technology and the linguistic sciences are confronted with a vast array of language resources, richly structured, large and diverse. Texts, recordings, dictionaries, annotations, software, protocols, data models, file formats, newsgroups and web indexes are just some of these resources. The resources are growing in size, in number, in diversity
>
> ... multiple communities depend on language resources, including linguists, engineers, teachers and actual speakers ... These communities are growing in size, in number, in diversity ... And today, we have unprecedented opportunities to connect these communities to the language resources they need. (Bird and Simons 2001)

One way of directly connecting resources to communities is to use them in the context of language training delivered to the community. Using multimedia such as *Spoken Karaim* in community-based training can also make connections *within* the community by bridging between generations, a typical prerequisite for the transmission of a language. In the summer schools, multimedia provided a very effective catalyst for interaction between youngsters and the older speakers. They complemented each other's abilities: the young were adept in exploring the language materials while the older speakers could help them to understand and use them. Similar observations have been made for multimedia language resources in Australian Aboriginal communities (Nathan 2000b).

Multimedia resources can be created in the course of fieldwork and training events. This provides a richer form of connection, because what is eventually delivered to the community has been created with and by them. Our activities over the one-week summer school in 2004 culminated in the production of a new language multimedia CD. Each lesson was attended by fluent elders who performed teaching and cultural roles. Starting with situations in which the children found it natural to use Karaim – such as greeting others in the Karaim street, asking for food or drink at home or the Karaim restaurant, commenting on the weather, praying in the *kenesa*, and singing traditional songs – the children undertook various activities to build up their confidence and ability. They also elicited and recorded language from the older speakers. These materials were used to create the *Trakai Summer School* CD.

The CD also included a multimedia crossword generator that incorporated some of the learning items and recordings that had taken place over the week. It has three types of crosswords:

- standard crossword. This crossword looks like a standard newspaper crossword. Students must input a Karaim word in response to an English clue. The crossword-generation program creates a different crossword for each

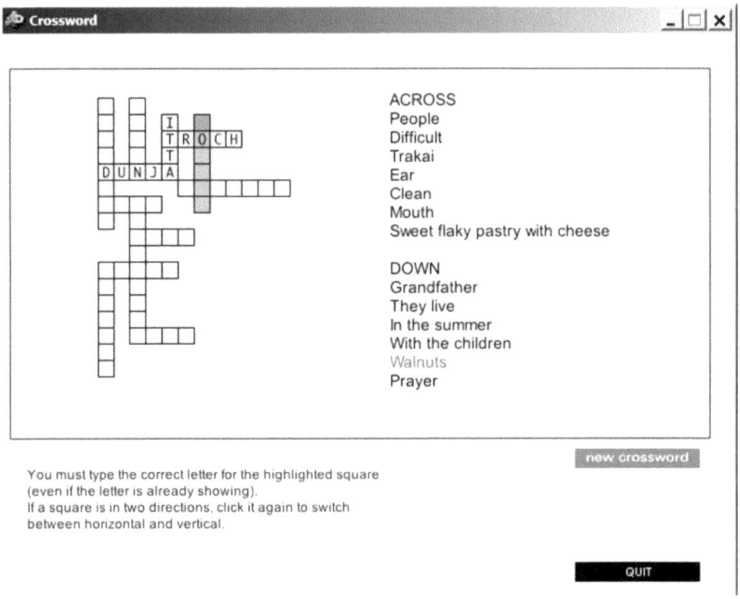

Figure 4. The Trakai summer school CD included this Karaim crossword game

game, drawing from a set of words, most, but not all of which the students had been exposed to (see Figure 4)
- talking crossword. This is a new kind of multimedia crossword. There are no clues: the students have to focus on the sound of the word, thereby drawing their attention to pronunciation and orthography
- picture crossword. Here, the students have to write the names of their classmates that they see pictured. This crossword is intended to mark the CD's relationship to the summer school participants (see Figure 5)

The crossword had an extraordinary impact on the students. They saw it evolving during the course of the summer school and became eager for its completion. Once completed, they enjoyed it as a game (since players get a "surprise" when they complete the crossword), especially when they could use two computers and compete in groups to see which group could complete their crossword first! We also used it as a vehicle for learning, and found that it provided a perfect context and motivation for them to explore and use *Spoken Karaim*. In particular, we found that they were looking up the CD's dictionary and, for the first time, using its "active morphology" inflection generator (Nathan 2000a). We also learned something important: that, if sufficiently motivated, students can handle complexities of orthographies without getting diverted or confused. Although the crosswords used a Polish-based orthography rather than the more linguistically detailed orthography used in *Spoken Karaim*, the students had no problem in translating between them (Nathan has made the same observation for Paakantyi students; see Nathan to appear).

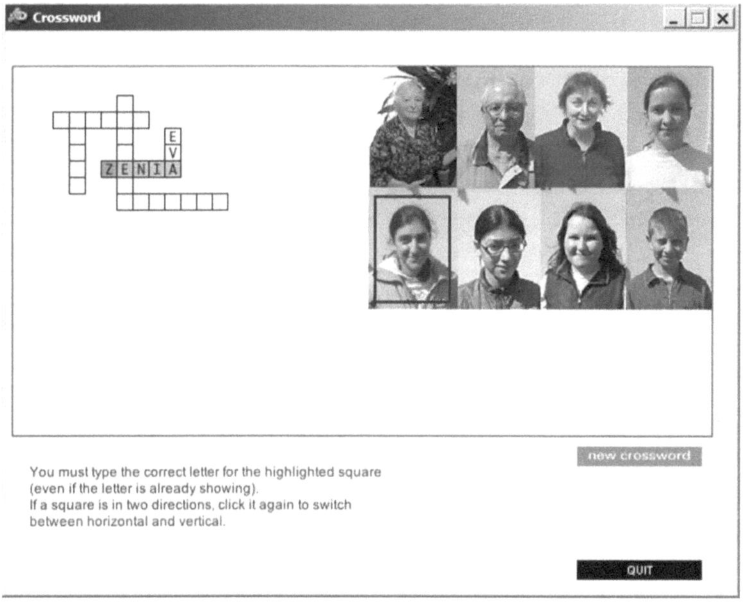

Figure 5. The Trakai summer school CD reflects the students – they are clues in the crossword!

At the end of the summer school, copies of the completed Trakai Summer School CD were distributed to each of the participants. The Karaim children were very excited to receive a new ICT product in the Karaim language, and were proud to have contributed to it. The CD contains:

- Expressions:
  - Greetings
  - Eating and drinking
- Recitations
  - Children's lullaby
  - Prayer
- Songs
  - Kïbïn song
  - Uzun kiunliar

- Game
  - Crosswords
- Images
  - An album of photos and video taken at cultural activities, and in classes

5.7. No boundaries: Multimedia and collateral activities

A multimedia product can provide a range of collateral resources, contexts, motivations, and inspirations for further language and cultural activities. In the second summer school, for example, students collected pictures and wrote Karaim text with the aim of turning them into further multimedia resources. They chose illustrations of objects that were relevant to their own life in the Karaim village – words such as 'lake', 'vegetable garden', 'fish', 'boat', 'cucumber', 'community school' and, together with the older speakers, formulated Karaim sentences for each picture. Next, recordings will be made and the materials will appear in the next community-based CD.

Spoken Karaim was involved in various indirect ways during the summer school, including the following:

- a printed "snakes and ladders" style game was created, based on the Trakai map (see Fig 2). This game was used as the basis of question and answer practice
- the CD's song player was used to teach Karaim songs
- students enjoyed exploring the CD as a "free time" activity
- the CD's dictionary was intensively used as a reference when playing the crossword games
- as a basis for discussion about an appropriate Karaim orthography

5.8. Cartoons

Nathan developed simple software for creating and playing comic-style cartoons, after the idea was suggested by Auntie Rose Fernando, of Collarenebri NSW (Australia), a Kamilaroi (Gamilaraay) elder, former teacher, and prominent and tireless promoter of language preservation in NSW. Comic books have been used for Aboriginal education before, notably the Streetwize series aimed at health promotion.[3] However, extending them to a computer platform has opened up new possibilities.

Cartoons provide an excellent environment for presenting and interacting with sound. It is difficult to find effective, "natural" interfaces for sound, because it is not only difficult to present on-screen (sound streams out in time, not space), but finds itself in competition with text, which becomes more seductively effective on-screen, even for people who are weakly literate in other domains (Nathan 1997, to appear). In the digital environment, text is nativised (Levinson 1999), because, like sound in the real world, it flows inseparably with its carrier.

The cartoon's signature "speech balloons" objectify utterances and serve as a transparent means of accessing them. Users know what to do (click on the balloon) and what to expect (the character associated with the balloon speaks). This makes cartoons one of the few rational, conventionalised interfaces for presenting the sounds of language without intermediation by written text.

In addition, cartoons can present a greater range of authentic language usage than most other forms of representation used in language documentation or teaching. Unlike dictionaries, they present words in context; unlike grammars, they present sentences with social meanings; unlike most stories, they show how everyday contemporary language is used in the context of real relationships. In the right circumstances, cartoons can even portray real people from the local community.

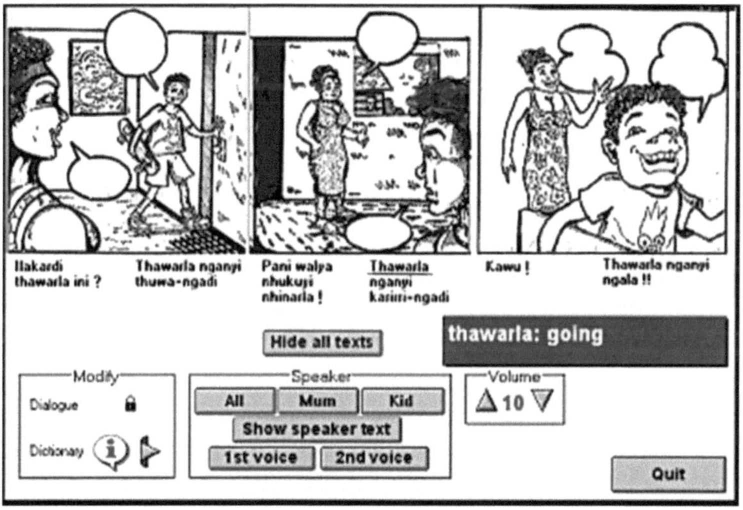

Figure 6. Yandrruwanda (SA/Qld Australia) cartoon. Produced with Greg McKellar and Muda Aboriginal Corporation

Cartoons are also a wonderful aid to elicitation. By objectifying utterances in social contexts, but without specifying their content, they help partial speakers, or "rememberers" to formulate expressions if asked, for example: "what would she say here?"

Furthermore, cartoons support the elicitation and presentation of the *kind of language* that speakers are likely to remember. Cartoons can portray familiar contexts and activities. Being less formal than other genres, they are better carriers of idiomatic and informal expressions that are otherwise often unwittingly censored from other linguistic materials. It is all too easy for linguists to fail to capture everyday, idiomatic expressions, the very expressions that can be crucial for maintaining identity. These expressions often continue to be used by the older generation, and, if passed on to younger people, can be markers of cultural identity. Expressions corresponding to English *poor me*, or *enough!*, or *ouch* can be used a dozen times a day (unlike, for example, constructed neologisms for "new" objects which might only be uttered on rare or artificial occasions). Such standard expressions can be easily elicited and presented using cartoons.

The value of cartoons for elicitation was well illustrated by a particular event during Nathan's fieldwork in preparing a CD-ROM. I worked with an elder on a cartoon story, sketching out the frames, participants, and the gist of the story. We were assisted by a linguist, who jotted down a version of the cartoon dialogue. On the day we planned to record the dialogue, the linguist arrived with a neatly typed up script based on the previous jottings, but with grammatical corrections. As we began recording, the elder started to read the written and corrected text; but she soon struggled, stumbled, and stopped, exclaiming in shame, "I'm not literate in my own language!" The value of cartoons as an elicitation methodology had been foiled by a quest for "accuracy", and a creative, authentic, spontaneous speech performance had been replaced by the reading of a prescriptively formulated text. It would be difficult to better capture, in a single event, the contrast between linguist-centred and community- centred approaches.

6. Resources on the internet

Up till now we have not mentioned the internet and, while there is no disputing its potential for supporting various communications, in general we do not regard its effectiveness for supporting endangered languages or for language teaching as highly as some others do (e.g. Crystal 2000). Fervent interest among language educators during the 1990s in the potential of email exchanges for language learning, for example, has not led to significant uptake of electronic communications in the language classroom.

Some Karaims do use formulaic Karaim greetings within emails (the emails are otherwise written in Lithuanian, Russian or Polish), but this illustrates the real problem: since the internet offers a more "genuine" communicative environment than staged multimedia resources, it also more genuinely reflects the reality of language shift situations.

The internet is more likely to be associated with written communications among younger, urban, professional, assimilated individuals living away from the main language community. Indeed, the main response to a small website representing the *Spoken Karaim* CD has been contact from a number of Karaims and Turkish people living in the diasporas of Europe, Turkey, and the USA. This does not mean that the internet is not useful for work *on* endangered languages, only that it has not so far been a fertile channel for communication *in* such languages (with some exceptions, this has been the case for endangered languages generally). It is, however, an excellent vehicle for organising activities, publicising language activities, providing download of digital resources etc.

Internet resources can complement, but not replace the kinds of ICT products and activities we have discussed in this paper. We do look forward to the day when Karaims in their far-flung locations can discuss in and about their language using the internet's "virtual spaces". But in the meantime, we cannot afford to lose the opportunities that multimedia can offer within a "deliver *to*" approach to endangered languages.

7. What communities want

A "deliver *to*" framework urges us to try to understand what communities want delivered to them. We have already argued that communities have not generally turned to the Internet as a means of supporting languages, and that multimedia, especially when constructed in and with the community, provides a better option. We have discussed above some examples, mainly for Karaim. Work in many Australian Aboriginal communities confirms the value of multimedia (Auld 2002; Nathan 2000b).

What then do endangered language communities want and expect from ICT? While there is no single voice or need that can provide a definitive answer to this question, here is a summary of wishes that we have heard many times in many places that might provide a partial answer:

- the sound of spoken language
- useful, everyday expressions
- product development processes that respect people's "ownership" of language
- products that represent the community's relationship to the language by implementing meaningful pathways between information providers and users

- a range of diverse and adaptable products from comprehensive linguistic and cultural multimedia documentations (such as *Spoken Karaim*) to learning resources, songs, games, and even spelling checkers (Manning and Parton 2001: 167)
- products that are easy to use. Goodall and Flick (1996) interpret this as requiring non text-based navigation; however, in practice this has *not* been observed nor suggested in community contexts. The *Paakantyi* CD (Hercus and Nathan 2001), for example, uses a contemporary, text-driven navigation system which has been well accepted and found easy to use

The "deliver *to*" framework outlined in the paper assigns the highest priority to delivering community members' language and cultural knowledge to other members of the community in order to assist in efforts to maintain and strengthen languages. We have discussed various examples of the development and usage of multimedia software within this framework. A complete implementation of the framework would also involve delivering multimedia skills to community members so that they can autonomously develop materials. We can best understand language endangerment through participating in practical efforts to deliver countermeasures to communities.

Notes

1. For OLAC, the Open Language Archive Community, see <http://www.language-archives.org/>. DoBeS (Dokumentation bedrohter Sprachen. Documentation of endangered languages) is funded by the Volkswagen Foundation; see <http://www.mpi.nl/DOBES/>.
2. A version of the *Spoken Karaim* CD accompanies Csató and Nathan (2003).
3. See also <http://www.streetwize.com.au/>.

References

AIATSIS
2000 *Guidelines for Ethical Research in Indigenous Studies*. <http://www.aiatsis.gov.au/corp/docs/EthicsGuideA4.pdf>. Accessed 23 December 2004.

Auld, G.
2002 What can we say about 112,000 taps on a Ndjebbana touch screen? *The Australian Journal of Indigenous Education* 30 (1): 1–7.

Austin, Peter K., and David Nathan
1996 *Gamilaraay Web Dictionary*. Online edition: <http://coombs.anu.edu.au/WWWVLPages/AborigPages/LANG/GAMDICT/GAMDICT.HTM>.

Bird, Steven
1999 Multidimensional exploration of online linguistic field data. In *Proceedings of the 29th Meeting of the North East Linguistic Society* (NELS29), vol. 1, P. Tamanji, M. Hirotani, and N. Hall (eds.), 33–47. GLSA, University of Massachusetts at Amherst.

Bird, Steven, and Gary Simons
2003 Seven dimensions of portability for language documentation and description". *Language* 79: 557–582.

Crystal, David
2000 *Language Death*. Cambridge: Cambridge University Press.

Csató, Éva Á.
1999a Should Karaim be 'purer' than other European languages? *Studia Turcologica Cracoviensia* 5: 81–89.
1999b Analyzing contact-induced phenomena in Karaim. In *Twenty-Fifth Annual Meeting of the Berkeley Linguistic Society. Special Session: Caucasian, Dravidian, and Turkic Linguistics*, S. S. Chang, L. Liaw, and J. Ruppenhofer (eds.), 54–62. Berkeley: University of California, Berkeley.
2000a Some typological features of the viewpoint aspect and tense system in spoken North-Western Karaim. In *Tense and Aspect in the Languages of Europe*, Östen Dahl (ed.), 671–699. Berlin: Mouton de Gruyter.
2000b Syntactic code-copying in Karaim. In *The Circum-Baltic Languages: Their Typology and Contacts*, Östen Dahl, and Masha Koptjevskaja-Tamm (eds.), 265–277. Amsterdam: John Benjamins.
2001a Karaim dictionary on CD-ROM. In *Uluslar Arası Sözlükbilim Sempozyumu bildirileri* [Proceedings of the International Conference on Lexicography], Nurettin Demir, and Emine Yılmaz (eds.), 35–40. Gazimagusa: Doğu Akdeniz Üniversitesi.
2001b Karaim. In *Minor Languages of Europe*, Stolz, Thomas (ed.), 1–24. Bochum: Brockmeyer. [Bochum-Essener Beiträge zur Sprachwandelforschung 30].
2002a Karaim: A high-copying language. In *Language Change. The Interplay of Internal, External and Extra-Linguistic Factors*, Mary C. Jones, and Edith Esch (eds.), 315–327. New York & Berlin: Mouton de Gruyter. [= Contributions to the Sociology of Language 86].
2002b The Karaim community in Lithuania. In *The Baltic Sea Region. Cultures, Politics, Societies*, W. Maciejewski (ed.), 272–275. Uppsala: Baltic University Press.

Csató, Éva Á., and David Nathan
2003a Multimedia and documentation of endangered languages. In *Language Documentation and Description*, vol. 1., Peter A. Austin (ed.), 73–84. London: Hans Rausing Endangered Languages Project.
2003b *Spoken Karaim* Version S. Institute for the Study of the Languages and Cultures of Asia and Africa, Tokyo University of Foreign Studies, and HRELP, School of Oriental and African Studies, University of London [Interactive multimedia CD-Rom].

Gibbon, Daffyd
n.d. Ega Web Archive. <http://coral.lili.uni-bielefeld.de/LangDoc/EGA/>.

Good, Jeff, and Ronald Sprouse
2001 Creating a database and query-tools for the TELL multi-speaker linguistic corpus. In *Proceedings of the IRCS Workshop on Linguistic Databases, IRCS University of Pennsylvania, December 2001*, Steven Bird, P. Buneman, and Mark Liberman (eds.), 82–91. Pennsylvania: University of Pennsylvania, Philadelphia.

Goodall, Heather, and Karen Flick
1996 Angledool Stories. AUC Conference *From Virtual to Reality*. The University of Queensland. Online edition: <http://auc.uow.edu.au/conf/Conf96/Papers/Goodall.html>.

Grinevald, Colette
2003 Speakers and documentation of endangered languages. In *Language Documentation and Description*, vol. 1, Peter A. Austin (ed.), 52–71. London: Hans Rausing Endangered Languages Project.

Hale, Kenneth
1992 Endangered languages. *Language* 68 (1): 1–3.

Hercus, Luise, and David Nathan
2001 *Paakantyi*. Canberra: ATSIC. [CD-ROM]

Himmelmann, Nicholas
1998 Documentary and descriptive linguistics. *Linguistics* 36: 161–95.

Johanson, Lars
2003 Smaller Turkic languages. In *Minor Languages. Approaches, Definitions, Controversies,* Joel Sherzer, and Thomas Stolz (eds.), 15–39. Bochum: Brockmeyer.

Levinson, Paul
1999 *Digital McLuhan*. London: Routledge. [Chapter 4].

Manning, Christopher, and Kristen Parton
2001 What's needed for lexical databases? Experiences with Kirrkirr. In *Proceedings of the IRCS Workshop on Linguistic Databases*, 167–173. Pennsylvania: University of Pennsylvania, Philadelphia.

Nathan, David
2000a The spoken Karaim CD: Sound, text, lexicon and 'Active Morphology' for language learning multimedia. In *Studies on Turkish and Turkic Languages*, Aslı Göksel, and Celia Kerslake (eds.), 405–413. Wiesbaden: Harrassowitz.
2000b Plugging in Indigenous knowledge – Connections and innovations. *Australian Aboriginal Studies* 2: 39–47.
2006 Thick interfaces: Mobilizing language documentation with multimedia. In *Essentials of Language Documentation*, Jost Gippert, Nicholas Himmelmann, and Ulrike Mosel (eds.), 363–379. Berlin: Mouton de Gruyter
to appear The Paakantyi Talking Dictionary: A case study. In *Indigenous People and Information Technology*, E. Dyson, M. Hendriks, and S. Grant (eds.). Sydney: Idea Group.

Simons, Gary, and Steven Bird
2000 Requirements on the infrastructure for open language archiving. <http://www.ldc.upenn.edu/exploration/expl2000/requirements.html#2.2.> Accessed 2 January 2005.

2001 OLAC overview. <http://www.languagearchives.org/documents/overview.html>.

Wittenburg, Peter
2003 The DOBES model of language documentation. In *Language Documentation and Description*. vol. 1, Peter A. Austin (ed.), 122–140. London: Hans Rausing Endangered Languages Project.

Woodbury, Anthony
2003 Defining documentary linguistics. In *Language Documentation and Description*. vol.1, Peter A. Austin (ed.), 35–51. London: Hans Rausing Endangered Languages Project.

Language survival kits

Jens Allwood

1. Should we save the languages of the world?

According to the survey of the world's languages published by *Ethnologue* (Grimes 2000), there are around 6800 languages in the world. Most of these are spoken by very few persons. Approximately 96% of the world's languages are spoken by less than 4% of its population, implying that 96% of the world's population speak only 4% of the world's languages. Since many of the speakers of "small languages" (in terms of available economic resources and number of speakers) often are older people, this means that languages are presently disappearing at a high rate. Crystal (2000) claims that the rate of language disappearance is as high as two languages each month. Most of the "small languages" are located in hot climate zones and are often spoken by people who are very poor in terms of economic and material resources. Because of the prevailing socio-political and economic conditions, a safe but sad prediction is that we will be losing between 1000 to 2000 languages over the next century. In contrast to that, around 100–200 "resource strong" languages are likely to maintain their position. Of these, around 10 languages, languages such as English, Chinese, Spanish, Arabic, Malay, Hindi, French, German, Russian, and Japanese are likely to strengthen their position. Of the other "strong" languages, more than half are spoken in Europe, within in economically affluent countries that have a comparatively high standard of living (see Matsumara 1998, Crystal 2000 and Romaine 1989 for a more general background).

Is the process leading to language loss good or bad? Although most linguists deplore language loss, there are those who are more positive to it. Let us therefore briefly summarize some of the arguments for and against linguistic diversity on Earth.

1.1. Arguments against linguistic diversity

1. The pressures of military, economic, technological, and social development push towards more integration in the world, with the use of fewer languages as means of communication. Languages which are tied to political,

economic, and military power are likely to survive if the trend continues. Trying to maintain linguistic diversity, in the end, may be a meaningless effort. It is thus an unwanted and hopeless quest to try to preserve "small human languages".

2. The diversity of languages on earth is a hindrance to global trade, scientific cooperation and communication in general. Large multinational companies need to use immense economic resources to localize products to fit with "local" languages and cultures. These resources could be used more productively if one could avoid these "local adjustments". From a commercial outsiders' perspective, linguistic diversity mainly benefits those who can use linguistic diversity to their own advantage, such as the "localization industry", the growing community of translators and interpreters, language teachers and language training institutes.

3. Linguistic diversity not only leads to a waste of resources, it also complicates interaction between different parts of the world and can lead to conflict through the misunderstandings that often arise when people do not share languages and cultures.

In other words, cooperation in the world would be considerably easier, more efficient, and less expensive if there were fewer languages (perhaps only one).

1.2. Arguments for linguistic diversity

1. Human languages are our greatest collective cognitive achievement . The development of human languages probably coincides with the development of human beings as a species. Through the mutual attunement of increased brain capacity and human languages, humans were able to develop not only individual information processing but also collective information processing and human cultures. Through languages, humans were able to coordinate thoughts and actions which enabled a survival capacity in diverse environments. Human languages maintain and, for new generations, preserve human conceptual and social development. Through human languages, we have access to diverse ways of classifying both the natural and the social environment, including artifacts as well as concepts having to do with cognitive states and abstract features of a world view. There is no better testimony to human intellectual effort over one hundred thousand years than human languages. Rather than letting go of this rich source of information, we should try to preserve as much as possible of it before it is too late.

2. The conceptual frameworks associated with different human languages have developed under mutual influence, interaction, and competition between languages and cultures for a long time. For millennia, human languages have provided a basis for a cultural and conceptual competition between different communities which probably has been for the good of mankind. If this competition were to be diminished or disappear, the risk is that the population of Earth would have to collectively suffer the drawbacks which are usually associated with a situation of monopoly.

3. Multilingual individuals make up a large part of the world's population. The majority are, and have always been, multilingual. We are perhaps not only genetically prepared for one language but for several. At any rate, research seems to show that multilingual people, by having to learn to live with the conceptual frameworks of several languages, become more creative (Allwood, Strömqvist and McDowall 1985; Skutnabb-Kangas 2000). Multilingualism is thus good for our cognitive development. Learning several languages increases mental capacity, flexibility, and creativity.

4. There are also ethical arguments in favor of a multilingual world. The first one is of a more general nature: Should we human beings really give in to military and economic pressures in the shaping of our future world? If we believe that linguistic and cultural diversity is beneficial, should we not be able to create social and organizational structures which would make it possible to preserve this diversity? Should we not use technology as a tool in maintaining linguistic diversity?

There is also an ethical argument favoring linguistic diversity at a more individual level. What happens to the generation of people whose language disappears? Their language loss, in many ways, makes them a lost generation. By losing their language, they also lose access to their conceptual, cultural, and social heritage. In most cases a language loss occurs through language shift, when speakers of a language start using another language in domains which they earlier used their own language for. This language shift brings with it the cultural values of the groups associated with the new language, both towards their own language (community) and to the outside world. So, the language shift means not only that speakers lose their mother tongue but also that they might lose their socio-cultural values and traditions. Thus, loss of languages is usually connected to social and psychological suffering for those whose languages is lost.

5. The final reason for wanting to preserve human languages is related to the first point mentioned above. Human languages can potentially provide insight not only about the nature of human languages *per se* but also about human

nature itself. Languages give us information about possible cognitive, social, and communicative structures both between and within human beings. If the diversity of human languages diminishes, this source of information for insights into human nature also diminishes.

Evaluating the arguments for and against linguistic diversity, I suspect that the situation is still not entirely clear. Depending on one's goals and orientations, one will favor one of the two positions outlined above. However, being a linguist with an interest in the nature of human languages and in the relation of language to human nature, communication and social organization, I tend to be more impressed by the arguments in favor of trying to maintain linguistic diversity than by the arguments for not doing so.

2. Levels of survival

Even if one is persuaded that maintaining linguistic diversity is important, one is confronted with the question of what one means by the "survival", "preservation", and "maintenance" of linguistic diversity, as these can be of many different types. Let us distinguish at least the following three levels (of survival, preservation and maintenance).

Level 1: Possible extinction
Level 2: Preservation for the record (museum)
Level 3: Maintenance of full functional viability

To reach the first level, it seems that we do not need to do anything. This seems to be the end result of the present development for many of today's threatened languages. The second level implies that we recognize the importance of linguistic diversity, and that we want to do something about it before a language completely vanishes from the face of this earth. This, at the second level, can, for example, be done by collecting written, spoken, and gesture data of these languages and trying to store the data in such a way that they will be accessible for future study. It also implies that we should try to describe and explain the data as fully as possible, by providing contextual and cultural background for the data we have collected. One reason for this is that languages are not self-explanatory. They require interpretation based on use in context. We can see this by examining now extinct languages like ancient Maya or Hittite. Even if we have been able to collect written data from these languages and even if we have been able, to some extent, to interpret some of the texts, our interpretation is limited by the fact that we do not have access to the speakers and cultures connected with these languages.

The third level (i.e., maintenance of full functional viability) is a level where the language is used for all communicative functions which an individual encounters in his or her private or public life. If we also include in this the need to communicate with people from other language communities than our own, or to interpret documents and other remnants of the historical past, perhaps no language has been able to provide full functional viability in this sense. However, in a world which was less interconnected through uses of transportation and information technology than the world of today, many more languages came very close to the goal of full functional viability. In today's world the number of languages which provide full functional viability is diminishing, with one language – most likely English – taking over functions (like scientific writing) that were previously more distributed between languages.

The general picture in the world today, could in fact be characterized as one of English plus (+). In the English-speaking countries only one language, English, needs to be learned. This will be sufficient both for internal domestic and most external communication. In a nation with a strong (official) dominant language, the situation is one of "English + a dominating language". This would be the case, for example, for speakers of Hindi in India but also for speakers of most of the European languages other than English in Western Europe. For speakers who come from a minority group in a region with a more dominant language, the situation is one of "English + a dominant language + a minority language" and in some cases "English + a dominant language + a subdominant language + a minority language". This is, for example, the case with a Kinnauri speaker in Northern India, who learns Himachali at an early age as it is the neighboring dominant language, and, then when this person starts going to school, learns Hindi, and then English. In other words, speakers of such minority languages often learn three, four, or more languages. Today, this is the situation for most speakers of minority languages. A main reason for this is the high penetration of dominant cultures into most areas of the world through the development of means of transportation and information technology.

Against this background, maintenance of functional viability means preserving a minority language in at least the functional domains it is used in today and, if possible, in regaining some of the functional domains that have been lost.

3. Means to prevent language extinction

We now turn to the question of what can be done to support languages that are threatened by extinction.

Depending on what our ambitions are, we can imagine different means to be used for different levels of language survival. The means can be, for example, conceptual clarification of goals, legislative, educational, and/or technological. We will now consider some of the possible types. Some of these could possibly become part of the "language survival kit" which will be discussed in section 4.

A first step would therefore be to discuss and clarify the kind of survival or functional viability that would be optimal, possible, and realistic in a particular language community. It should be fairly clear that what is possible might not be realistic and that what would be optimal might be neither probable nor realistic. At any rate, the contrast of the three concepts might help to clarify the situation. As we have already noted, in most cases, in order to be realistic, this will probably mean support both for some type of individual multilingualism and for some type of societal multilingualism.

A second step might therefore be to investigate what legislation exists in relation to both individual and societal multilingualism, what "individual linguistic rights" exist in the society, and what "community linguistic rights" exist.

Three other important areas of support are education, the media, and increasingly the internet. When it comes to education, perhaps the most basic question is: "what kind of education is available in the language?" Since education is one of the main ways in which children gain access to participation in a particular society, and this participation usually involves use of language, providing education in a particular language is one of the key means to maintaining a language. If for economic or practical reasons only some topics or subjects can be taught in a local language, it is useful to make an analysis of what topics would make most sense given the needs, interests, and beliefs of the local population. In fact, local participation in the educational process (based on local competence) will probably almost by necessity be the decisive criterion for what can, at least initially, be taught in a local language. This could, for example, be information about the culture and customs which the language encodes. Over time, however, also other topics should be taught locally.

In fact, irrespective of whether local education takes place in the local language or not, it is usually preferable to arrange the education locally. There is otherwise a considerable risk that education might be a main cause of a "brain drain" out of the local community. People go elsewhere for education and since conditions are often better than in their original home, they do not return to bring the benefits of their education to bear on local problems so that, in the worst case, the local community is deprived both of people with an entrepreneurial spirit and of native speakers using their languages in a variety of functional domains.

Besides education, the mass media are a second strategic area for use of a language. If a language is used in the media, it automatically acquires a sort of public/official status. The language becomes publicly recognized and its speakers usually experience this as a boost to their self-esteem and prestige. Access to as many media as possible is probably helpful. But there should be an analysis of which medium, (e.g., books, newspapers/magazines, TV or radio) is most beneficial in a given situation. At the present stage of language technology, if there is no written language, only TV or radio can be considered. Further, for economic reasons, radio will probably often turn out to be most cost-effective. Local radio stations broadcasting in the world's non-written languages should therefore be high on the list of contents in a "language survival kit". It is also desirable that educational initiatives be combined with media use. In an "oral community", radio (and TV) combined with face-to-face interaction are the most straightforward ways of providing education in a threatened language.

A third means that can be enlisted in empowering endangered languages is the internet. Since so far most of the information on the internet exists in written form, its use is mainly restricted to languages that have a writing system. A strong *desideratum* is therefore that we develop multimedial uses of the web (including sound and film) which are easy to use and are widely available. The internet can, in this way, be used, for example, to provide multimodal literacy training, perhaps starting with illiterate speakers of a language that already has a writing system and later continuing with speakers of languages that are acquiring writing systems. In practice, not all languages that have writing systems can make use of the internet since most uses of the web are based on the Latin alphabet and the ASCII code. Again, a strong *desideratum* is that standardized uses of Unicode for non-Latin writing systems be made more widely available. This would greatly improve the possibilities for these languages to make use of the internet to provide increased functional viability. (For some of the problems that need to be solved, see the contribution by Hardie et al. in this volume.)

We now turn to a discussion of some of the ways in which language technology can be used to support the threatened languages of the world, cf. also Allwood 2001 and McEnery and Ostler 2000.

4. Language survival kits

A very basic level or goal of language survival is the maintenance or preservation of a language "for the record" on a "museum level". However, this goal is not incompatible with the more ambitious goal of functional viability. In fact, meeting the requirements connected with documenting a language for the record is often a prerequisite for realizing the more ambitious goals. In what follows, I will briefly discuss some ways in which language technology can be used to help make up the contents of a "language survival kit" which is both meant to be a means of preserving a language for the record and of providing a basis for more active and functional use of the language. The contents of the kit correspond to three kinds of *desiderata*:

(1) general *desiderata*
(2) creation of language resources
(3) creation of useful applications

Below, we will now discuss these three kinds of *desiderata*.

4.1 General desiderata

Some of the general *desiderata* for a language survival kit might be the following:

- Both languages with and without writing systems should be covered
- Low-cost or free ware
- Open source and general standards
- Enable automatic computer-based analysis
- Enable reuse of technology and linguistic analyses

The ambition is to be useful both in relation to languages with a writing system and to languages that lack a writing system. Since speakers of "small languages" usually also have very small economic resources, another concern is that everything that is suggested should preferably be low-cost or free ware. Since we also want to invite participation from sympathizers around the world, open source programs like Linux are to be preferred. Similarly, to facilitate cooperation, general standards that can handle a variety of linguistic phenomena should be used. For languages with writing systems, perhaps the best alternative at the moment is Unicode.

A further *desideratum* is that we should be able to make use of automatic computer-based analysis if possible. One example of this is to use machine-learning techniques, i.e. algorithms which automatically learn linguistic generalizations from raw or annotated language data. We will need such techniques considering the fact that the number of endangered languages is high, time is short, and economic resources are limited.

Finally, it is desirable that language technology tools are resuable. Languages have a great deal in common and the components of a language survival kit should be sufficiently generic, standardized, and robust to be reusable when moving from one language to another. Similarly, in many cases, especially when languages are typologically and historically related, many of the features of an analysis for one language probably are reusable in the analysis of a related language. Thus, the kit should facilitate structurally similar analyses of the phonology, morphology, lexicon, phraseology, syntax, and even pragmatics for related languages.

4.2 Creating linguistic resources

The first task of a language survival kit will be to establish what is often called "linguistic resources" for the language community. The following are some of the resources that should be created.

Basic resources

- A multimodal spoken or written database (corpus)
- Routines for recording, storing, and analyzing the data
- Transcription standards and transcriptions
- Standards for the creation of writing systems and their digital encoding (e.g. ASCII, Unicode, SAMPA)
- Annotation standards of various types
- Routines for automatic linguistic analysis (e.g. by machine learning)
- Guidelines for descriptions (and explanations) of the language, e.g. grammars and lexica

Thus, the first goal will be to establish multimodal spoken or written corpora for any given language. What this means is that a database for the language containing written, spoken, or gestural data in combination with cultural information in digital form should be created. The "raw data" will be texts, audio recordings, and video recordings. Everything should preferably be in digital form to facilitate later processing. The survival kit, thus, has to include good robust low-cost digital audio and video recording equipment or information about such equipment. It has to include the means of storing recorded data (tapes or digital storage). In cases where digital print files are available, there should be routines for organizing them into a corpus. In the event that printed digital data is not available but there is printed material on paper, scanning equipment has to be included.

Creating the corpora should be seen as an incremental process, where multimodality might not be reached initially or goals of size might only be reached after some period of time. For this reason, it is not meaningful to give absolute quantitative goals for the size of the corpus. However, a reasonable ultimate goal for a spoken language corpus might be between 50 and 100 hours of recording, which depending on how analytical the language is typologically speaking, corresponds to between 500 000 and 1 000 000 words. For written language, the initial goal might be a corpus of between 1 000 000 and 3 000 000 words.

In collecting the corpora, the recordings which are made should strive for optimal sound and video quality, and at the same time also strive for "ecological validity". This can be achieved by obtaining recordings of a representative sample of societal activities and speakers, that is, drawing from the life of the community. Since the goals of optimal sound quality and "ecological validity" are not always compatible, deciding what to record will sometimes involve a process of "suboptimization". Another important concern, related to recording, is that the video recordings provide a constant view of the interaction between as many of the speakers as possible. Only this type of recording allows for a study of communication as an interactive process. Focusing and cuts are, thus, to be avoided since they remove a clear view of the interaction.

Once records have been made, it is essential to agree on standards for storage and data classification (metadata). In the worst case, large amounts of data are recorded which can never be used, since it was not classified and stored in such a way that it can be retrieved. The language survival kits should therefore include suggestions for a system of data classification and storage. The system should be rich enough to allow searches on, for example, the date of recording, what was recorded, who was recorded, what transcriptions were made, and what kinds of analysis are connected to the recording. For more detail, cf. Allwood et al. 2000.

The next step will probably consist in some form of annotation (coding) and/or transcription (if there is a writing system). If the data are not transcribed and annotated in a standardized way, no consistent patterns will be found, even if they exist. Standardization is, thus, a clear prerequisite for further analysis, especially when automatic analysis is used. Since some languages are connected with several written variants and there is often a fairly large distance between spoken and written language, it is essential to agree on a standard for annotation and/or transcription, in order to avoid much effort at a later date.

Often it is desirable that the standard be such that more features of spoken language can be included than are reflected in normal standard written orthography. The spoken language features can be divided into two types. Some of the features, like overlap, stress, pausing, and the use of gestures, occur in all spoken languages and can thus be standardized independently of language. Other features of spoken language concern pronunciation. For these features, one alternative would be phonetic or phonemic transcription. Since this is rather labor-intensive, it might be more desirable to use standard orthography with some modifications for spoken language (cf. Nivre 1999). The corpus will also be more valuable if audio recordings, video recordings, and transcriptions are aligned temporally, so that it is possible to simultaneously read the transcription and to hear and view the relevant recorded passages on which it is based.

A special and intriguing problem is connected to languages that have no system of writing. Either these languages must somehow be processed and analyzed directly on the basis of audio and video files, or they have to be provided with a system of writing.

Regarding the first option, there is hardly any available language or speech technology that does not require written language as input or output. For example, writing is the normal input for speech synthesis and the normal output of speech recognition. One could imagine having pictures or diagrams as input or output instead but, by and large, we are still lacking language (or speech) technology that can work without a writing system. To aid in the preservation and analysis of languages without a writing system, it would therefore be important to develop such technology. Some examples of what could be done are use of recorded samples, use of concatenated synthesis and use of multimodal interfaces with graphics, photorealism, streaming etc. A future goal would be to use speech recognition to build interactive dialogue systems of different types, e.g. public information and educational systems.

The second option is therefore perhaps equally or even more realistic – providing a language with a writing system. One *desideratum* here would be to combine automatic analysis of speech with phonemic manual analysis. Automatic analysis based on speech recognition would provide suggestions for sound units, which would then be subjected to manual phonemic analysis. The end result might be a writing system based on the International Phonetic Alphabet (IPA), e.g. in its ASCII compatible form SAMPA. Since IPA is an extension of the Latin alphabet, this solution might, for cultural and historical reasons, be undesirable in some parts of the world (e.g. India). Here, instead, a writing system based on the local tradition of writing, e.g. some extension of Devanagari or Arabic script, might be a better alternative.

Once a corpus has been established (or even before), it should be described and explained. The description should include the standard linguistic types of analysis, i.e. phonological, morphological, lexical, phraseological, syntactic, semantic, and pragmatic analysis. In the interest of time and resources, as much as possible of the analysis underlying the descriptions should be done automatically. If the language has a writing system, or if a writing system can be established for the language, one of the earliest products of analysis could be a frequency dictionary of morphemes, words, or collocations. This can later be elaborated and refined through the addition of manual or automatic analysis yielding outputs like a lemmatized word frequency list, concordances, part of speech tagging, or morphological analysis. In performing such analyses, results from related languages should be reused and attempts should be made to gradually convert manual routines into automatized routines, for example, by including them in a machine learning program.

4.3. Useful applications

In order to have functional viability rather than mere preservation for the record, the language must be usable. From a language technology point of view, this means that it must have some useful applications.

Some of the applications that might be useful are the following. As above, the list is not to be taken in an "all or nothing" sense, but can be gradually extended as resources develop.

Useful applications
- Multimodal and text interfaces
- Speech synthesis
- Authoring support (word processing)
- Multimodal tutoring support
- Support for internet use
- Support for information retrieval
- Support for translation
- Support for generic dialogue tools
- Speech recognition

For languages without a writing system, multimodal interfaces with icons and recorded speech should be created. For these languages, good handling of recorded speech and speech synthesis based on concatenated phones, diphones, or triphones is essential. Such multimodal interfaces (containing combinations of icons, animated cartoons, and recorded or synthesized speech) can then be used to create multimodal tutoring systems that can be used to distribute information about health care, agriculture, or low-cost (e.g.

solar) energy production. Multimodal interfaces could also be used to develop voice mail systems that can serve as alternatives to e-mail based on writing. Future developments of speech recognition might even make an integration of voice-based and writing-based email possible. Further, many household appliances could be equipped with systems for speech control, the use of which perhaps could be extended to some of the "small languages" of the world.

Similarly, for languages that have a writing system, as already mentioned, one of the most basic things is to provide users with interface texts to the computer. Since levels of literacy might not be high, a combination with a multimodal interface using icons and recorded speech might also be useful in this case. Given a writing system, perhaps the most basic application is a tool providing authoring support (a word processing system). This can be incrementally extended as components become available. In other words, basic functions, like delete, copy and paste, are perhaps more needed than hyphenation, spelling, and grammar correction. A special challenge might here be the construction of a speech-based word processing system, where an example of a question requiring an answer might be: What kind of (graphical) means could be used in a speech-based system to give the kind of overview that is normally associated with the reading and editing of a text?

As we have also briefly mentioned above, another area of functional use will be the internet. Just as with radio broadcasting, speakers of a language will receive a boost in self-esteem and will experience a heightened prestige for their language if it can be used on the internet. There should therefore be support for internet use, in the form of email programs and tools for the creation of homepages. But besides being used for private personal communication, the internet should of course also be used to provide public information and to create the educational programs discussed above.

Further development should lead to support for information retrieval, for classification in a database, and for translation. Finally, various interactive applications utilizing generic tools for dialogue can be created. At first, these will probably involve written language, but over time attempts should be made to create systems for speech recognition, so that interactive systems for illiterate users can also be established.

5. Conclusions

Many of the world's languages are threatened by extinction. After having discussed arguments for and against interfering in this process, I conclude that it would be desirable to interfere. I then go on to briefly discuss some means to prevent language extinction, and suggest that one of the ways to do this would be to use present day language and speech technology to create a "language survival kit". Such a kit could be used not only to preserve samples, descriptions, and explanations of the language for future generations, but also to make the language more functionally useable for its present speakers. The paper provides a discussion of what the contents of such a kit should be, and of some of the challenges that have to be faced in putting it to use.

References

Allwood, Jens
2001 Language Technology as an aid in preserving linguistic diversity. *ELSNews* 10.1. <http://www.elsnet.org>.

Allwood, Jens, Sven Strömqvist, and Monica MacDowall
1982 Barn, språkutveckling och flerspråkighet [Children, language development, and multilingualism]. Government Report SOU 1982: 43. Also in *Gothenburg Papers in Theoretical Linguistics S6*, 1–290. Department of Linguistics, University of Göteborg.

Allwood, Jens, Maria Björnberg, Leif Grönqvist, Elisabeth Ahlsén, and Cajsa Ottesjö
2000 The spoken language corpus at the Department of Linguistics, Göteborg University. *FQS – Forum Qualitative Social Research* 1(3): 1–22.

Crystal, David
2000 *Language Death*. Cambridge: Cambridge University Press.

Grimes, Barbara (ed.)
2000 *Ethnologue: Languages of the World*. Dallas: SIL International.

Matsumara, Kazuto (ed.)
1998 *Studies in Endangered Languages. Papers for the International Symposium on Endangered Languages, Tokyo 18–20 November 1995*. Tokyo: Ititzu Syobo.

McEnery, Tony, and Nicholas Ostler
2000 A new agenda for Corpus Linguistics – Working with all of the World's languages. *Literary and Linguistic Computing* 15: 401–418.

Nivre, Joakim
1999 Transcription standard version 6. Department of Linguistics, Göteborg University.

Romaine, Susan
1989 Pidgins, creoles, immigrant and dying languages. In *Investigating Obsolescence: Studies in Language Contraction and Death*, Nancy Dorian (ed.), 369–383. Cambridge: Cambridge University Press.

Skutnabb-Kangas, Tove
2000 *Linguistic Genocide in Education – or Worldwide Diversity and Human Rights?* Mahwah, NJ & London: Lawrence Erlbaum.

Grammatically based language technology for minority languages

Trond Trosterud

1. Introduction

Languages may live on without an orthography. But no language will be able to function as an administrative language in a modern society without a developed language technology. As long as work on language technology is restricted to languages for which the demand for products in a market will fund the necessary work, the number of languages capable of functioning as administrative languages will shrink drastically as literacy becomes increasingly digitalised.

The present article argues for a wider view on language technology. If seen as a central means of achieving basic knowledge of the linguistic structure of the languages involved, work on language technology can combine the efforts for descriptive adequacy and basic linguistic research with efforts for developing practical tools for the languages involved. In this way, the number of languages for which basic language technology work is conducted will rise drastically.

The article only deals with a small part of language technology, with lexical and grammatical analysis of written language. I will not go into machine translation or speech-based language technology, as such work is at least partly based upon basic grammatical analyses. I will look at ways of building linguistically based language technology applications for minority languages, and at what benefits there are in such applications. Finally, I will discuss the limitations of technology: What is it that languages cannot get via computerisation and the Internet. The focus will be on South Asian languages, with occasional examples drawn from the Sámi languages of Northern Scandinavia, but the issues will be relevant for any multilingual society.

2. Language technology for minority languages

2.1. Which languages get language technology applications?

For languages with a large number of monolingual and wealthy speakers, there is a commercial market for language technology products. Costumers need applications to handle their language, and they are sufficiently numerous to pay for the developmental work. At the moment, it seems that the number of languages for which substantial work is being done on a commercial basis is between 50 and 100.[1] For other languages, there are not enough buyers willing to pay for developmental work.

If language technology is done within a grammatical framework, this issue may be reconsidered. When work on language technology is conducted partly so as to increase insight into the linguistic structure of the language in question, the size of the language community becomes irrelevant. Any language may get language technology applications, simply because they are needed for the descriptive work on the language in question. Many types of linguistic problems may best be solved via the construction and subsequent use of basic language technology tools. Seen in this perspective, any language for which descriptive work has been done, may be subject to development work within language technology, and hence also as a side effect produce different types of practical applications.

2.2. Language and orthography in South Asia

South Asian languages are poorly represented in localised versions of computer programs, despite India's indisputable strength as an IT nation. Thus, the first version of Microsoft multi-lingual XP was translated into 33 languages, the smallest of these languages, Estonian, is spoken by fewer than one million. In contrast, none of the languages of India were among the ones listed in 2001 as languages with a translated user interface. When turning to languages where keyboards and date-time format are included, Microsoft listed 5 Indian languages,[2] Hindi, Konkani, Marathi, Sanskrit and Tamil, they hence included keyboard support for two Indic scripts, Devanagari and Tamil (Konkani may be written both with the Latin and the Devanagari alphabet). At the same time, Macintosh had a slightly different set of scripts, Devanagari (for Sanskrit, Hindi, Marathi), Gujarati (for Punjabi) and Gurmukhi (for Punjabi). India and South Asia in general of course have a domestic computer industry, where the Brahmi scripts are encoded, software is written and being localised, as exemplified by Indix, a Hindi version of Linux. But compared both to smaller European countries and to other large Asian countries, the international computer localisation of Indian languages for major software lags behind. The situation is more parallel to that of Swahili than to that of Japanese, Chinese or Korean.

There may be several reasons for this state of affairs. One is no doubt the central position of English in South Asia, as compared to other parts of Asia.

It seems there is a diglossia situation, with English dominating the arenas where computers are used, especially on the internet, but probably also in off-line computing. Software producers thus find customers willing to buy programs that are not localised in their native language. Needless to say, this put the 95% of the population of India that does not master English at a great disadvantage (the phenomenon is referred to as the "digital divide" in official documents).

Another reason may be connected to controversies over character encoding. The Unicode standard is by no means a de facto standard in South Asia, instead national standards and proprietary encoded fonts still dominate, even on the net. Internet sites, including governmental ones, are thus mostly in English. A telling exception is the presidential office of the Maldives, which publishes bilingually, but with the Maldivian version rendered as a picture file rather than as Thaana text.

Even though English has a strong position in South Asia, especially in business, higher education and government, the national languages are by no means marginalised. National and regional languages are being widely used in the press, and in the administration.

The number of languages in South Asia is an open question. The sources arrive at widely different numbers of languages, one estimate is given by the Ethnologue,[3] which lists 398 languages for India (470 total for South Asia). Very few of the languages in South Asia have a standard orthography. In a recent survey, Colin Masica (Masica 1991) claims there to be as few as 78 Indo-Aryan languages in India (the rest he sees as dialects of these 78), of these he counts 12 written languages and 7 "aspiring written languages" (Doori, Kashmiri, Khowar, Konkani, Maithili, Marwari, Siraiki). This gives a rate of languages with standardised orthographies at ¼, at the very best (more like 10%, if we keep the number of standardised orthographies constant and raise the number of actual languages to the level suggested by the Ethnologue. This is surprisingly low. The global rate of languages with a standardised orthography is not known. 3500 languages, or 2/3 of the languages of the world, have at least some standardised orthography, since at least parts of the Bible is available in so many languages (Barbara Grimes, p.c.). This number does not tell for how many languages there is a literate population actually reading and writing their language

With the advantages of parallel engineering, making language technology applications for large number of South Asian languages has become a manageable task. Since the relevant medium is digital storage space, the only factor limiting the number of language technology application for even the smallest languages, is the availability of skilled linguists and computer scientists.

2.3. Language technology in India

A good overview of status quo for language technology in India is given by the *Technology Development for Indian Languages*, <http://tdil.mit.gov.in> (cf. also Jha 2003). There is developmental work for 18 languages in 13 different regional research centres. The centres have concentrated upon basic localisation: character conversion, translation of graphical user interfaces, and OCR applications, but also upon content analysis such as corpus building, wordnet tools, and, in some cases, machine translation. For approximately ten of the official languages spell checkers are being developed. Several of these are modelled as automata, especially for the morphologically complex Dravidian languages (cf. section 3 below for a discussion on automata). Judging from the documentation available, the morphological analysers are still not built as transducers, or in other ways provided with grammatical information.

2.4. Grammatically or statistically based language technology?

Many language technology applications for English have been done with a minimum amount of grammatical analysis. Spell checkers have been made based upon a compressed list of all wordforms found in a large corpus of proofread running text. Grammatical analyses have been assigned to word forms from a list, and thereafter disambiguated on a statistical basis. Two factors have contributed to the success of this approach: First, there are very many texts electronically available for English. For the majority of languages of the world, we do not have such resources. Second, English has a very poor morphological structure. Each lexeme is represented by at most a handful of word-forms, with no forms significantly less frequent than the others. Most of the languages of the world have a more elaborate word structure than English, with large paradigms for each lexeme, containing tens or hundreds of word-forms. In these languages, some of the paradigm members (also the ones of common lexemes) have a very low text frequency. Building spell-checkers or other language technology applications from compiled word-lists therefore runs the risk of omitting possible wordforms rarely used in corpus texts. Confining ourselves to the South Asian languages, we see that they all have a morphological structure more complicated than that found in English. Even though Indo-Aryan languages show more modest inflectional paradigms than the Dravidian ones, especially when it comes to nominal morphology, they still have dynamic compounding of words.

Statistical approaches work best for languages with huge amounts of text electronically available, and a minimal amount of morphology. Languages with fewer available texts and more productive morphology should choose linguistically based methods. In order to make basic language technology tools one needs a comprehensive grammar and a large dictionary, with grammatical information linked to the lexemes.

3. Finite-state automata and transducers

3.1. What are automata and transducers, and how can they be built?

A *finite state automaton* is a machine, with one initial state and one or more final states (marked with hooked circle and double circle, respectively), and a number of paths to travel from the former to the latter, via a *finite* number of intermediate states (hence the name). Figure 1a shows an automaton that generates, among other, the strings *b, ab, aab, aaab, …* Automata can obviously be used for many purposes, relevant in our context is their ability to model morphological structure. In a morphological automaton, each path represents from the start state to a final state one wordform. An automaton generating the wordforms *boy*, *boys*, *girl*, *girls*, *child*, *children*, *sheep* is shown in Figure 1b.

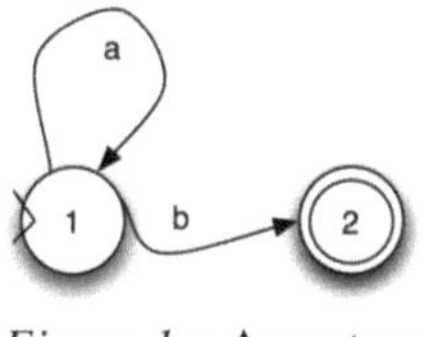

Figure 1a. An automaton

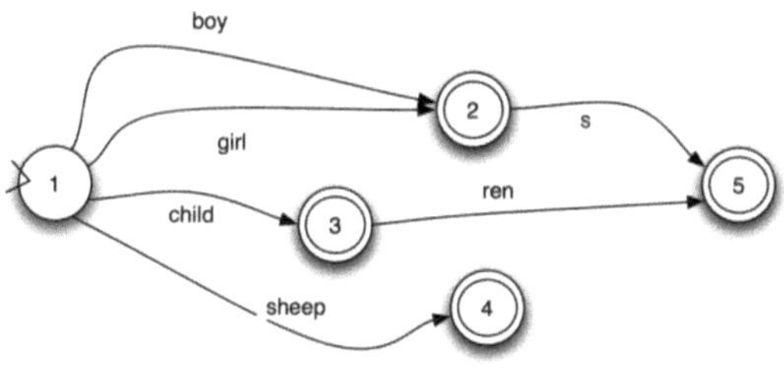

Figure 1b. Generating 7 wordforms

In order to model morphological processes, we need a special kind of automaton, called a *transducer*. A transducer is an automaton where each move from one state to the next is represented by a symbol pair, and not by a single symbol. The symbol pair may be seen as a transition, so that travelling from one state to the other via the symbol *a:b* implies changing the symbol *a* into *b*. A simple transducer that converts nouns to grammatical representations is shown in Figure 2. Morphological transducers typically convert a grammatical form to a word form and vice versa. Morphophonological processes, such as Umlaut, Sandhi phenomena, etc., may be handled in a separate component.

In order to cope with morphophonological processes we may build two transducers, one from a grammatical form to an intermediate representation, and one via morphophonological processes to a surface realisation. In some

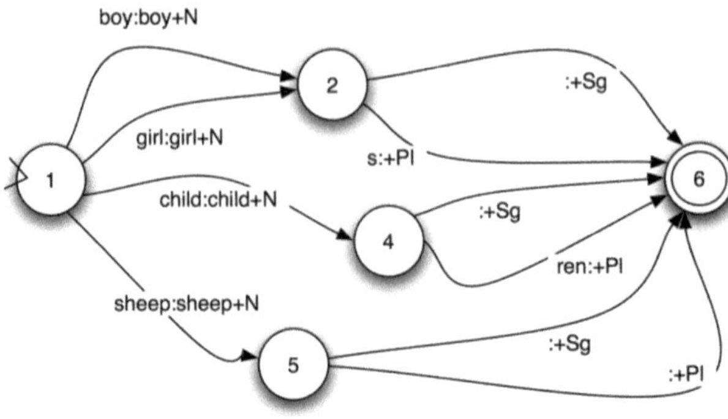

Figure 2. A simple transducer

cases, the process is purely phonological. Thus, in Tamil the final *-u* of bisyllabic nouns is deleted in oblique cases whenever the root syllable is short, but retained when it is long. This may of course be hand encoded in the lexicon (as it should if it were to turn out that this rule had become unproductive), but it may also be handled with a phonological rule encoded as a transducer taking the morphological transducer as input and giving a surface form as output (the rule formalism is the one of Beesley and Karttunen 2003, it should be fairly transparent: "||" means 'in the context of', and "A|B" means 'A or B').

(1) u -> 0 || Vow (Vow|Cons) Cons _ Case-suffix ;

As can be seen from the literature (cf. e.g. the references cited in Beesley and Karttunen 2003), both morphophonological processes and suprasegmental morphology may be handled as additional transducers modifying the basic lexical segmental transducers, thereby keeping the lexical component simple and manageable, and being able to express the grammatical properties of the wordforms involved.

3.2. Transducers in language technology applications

Many language technology applications are built like simple automata, and not like transducers, either automatically via compressing fullform lists, or explicitly, for example with the popular *ispell* framework for spell checkers,[4] or simply in Perl. While transducers imply more complicated source code, they come with two clear advantages. First, by being able to separate segmental and suprasegmental morphology, transducers provide simpler morphological representations. Ispell-type automata need one continuation class for each morphophonological process. Umlaut nouns will need one entry for each stem, and paradigms with sandhi phenomena require separate treatment, even when the process in question is phonologically regular. Second, since transducers allow for several levels of representation, it is also possible to include linguistic information with each word form. In addition to recognising a complex form like *men* as "man+N+Pl", it is also able to link it to the dictionary form and its morphological analysis.

Many languages, including the Indo-Aryan ones, have productive derivational and compounding processes. Even though a compound form in itself may be built as a simple automaton, there are always restrictions on such processes, and such processes are dependent upon grammatical properties of the compound parts. A language may for example allow N + N compounds, but not N + V compounds. In addition there may be a productive suffixation process turning verbs into nouns. Now, we want to allow N + $[\text{V–suf}]_{\text{N}}$, but disallow N + V compounds. In order to do this without doubling the whole set of verbal stems, the grammatical information provided by the multiple levels of the transducer is needed.

Morphological transducers form the core of a large range of applications, both theoretical and practical. A machine that accepts grammatical wordforms and rejects ungrammatical ones, may constitute a spell-checker; it may offer a way of registering neologisms and loanwords for lexicographical purposes, or become the core of other text editing programs. The ability to analyse and generate word-forms makes it possible to make interactive pedagogical applications.

3.4. Transducers as language documentation

Building a transducer for generating word-forms gives as a side effect a comprehensive description of the morphology of the language in question. A paper-based reference grammar may overlook details on the morphological system, this is immediately revealed in a generating transducer. Chomsky (1965: 4) claims that a generative grammar is the same as an explicit grammar. The grammars he had in mind dealt primarily with syntax, and they were quite different from the ones advocated here, but his point is valid for morphology as well. In order to build a transducer that is able to generate all possible word-forms, and reject all the impossible ones, one must have an explicit grammar for the language in question. A morphological description of any language runs the risk of being incomplete. The grammarian may leave conclusions implicit, deducible only via the combination of scattered information, and s/he may also unintentionally leave out information on different parts of the grammar, of lexically conditioned morphological patterns.

From this point of view, making a morphological transducer represents a unique opportunity to build a really comprehensive grammar. Inaccurate descriptions of the grammar will be revealed during a good testing procedure for the transducer (analysis of real corpora, perhaps containing millions of words). The vast majority of the languages of the world are poorly documented, and thorough empirical work based upon the current linguistic diversity will, if undertaken, no doubt be the main contribution to the history of linguistics provided by the linguists of the 21st century. Collecting texts from smaller languages, building comprehensive lexica and combining them with morphological transducers should thus come high on any research agenda.

In practice, empirical work on languages with a written tradition never begins from scratch. Typically, there will be a grammar, a dictionary, and a corpus. These are exactly the tools that are needed in order to build a transducer for a given language. Unless the grammatical structure is very simple, or the grammar exceptionally good, the work on the transducer will quickly reveal the omissions of the reference grammar. The result will be a dialectic process, where the transducer and the basic linguistic description will complete and support each other as the work progresses.

Among some linguists, language technology has a reputation of (even deliberately) focusing upon "getting it to work", as compared to both descriptive and formal linguists, who want to "describe it in the best possible way". In some cases, this is a real conflict. A formal linguist may want to find the most economical rule set generating a given paradigm, where economy is measured as using the minimal number of symbols and states. When writing a transducer, a conflicting priority may be to get a source-code that is easy to read, understand and correct by the linguist, even though it implies more symbols and states. As soon as the transducer is not seen as a goal in itself, but as a tool for expressing the grammar, such conflicts should be solvable.

When the transducer reaches a stage when it can be used for analysing running text, it may also contribute to basic research by providing morphologically annotated corpora. Cf. section 6 below for a discussion.

4. Some case studies

4.1. Phonematic representation vs. official scripts in morphological transducers

The transducers made for analysing written text represent generalisations not over morphophonology, but over morphographemics. Since they aim at analysing written text, rules are written for the original orthographies, for letter sequences. Technically, there is no problem in writing transducers for phonematic representations. This is seldom done, since one important aim of the transducers is to analyse written text. In cases where there really is no written standard, a natural choice is of course to use phonematic representations, but in most cases there will be a written standard to relate to. Even if, in certain cases, it may feel linguistically irrelevant to hassle with getting purely orthographic alternations right, the possibility of being able to analyse written texts will motivate the construction of a transducer for the written standard.[5]

A further question will be whether to represent the languages with their native orthography, or to reduce it to some ASCII representation. Language technology applications have often used such ASCII representations, due to limitations of the software involved. Today, the scripts of all living languages, and indeed of all South Asian ones, are in principle included in the Universal Character Set, UCS, known as Unicode (possible missing scripts will be added as they are detected). More and more development tools are able to handle the 8-bit multi-byte format of the Unicode encoding standard, UTF-8, thus making it possible to write all scripts directly into the source code. There are still environments where plain 7-bit ASCII or single-byte 8-bit Latin 1 will work better than UTF-8, but a migration from ad hoc digraph representation to UTF-8 should always be considered. Especially when it comes to representing huge amounts of lexical data, the benefits of representing them with the letters the users know from everyday life will make both addition of new data and proofreading of the source code easier.

For South Asian languages, using the standard orthography will in the majority of cases mean one of the descendants of the Brahmi scripts. Other, less common scripts are the Arabic and Latin scripts, they will not be dealt with here.[6]

The consonants and vowels of the Brahmi scripts are given separate characters in the Unicode standard. Also when the vowel glyph in CV sequences is placed to the left of the consonant glyph, the underlying character sequence is CV, not VC. This means that making transducers for South Asian languages is straightforward, as long as the software accepts Unicode, for example in multi byte (UTF-8) encoding. This encoding method is not the only possible, and also not the only one in use. An alternative would have been to encode each syllabic glyph, thus each glyph representing a CV sequence would have corresponded to one character, instead of two or even more characters. Such an encoding would have offered a one-to-one correspondence between character and glyph, and made glyph composition unnecessary. For Tamil there are glyph-encoding character sets in use, and the Association for

Tamil Computing[7] argues for such a solution for Tamil in Unicode as well. Now, text can always be transposed from one code table to another. As far as the ordinary user is concerned, character encoding is invisible as long as the conversion works. The Association for Tamil computing claims Tamil would have been better off with a single-character encoding of the Tamil glyphs, arguing that one-to-one encoding between characters and glyphs would require smaller storage space and simpler ways of rendering the glyphs.

Seen from the perspective of transducer construction, the multi-character (phonemic) glyph encoding is clearly superior to the single-character (syllabic) one. The reason for this is that morphological processes frequently generalise over processes smaller than the syllable. Tamil inflectional paradigms, such as the one of consonant-final nouns (e.g. மனிதன் *maṇitaṇ* 'man') do add vowel-initial case suffixes to the consonant-final stem. As long as vowels and consonants are encoded as separate characters, this is straightforward, but with a syllabic script, where each syllable is represented by one character in the digital representation, the rule would have to include an intermediate phonological representation for the syllable symbols. Rather than adding *-ai*, *-aal,* etc. for each case, one would have had to reformat the stem, and exchange *-na-* with *-la-*, *-ta-*, etc. Thus, there are indeed advantages with a multi-character phonemic encoding of the Brahmi glyphs when it comes to language technology.

4.2. Making transducers

Here, I will sketch some examples of how transducers may be built for Indo-Aryan languages. When making morphological transducers, one is faced with the choice of software. I use the Xerox tools *lexc* and *xfst* . I will refer to these tools here, since they accept Unicode UTF-8-input, but the crucial point concerning the transducers described here is their formal properties, they may potentially be implemented with other tools as well.[8] It is also important to notice that the mathematical properties of transducers make them independent of the particular software chosen. In practice, when writing source code, the notational conventions of course differ between the software providers, but since the basic needs are the same for all the programs, it should be possible to convert the source code from application to application with a mixture of Perl programming and manual adjustment.

4.3. An Indo-Aryan example: Hindi[9]

Like most Indo-Aryan languages, Hindi has a predominately agglutinative morphological structure including both inflection, derivation and compounding. Seen from a morphological point of view, the nouns inflect for 3 cases (nominative, oblique, vocative) and 2 numbers (singular, plural). The oblique form never occur without a case postposition (ergative, accusative/dative, instrumental/ablative, locative or genitive), the oblique form and the postpositions are still analysed separately, as they are written separately, and the transducers generalise over written input.

Hindi nouns divide into two genders, each with two stem classes with subgroups. Here, as for any other language, the grammar writer has the choice between a regular lexicon and a regular morphology. For Hindi, the -ा (*-ā*) and -ाँ (-ā̦) stem paradigms do not have the same number of affixes, thus, keeping them apart gives more transparent rules. The ā- and -ā̦ stems illustrate the case where the stem differs from the nominative singular form. The automaton may be construed so that it takes the base form as input (गधा *gadhā* 'donkey'), or it may be the stem (गध *gadh*). In the former case, an adjustment rule is needed in order to delete the superfluous ा (*ā*) symbol in vocative (*गधो **gadhāē*). In feminine ि (*-ī*) stem nouns, on the other hand, basic form and stem overlap, here writing rules for different shapes is a trivial matter. Let us write the transducer with a stem form input.

(2) गधा+N:गध → +Nom+Sg:ा; +Obl+Sg: ; +Voc+Sg: ; (Devanagari)
+Nom+Pl:े ; +Obl+Pl:ों; +Voc+Pl:ो;
gadhā+N:gadh → +Nom+Sg:ā ; +Obl+Sg:ē; +Voc+Sg:ē; (Latin)
+Nom+Pl:ē; +Obl+Pl:ō̦; +Voc+Pl:ō;

Upper and lower forms are separated by a colon, and the path concatenates upper with upper and lower with lower, so that the oblique singular of गधे *gadhē* will be represented as गधा+N+Sg+Obl:गधे gadhā+N+Sg+Obl:gadhē. Similar lexica and sublexica may be made for masculine -ā̦-stems (कुआँ *kuā* 'well'), and non-ā-stems, such as मुनि *muni* 'hermit' and दिन *din* 'day'. There will then be similar affix series for the feminine forms.

Derivation and compounding are the areas where transducers prove their advantage over list-based approaches. Derivational processes are handled by pointing the stem to a set of derivational suffixes, the result is then directed to the same set of inflectional affixes as in the example above. Here, the distinction between productive and non-productive derivation becomes crucial. If a given suffix can be added only to a small number of roots, the resulting stems may as well be treated as unanalysable forms, whereas productive derivation calls for separate lexica for the suffixes in question. A standard example of productive derivation is the treatment of participles as adjectives. An example from Hindi is valence-changing derivation, passive and causative derivation. A large subset of the verbal stems should be directed to sublexica for causative and passive formation, and thereafter to the ordinary verbal morphology.

Hindi valence-changing derivation also illustrates how non-segmental morphological processes may be handled with cascaded transducers. For a large subset of Hindi verbs, root vowels *-ā-*, *-o-*. *-e-*, *-ī-*, *-ō-* in transitive stems correspond to *-a-*, *-u-*, *-i-*, *-ū-*, respectively, in intransitive and causative stems (with the addition of suffix *-(v)ā-* for the causatives). The continuation lexicon for basic verb roots may lead to an intermediate lexicon adding an arbitrary diacritic ^ITC (here a mnemonic for intransitive-causative). The resulting enriched stem may then be directed to the usual sets of inflectional affixes.

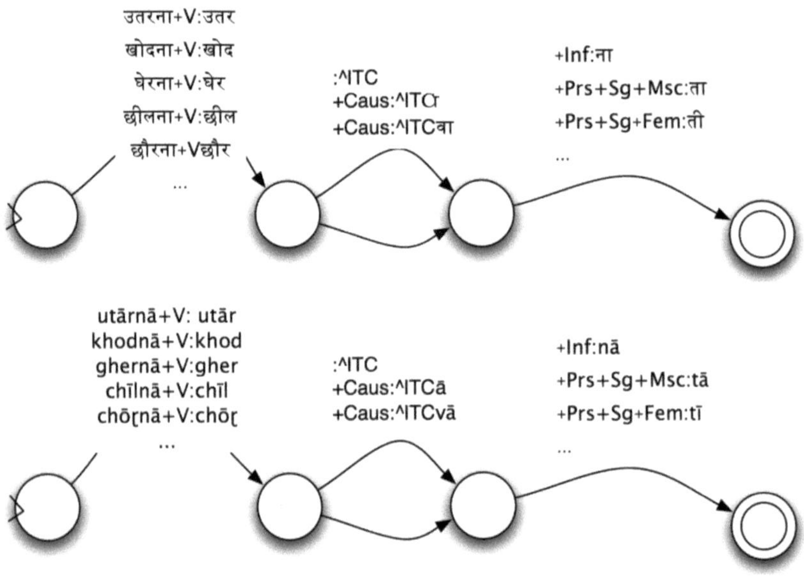

Figure 3. A transducer for causative and intransitive formation

The result will be an intermediate string, e.g. like the one in (3):

(3) utārnā+V+Caus+Prs+Sg+Msc:utār^ITCvātā

The lower string (the one following the colon) will then be given to a morphophonological transducer, containing, among others, the following ordered rules:

(4) ā -> a, o -> u, e -> i, ī -> i, ō -> ū || _ Cns* ^ITC ;
^ITC -> 0 ;

The morphological and morphophonological transducers will then be combined, and the intermediate level will be invisible.

A proper treatment of compounds is crucial for all languages where they are written as single words. This is the case in Indo-Aryan languages, as it is in all Germanic languages but English. Even with enough storage space to allow for a list representation of all the potential compounds, the resulting lexicon would not be human-readable. In contrast, a transducer will redirect the noun lexicon in a loop back to itself, and thereafter to the inflectional component. The result is a lexicon that can be manually maintained, with each non-compounded stem listed once. It will also be possible to introduce grammatical restrictions upon the compounds. Thus, in A + N compounds, adjectives have the gender form of their respective head nouns (कालापानी *kālā-pānī* 'penal servitude' < *kālā* m. 'black' + *pānī* m. 'water', कालकोठरी *kālkōṭhrī* 'solitary confinement' < *kālī* f. 'black' + *kōṭhrī* f. 'room'). In a morphological transducer, this may be ensured by directing the different adjectival forms to the corresponding masculine and feminine noun lexica.

The transducers described above have been compiled with the Xerox tools *lexc* and *xfst* with the Devanagari script written directly into the source code, in Unicode (UTF-8), as indicated above. The possibility of using the ordinary orthographical forms directly in the source code allows for a direct integration of lexical and morphological resources. The time when lexicographical projects and morphological transducers are construed independently of each other should now indeed be over.

4.4. Mising

As a further example, let us take Mising, or Miri, a language belonging to the North Assam group of Tibeto-Burman languages spoken in Assam, by 136 698 speakers (1961 census). Mising is written both with the Devanagari and the Bangla (Bengali) scripts, text can be converted between the two with no information loss. The basic grammatical structure of Mising is documented in the concise 130-page exposition of Bal Ram Prasad (Prasad 1991). The grammar uses a phonological representation, so in order to make a transducer for written text, a grammar using Devanagari or Bangla will be needed. The grammar still makes it possible to outline the transducer.

Mising nouns are inflected for number (sg, pl), case (10 grammatical and adverbial cases). Verbs are inflected for tense (6 tenses), aspect (3 aspects) and mood (4 moods). In addition, interrogatives and negation are morphologically expressed. Just as for Hindi, Mising transitive and intransitive verbs may change valency, and form causative, reflexive and even reciprocal verbs via derivational processes. In addition, all lexical categories may form compounds. Compounding in Mising is a non-trivial process, with elaborate morphophonological and probably also morpholexical processes in addition to simple concatenation. The conditions governing these processes are only partly spelled out in the grammar, and it is also an open issue to what extent there is lexical variation , or whether the processes are phonologically regular. A morphological transducer and a text corpus make it possible to find out: The transducer may allow for both concatenative and truncating compounding, and the linguist may inspect the result and draw conclusions on the distribution of the different compounding mechanisms.

Although genetically not related to the Indo-Aryan and Dravidian languages, Mising has a morphological structure that is not too different from the other South Asian languages. Mising is a tone language, but as long as this is not reflected in script, and not part of suprasegmental morphological processes visible in script, this may be ignored in a transducer based on standard orthography. The construction of the transducer itself is not a time-consuming task, especially not if a transducer made for some other language using Brahmi script has already been through the problems of setting up the software infrastructure. The crucial part is whether there is a dictionary available in electronic format. If that is the case, the transducer and the dictionary may be integrated directly. If the dictionary exists only on paper, in a format that cannot be scanned, then more resources are needed. In any case, much of the work may be done by people already engaged in work on Mising grammar and lexicography. The result will be tools with application possibilities as indicated earlier in this article.

4.5. Transducers for the Sámi languages in Scandinavia

As an example of how transducer construction may be carried out, consider the author's project for building transducers for the Sámi languages in Scandinavia. The Sámi languages belong to the Uralic language family (together with Finnish, Hungarian, Estonian, and some 30–40 languages spoken in Russia). There are approximately 10 Sámi languages, 6 of which have a standard orthography. North Sámi, the largest language, has 17 000 speakers, Lule Sámi has fewer than 2000 speakers, and the other languages fewer than 500 speakers each.

The starting point for the project was an electronically available dictionary for North Sámi, and decent morphological descriptions for North and Lule

Sámi. Building a prototype transducer for North Sámi took approximately 2 ½ man-years, including the development of a computer infrastructure. With the infrastructure and the North Sámi transducers available, the development of the transducer prototype for the closely related Lule Sámi took approximately 6 man-months.[10] Addition of the other Sámi languages is anticipated to take longer, as they are more complicated. The work on the transducers has revealed the morphological and morphophonological omissions of the standard grammars, among them sandhi phenomena, especially linked to compounding, lexical idiosyncrasies in the declension of adjectives, and variation in the treatment of recent loan words.

The resulting transducers will now form the core of several planned applications, among them both practical applications such as spell checkers, intelligent content search and pedagogical programs, but also applications geared toward basic research, such as tagged corpora, frequency dictionaries, etc.

4.6. Resources needed for morphological transducers

The most time-consuming part of a morphological transducer is its lexicon. Fortunately, lexica are made for other reasons, and for most languages there will be available dictionaries that can form the basic lexicon of the morphological transducer. In the lucky cases, the dictionary is even available electronically, otherwise, scanning or hand typing it will be a time-consuming task.

A list of lexemes in itself is not enough. For each lexeme, there must be information on inflection class, gender, and other relevant features. Later applications will also require more information, such as valency, animacy, etc. If there is no dictionary available, but there is a text corpus, then a transducer may be used to extract stems in order to build a lexicon. Especially for agglutinative languages this is a time saving way of building lexica.

5. Disambiguation

Within morphological paradigms, there is massive homonymy, as seen e.g. in the previous section on Hindi morphology. Many languages also have homonymy across major parts of speech, such as when English *walks* may be either a noun of a verb. Only when this ambiguity is resolved do we have reliable information on the morphological properties of running text.

5.1. How to disambiguate

One way of disambiguating morphological homonymy in running text is via *constraint grammar*, a framework initiated by Fred Karlsson (Karlsson 1990; Karlsson et al. 1995), and further developed by e.g. Pasi Tapanainen (Tapanainen 1996). The intuitive idea behind this framework is the insight that we as humans are able to resolve ambiguity by looking at the linguistic context of the polysemous word form. As linguists we then write rules that simulate this ability.

As a concrete example, let us take the case homonymy of Hindi singular masculine non-ā-stems and feminine ī-stems. For these stems, nominative, oblique[11] and vocative are all identical, whereas *ā*- and *ą̄*-stems have a separate nominative form. Syntactically, the nominative is the case of the grammatical subject, the predicative, and in certain cases the direct object. The oblique case form is used in front of postpositions, in adverbial expressions associated with time, direction or manner, and in certain idioms. The vocative is used to address the hearer. The task of the disambiguator is now to identify these contexts.

The clearest cases are the one of postposition complements. Thus, whenever a postposition follows a possible oblique case form, the oblique reading should be chosen. If the clause contains two conjoined NPs followed by a postposition, both NPs carry oblique case. A typical vocative context will be a sentence fragment terminated by an exclamation mark, or by a comma and an imperative sentence. (Tikkanen op. cit. p. 44)

(5) गाड़ी की खिड़की और दरवाज़ों से धुआँ निकला
gāṛī kī khiṛkī aur darvāzõ se dhuā niklā.
train GEN *window and door* ABL *smoke came*
'Smoke came through the windows and doors of the train'
बहनो और भाइयो!
bahno aur bhāiyo!
sister.PL *and brother.PL*
'sisters and brothers'
हे कुकर्म करनेवालो, मेरे पास से चले जाओ!
he kukarma karnevālo, mere pās se cale jāo!
oh bad doer.PL, *me away* ABL *go* IMP
'Oh, bad doers, go away from me!'

In the CG-2 formalism of Tapanainen 1996, the appropriate rules would be formulated approximately as follows ("0" indicates the position of the word to be disambiguated, "1" is the position to the right, "-2" two positions to the left, and "*1" one or more positions to the right, etc. The "C" in "*2C" below signals 'careful mode', it matches only when the reading in question, here "Obl", is the only one available. "Y BARRIER X" means 'unless some of the cohorts between the starting point and Y contains X':

(6) SELECT Obl IF (*1 Post BARRIER Obl);
*SELECT Obl IF (1 Conj) (*2C Obl BARRIER Noun-not-Obl) ;*
*SELECT Voc IF (0 Animate) (*1 Exclamation-mark BARRIER Verb-not-imperative);*

The first rule selects the oblique case reading of the cohort if somewhere to the right there is a postposition, with no intervening oblique nouns. Thus, assigning oblique to a subject NP followed by oblique NP plus postposition is blocked. The second rule takes coordination into account. It selects oblique case for a noun followed by a conjunction that links to another oblique nouns. This rule will come late in an ordered rule set, and it will thus work only after the noun following the conjunction has been disambiguated as an oblique noun. The third rule is triggered only for animates, and it selects vocative if there is an exclamation mark to the right, and no non-imperative verb intervenes. As can be seen, the rules may refer both to tags (like "Obl") and tag sets (like "Noun-not-Obl", which we here suppose has been defined as the union of "N Nom" and "N Voc"). Needless to say, these are simplified rules for expository purposes, but they still may give a general impression of the framework.

5.2. Disambiguation in Sámi

In English, homonymy is often found across POS boundaries, so that some word-form may be a verb or a noun, but if you know which one it is, you also know the placement within its respective paradigm (we know that *walks* is plural if it is a noun). In languages with richer morphology, homonymy is often different. Here, derivation is not done via conversion, but via suffixation, and the homonymy is found within the same POS, and only marginally across POS borders. Whereas disambiguation in English thus may be seen as the task of finding some starting point ("if I am a verb then you must be the noun"), homonymy in languages with richer morphology is more dependent upon the morphosyntactic properties than upon the basic POS classification of its neighbours. As an example, let us again take North Sámi. Here, the verb for 'to throw' is *bálkestit* and 'a throw' is *bálkesteapmi*. There is no cross-POS homonymy, as the former must be a verb, and the latter must be a noun, as can be seen from the affixes attached to the stem *bálkest-*. But in addition to being an infinitive, *bálkestit* may also represent indicative first and third person plural, and indicative past tense second person singular. Distinguishing verb forms from each other differs from distinguishing verbs from nouns, since the contextual differences are smaller in the former case. As an example of disambiguation of different verb forms, let us take the North Sámi clause *Mii eat leat dan muitalan* 'We haven't told it', with the verbs *leat* 'to be' and *muitalit* 'to tell'. By the transducer, the sentence is given the following morphological analysis, prior to disambiguation:

(7) "<Mii>"
    "mun" Pron Pers Pl1 Nom
    "mii" Pron Interr Sg Nom
  "<eat>"
    "ii" V Neg Ind Pl1
  "<leat>"
    "leat" V Ind Prs Pl1
    "leat" V Ind Prs Pl3
    "leat" V Ind Prs Sg2
    "leat" V Inf
    "leat" V Ind Prs ConNeg
  "<dan>"
    "dat" Pron Dem Sg Acc
    "dat" Pron Dem Sg Gen
  "<muitalan>"
    "muitalit" V PrfPrc
    "muitalit" V Act
    "muitalit" V Ind Prs Sg1

The only unambiguous word is *eat*, first person plural of the negation verb. In reality the sentence does not have 60 readings (2 x 5 x 2 x 3), but one:

(8) "<Mii>" "mun" Pron Pers Pl1 Nom
  "<eat>" "ii" V Neg Ind Pl1
  "<leat>" "leat" V Ind Prs ConNeg
  "<dan>" "dat" Pron Dem Sg Acc
  "<muitalan>" "muitalit" V PrfPrc

Here are the disambiguation rules that were used to arrive at the correct reading:

```
(9) SELECT Pers IF (0 ("mii")) (*1 V-PL1 BARRIER NON-ADV);
    SELECT ConNeg IF (*-1 Neg BARRIER VFIN);
    SELECT Acc IF (*-1 LEAT-FIN-NON-IMP BARRIER NON-PRE-N)
            (1 PrfPrc);
    SELECT PrfPrc IF (*-1 Neg BARRIER CONTRA);
```

The form *mii* may be a personal or interrogative pronoun. The rule states that if there is a PL1 verb to the left, with no other words than adverbs between the two, then the personal pronoun reading is selected. In order to get the correct reading for copula, the ConNeg form (the form connected to negative verbs) is chosen if a negation verb may be found somewhere to the left, before we find any other finite verb. The rule for perfect participles is similar, but here the barrier is a set of words cancelling negation, like the word *muhto* 'but'. This set has been listed earlier, and is labelled CONTRA. The rule for accusative demands a finite copula to the left, and with nothing but NP-internal pre-modifiers intervening, and a perfect participle to the right.

In order to get a close to perfect disambiguation of running text, approximately 1500 to 2500 such rules are needed, but already after a couple of hundred rules, the performance creeps above 80%.

5.3. Other disambiguation approaches

Constraint grammar is not the only approach to disambiguation. One alternative is presented by Koskenniemi 1997 (the so-called FSIG, Finite State Intersection Grammar). Rather than removing erroneous readings one by one, until only the correct one(s) is/are left, this approach models the syntax as one finite-state automaton, and sees sentence analysis as finding one reading corresponding to the grammatical model among all the possible sets of readings for any given sentence. So, rather than reducing the number of erroneous readings in a stepwise fashion, this method tries to accept the only one right on. Since no working versions of FSIG have yet been presented, its merits remain to be evaluated.

There are also machine-learnable disambiguation methods. In this context, they suffer from two disadvantages: They require large amounts of correctly tagged text as a training corpus, and the methods tend to result in a black-box type disambiguator, working to a certain extent, but with a rule set that tend to be difficult to update manually. Comparison also show that grammar-based disambiguation performs better that statistically-based ones (cf. Tapanainen and Voutilainen 1994; Samuelsson and Voutilainen 1997).

5.4. Applications

The list of applications presented in section 3.2 shows that morphological transducers may be useful, also without a disambiguator. When a disambiguator is added, it is possible to analyse running text, and state not only that some form *may* be a vocative (or a verb), but also that it actually *is* a vocative (or a verb) in this particular case. Equipped with disambiguation tools, the usefulness of morphological transducers increases.

Intelligent content search is dependent upon disambiguated morphological information. For a language without such tools, information saved in digital form will be much less accessible. Multilingual software and administration will rely heavily on machine translation, especially since much of the textual infrastructure (schemes, interactive menus, etc.) is standardised, and therefore well suited for machine translation. Languages without the necessary tools will not be able to enter these important arenas.

6. Limitations of language technology

It is worth stressing once again that language technology cannot save a language from dying. A language dies when it is not passed on to the next generation. Also, in cases when a language is dependent upon the education of younger semi-speakers or non-speakers, the best formal way of learning a language is still with the help of a teacher, and a good textbook, grammar and dictionary. Language technology such as the one sketched here may provide interactive pedagogical programs, but they will always supplement the printed material and human interaction.

Although language technology-based applications as described in this article thus become increasingly important for the use of written languages in formal settings, they can and should by no means replace the oral transmitting of languages from generation to generation, and in language documentation and teaching, the basic grammars and dictionaries.

7. Conclusion

The present article has focused upon a small, but fundamental part of current language technology, the construction of morphological transducers and disambiguators. It was argued that such tools are needed for all languages in order to make it possible to use them in administrative and literary contexts in a modern society. Given both the paucity of multi-million word text corpora and the complex morphological structure of most South Asian languages, such transducers and disambiguators should be built as grammar-based applications. If this work is conducted in parallel with language documentation work, language technology could function as a much needed supplement to basic grammatic research on poorly known languages, thereby both increasing the knowledge base of linguistic theory, and enabling the speakers of the languages in question to enjoy the same linguistic rights as speakers of dominant languages.

Notes

1. For the 191 UN member states (2004), there are less than 100 languages de facto used in the central civil administration of the country in question. Including large regional languages, such as the 18 official languages of India, the number rises somewhat, but on the other hand, several of the official languages have a marginal position within language technology. Most of the rest of the world's approximately 3500 languages with standardised orthography enjoy basic localisation support (encoded characters, via Unicode, and perhaps keyboard resources), but only marginal language technology support.
2. <http://www.microsoft.com/technet/prodtechnol/winxppro/evaluate/ muiovw.mspx>.
3. <http://www.ethnologue.com>.
4. <http://www.ispell.org/ >.
5. The situation is of course different when it comes to speech-based technology. Such applications will not be discussed here, though.
6. See, for example, Beesley (1998) for a presentation of a morphological transducer for Arabic, facing problems similar to the ones relevant to Urdu and Thaana.
7. <http://www.kanithamizh.org/>, cf. especially <http://www.kanithamizh.org/16-bit-encoding.htm> for their view on the encoding issue.
8. The Xerox tools are described and included (on a CD-ROM) in Beesley and Karttunen (2003), (cf. <http://www.fsmbook.com>), a book that also contains a general discussion of the use of transducers in computational linguistics. An open-source finite-state utility similar to the Xerox one is the Stuttgart Finite State Transducer SFST, cf. <http://www.ims.uni-stuttgart.de/tcl/resources.html>. Other tools include the finite-state utilities of Jan Daciuk (<http://www.eti.pg.gda.pl/ ~jandac/fsa.html>), the PC-KIMMO software of Evan Antworth, (<http: //www.sil.org/computing/catalog/pc-kimmo.html>, cf. also Antworth (1990). based upon Kimmo Koskenniemi's work), Gertjan van Noords software (<http: //odur.let.rug.nl/~vannoord/Fsa/ fsa.html>), and Richard Sproat's lextools software (<http://www.research.att.com/ sw/tools/lextools/>),
9. Hindi is of course not a minority language. It was chosen as an example language due to the availability of grammatical descriptions. Indo-Aryan languages differ from each other in many ways, but Hindi is as good as any Indo-Aryan language to serve as an illustration. Indo-Aryan and Dravidian languages have a lot in com-

mon, both typologically and typographically, so much of what holds for Hindi will be directly relevant for Dravidian languages as well. Here and below the data on Hindi are drawn from Tikkanen (1991) and Shukla (2003).

10. The transducers are available at <http://giellatekno.uit.no>.
11. The term *oblique* refers in Hindi grammar not to all other case forms than nominative, but to a distinct case form.

References

Antworth, Evan L.
1990 PC-KIMMO: A two-level processor for morphological analysis. Occasional Publications in Academic Computing 16. Dallas, TX: Summer Institute of Linguistics.

Beesley, Kenneth R.
1998 Arabic morphological analysis on the Internet. *Proceedings of the 6th International Conference and Exhibition on Multi-lingual Computing, Cambridge, 17–18 April 1998.* <http://www.xrce.xerox.com/publis/mltt/mltt-97-03.ps>.

Beesley, Kenneth R., and Lauri Karttunen
2003 *Finite State Morphology. Studies in Computational Linguistics.* Stanford, CA: CSLI Publications.

Chomsky, Noam
1965 *Aspects of the Theory of Syntax.* Cambridge, Mass: MIT Press.

Jha, Girish Nath
2003 Current trends in Indian Languages Technology. *Language in India* 3: 12. <http://www.languageinindia.com/dec2003/langtech.html>

Karlsson, Fred
1990 Constraint Grammar as a framework for parsing running text. In *13th International Conference on Computational Linguistics (COLING-90)*, vol. 3, H. Karlgren (ed.), 168–173. Helsinki : Helsingin yliopisto.

Karlsson, Fred, Atro Voutilainen, Juha Heikkilä, and Atro Anttila (eds.)
1995 *Constraint Grammar. A Language-Independent System for Parsing Unrestricted Text.* Berlin: Mouton de Gruyter.

Koskenniemi, Kimmo
1997 Representations and finite-state components in natural language. In *Finite-State Language Processing*, Emmanuel Roche, and Yves Schabes (eds.), 99–116. Cambridge, Mass: MIT Press.

Masica, Colin P.
1991 *The Indo-Aryan Languages.* Cambridge: Cambridge University Press.

Prasad, Bal Ram
1991 *Mising Grammar.* CIIL Grammar series 17. Mysore. Central Institute of Indian Languages.

Samuelsson, Christer, and Atro Voutilainen
1997 Comparing a linguistic and a stochastic tagger. *Proceedings of the 35th Annual Meeting of the Association for Computational Linguistics*, 246–253. Madrid: Morgan Kaufmann Publishers/ACL.

Shukla, Shaligram
2003 *Hindi Morphology*. LINCOM Studies in Indo-European Linguistics 15. München : Lincom Europa.

Tapanainen, Pasi
1996 The Constraint Grammar Parser CG-2. Publications of the Department of General Linguistics 27, University of Helsinki.

Tapanainen, Pasi, and Atro Voutilainen
1994 Tagging accurately – Don't guess if you know. *Proceedings of the Fourth Conference on Applied Natural Language Processing (ANLP'94)*, 47–52. Stuttgart, Germany.

Tikkanen, Bertil
1991 *Hindin kielioppi* [Hindi Grammar]. Suomen itämaisen seuran suomenkielisiä julkaisuja 23. Helsinki: Suomen itämainen seura.

Supporting lesser-known languages: The promise of language technology

Lars Borin

> The old quip attributed to Uriel Weinreich, that a language is a dialect with an army and a navy, is being replaced in these progressive days: a language is a dialect with a dictionary, grammar, parser and a multi-million-word corpus of texts – and they'd better all be computer tractable. When you've got all of those, get yourself a speech database, and your language will be poised to compete on terms of equality in the new Information Society. (Ostler n.d.)

1. Introduction

According to Nicholas Ostler, a language will not get by in the world of today unless it is equipped with a "parser and a multi-million-word corpus of texts" (see the quote above).[1] The parser, and also arguably the computer-tractable dictionary and grammar and the text corpus, are all examples of *language technology*. Ostler's statement reflects the fact that, recently, there has been a good deal of concern about the availability of language technology resources for other languages than English and a few others.[2] This of course is part of a larger concern about diminishing linguistic diversity and what can be done to counter this trend (see Saxena's introduction to this volume). For good arguments as to why language technology is a desideratum for any language in the modern world, I refer the reader to Trosterud's paper in this volume.

To those outside the field, it may not always be completely clear what language technology is all about. The term itself is fairly recent,[3] and there seems to be a tendency (in itself entirely natural and understandable) to refer to any computer application which deals with language in any form as "language technology". Thus, a multimedia program for language learning may be referred to as a piece of "language technology" simply because the purpose of the program is that of assisting language learning.[4] This usage is misleading, however, and should be avoided. The essence of language technology lies not in the circumstance *that* it deals with language in some way – many kinds of technology do that, including ordinary telephones – but in *how* it deals with language. Language technology is the common name for a spectrum of technologies, tools, algorithms, etc., which enable computers to deal with human language in all its manifestations – speech, writing and sign – as an open-ended system, rather than merely as a closed set of linguistic products (such as predetermined answers to vocabulary questions or fixed replies retrieved on the basis of of combinations of keywords found in the input).[5]

Language technology is a kind of information and communication technology (ICT), and it is obviously potentially applicable whenever and wherever people interact with machines, but also, and perhaps less obviously, in many cases where people interact with other people, in the form of various assistive technologies, e.g. online machine translation, speech-to-text applications for use with text telephones, etc. Historically, language technology includes the older fields of computational linguistics/ natural language processing and speech technology. Hence, language technology is a strongly interdisciplinary field, with linguistics and computer science being the traditional ingredients of the more written-language-oriented computational linguistics/ natural language processing branch, whereas phonetics and electrical engineering/ signal processing make up the speech technology branch of the field. The potential for other cross-disciplinary combinations is immense, considering that there is hardly any branch of scientific inquiry (especially in the humanities and social sciences) which does not at some level involve language.

2. Language technology resources for lesser-known languages

In the terminology of language technology, a *resource* is the term used about a collection of data or a piece of software, called *data resources* and *processing resources* (or *algorithmic resources*), respectively, depending on whether they are best thought of as static, declarative knowledge, or dynamic, processual, computer program-like knowledge. The data resources are often called *linguistic resources*, since they consist of samples of language (texts, recordings, etc.) or formalized knowledge about language (dictionaries, grammars, etc.). The processing resources use linguistic resources in order to analyze linguistic input or produce linguistic output for some purpose, e.g. for translating from one language to another, for finding grammatical errors in text, etc. This compartmentalization into linguistic and processing resources – the notion that grammatical generalizations about a particular language should be stated separately from whatever knowledge is needed for processing language[6] – means that in the ideal case only the linguistic resources need to be changed, but not the processor, when we wish to adapt a particular kind of language technology application to a new language.

What, then, are these linguistic and processing resources, and how do you go about creating them for a language which does not have them? In the next two sections, I will discuss some resources that have been mentioned in this connection. As my main point of departure I will use the very useful overview made by Kepa Sarasola, a well-known Basque language technology researcher, for the workshop on "Developing language resources for minority languages: reusability and strategic priorities" at the Second international conference on language resources and evaluation (Sarasola 2000; see also the Atlantis Project web pages <http://www.uoc.edu/in3/atlantis/>).

More recently, the US Linguistic Data Consortium (LDC) has undertaken a "Low Density Languages" (LoDL) survey, where they have surveyed the largest (in terms of number of native speakers) 300 languages of the world with respect to how well they are equipped with language technology and prerequisite resources,[7] including an in-depth survey of 8 of these languages: Bengali, Chechen, Hindi, Panjabi, Tagalog, Tamil, Tigrinya, and Uzbek (note that half of these are South Asian languages; Strassel et al. 2003). The LDC survey criteria deal mainly with prerequisites for creating language technology resources, whereas Sarasola discusses the resources themselves; hence the two overviews are complementary. A summary listing can be found in the appendix to this article. In the following two sections, I will endeavor to highlight some of the prerequisites and language technology resources mentioned in the listing.

2.1. Prerequisites for creating language technology

The LDC LoDL survey criteria implicitly identify some necessary prerequisites for the (rapid) creation of language technology resources for a language, using the current state of the art of language technology.

It is quite clear from their survey that basic literacy is seen as a necessary component of language technology, meaning that not only should there be a standard orthography for the language, but also that the language should actually be used in writing on a regular basis. However, this is not a logical requirement, only a pragmatic one. Current language technology was developed within a written-language framework, and consequently it fits most comfortably with well-developed literacy. It is a very interesting question – but well beyond the scope of this article – to consider whether a purely or predominantly oral language technology would be feasible.[8]

Current language technology tools are also aided by certain linguistic and orthographic characteristics, which we can sum up as those that tend to formally delimit and distinguish units of (linguistic) interest, namely lexical word and sentence sized units. In the LDC survey these characteristics are captured by the criteria "words separated in writing", "simple orthography", "sentence punctuation", and "simple morphology". Arguably, this means that the technology is biased toward English-like linguistic systems, a point to which I will return below.

2.2. Language technology resources

It turns out that the most central linguistic resources for language technology are text corpora proper and speech databases (and now also increasingly the products of modern linguistic documentation, i.e. video recordings of culturally and linguistically significant activities). The former are generally the basis for all kinds of language technology dealing with written language, whereas the latter form the empirical basis for speech technology applications. In this article, I will focus on the issues of corpus collection and the creation of tools for linguistic annotation of corpora.

Although they make up important basic resources for language technology, not all uses of text corpora fall within its purview. In fact, most work conducted under the heading of *corpus linguistics* (see McEnery and Wilson 2001 for a general introduction) has very little to do with language technology. Rather, this is traditional descriptive linguistics using fairly simple computer tools for searching, counting, and reordering large amounts of normally unannotated (see below) digital text. This kind of corpus use is still relevant for the purposes of this discussion, since it provides important input for the descriptive grammars and above all the dictionaries mentioned among the foundational resources.

A text corpus proper, as opposed to a "mere" collection of text – e.g. newstext downloaded from the WWW – is a selection of text intended to be representative for some particular purpose. E.g. a so-called balanced corpus of (published) written language is compiled from the main genres/ text types of published written language, in roughly the correct proportions, on the assumption that linguistic investigations of this material will yield results that generalize to written language in general. Compiling a proper text corpus entails a much greater amount of work than merely collecting any kind of text that you can lay your hands on, especially when other text types than newstext are difficult or impossible to acquire in electronic form (see Hardie et al., this volume). A *parallel corpus* is a bi- or multilingual text material containing original texts in one language and their translations into another language or other languages. Often, parallel corpora are aligned, meaning that corresponding units (sentences, phrases, even words) from the different language versions are explicitly linked together (see Borin 2002a).

Raw text collections and – even better – proper basic corpora are arguably today the single most valuable resources for language technology development, however, especially for those researchers who see automatic or semi-automatic acquisition of linguistic knowledge as the preferred route to quickly developing language technology tools for new languages.

Among the most basic tools, we find those used for fundamental statistical modeling of linguistic phenomena. Another basic tool is the so-called *part-of-speech* (POS) *tagger*, which actually normally (automatically) assigns not only part-of-speech labels in the traditional sense, but full morphosyntactic descriptions, i.e. part of speech and inflectional categories, to all tokens (words or punctuation signs) in a text (although they will not assign lemmas, i.e. basic, or citation, forms), in the form of tags, or labels, attached to the words, as in (1), taken from an automatically POS tagged learner corpus (Borin and Prütz 2004). POS tagging is a special case of what is generally referred to in the context of language technology as (automatic linguistic) *annotation* of text.

(1) Another/DD1 great/JJ fear/NN1 was/VBDZ that/CST wilderness/NN1 would/VM force/VVI civilised/JJ men/NN2 to/TO act/VVI like/II savages/NN2 ./YSTP

POS taggers normally assign only one POS label to each word, generally the most probable one for the word form, given the local context (defined as the one, two, or three immediately preceding words with their POS tags). A POS tagger is normally designed so that it will assign a tag to unknown words as well, i.e. words not in its lexicon – also called *out of vocabulary* (OOV) words.

Currently, there are very good POS taggers for English and other similar languages, which are *trained*, i.e. they work with algorithms which are capable of "learning" from correctly tagged corpora the linguistic knowledge needed for tagging new, previously unseen text. This of course means that there must be some manually annotated corpus resources available, the larger and the more varied, the better.

As the annotations grow more sophisticated and refined, the annotation work becomes correspondingly harder. At the moment, treebanks – syntactically annotated corpora – are among the hottest topics in language technology. There are innumerable treebank projects being pursued all over the world, and quite a lot of research on ways of making the treebanking more resource-effective; currently it seems that there is a constant labor cost of about one person-year per 50 000 words of tree bank, almost regardless of the language.[9]

Word and sentence level analysis of form (i.e., grammar in the traditional sense of morphology and syntax) are the areas where the bulk of mainstream language technology resource and tool development is being pursued in many language communities, including the smaller Western European languages, such as Finnish, Swedish, Norwegian, Portuguese, etc., some East and Central European languages, such as Russian, Hungarian, Czech, etc., and some South Asian languages, e.g., Hindi.

On the research front we find integration of (lexical and sentence) semantics, text-level phenomena (including dialogue structure), and general world-knowledge into applications, and also multilingual language technology. On this level, we find mainly large Western European languages: English, but also French, German, Italian, and Spanish. Outside Western Europe and North America, it is mainly Japanese and Chinese which boast at least some of these resources. Multilingual language technology has long been an important item on the research agenda in the European Union, which is founded on the principle that all the official EU languages have equal status within the union, and furthermore that the so-called lesser-used languages of the union also receive considerable support. Language technology, being a field dominated by US researchers, has been a bit slower to adopt this view,[10] but there the intelligence community and the military have pressed for language technology support for multilingual information-processing capacity of a slightly different sort. An analogous situation to that of the EU occurs in several places in South Asia, notably in India.

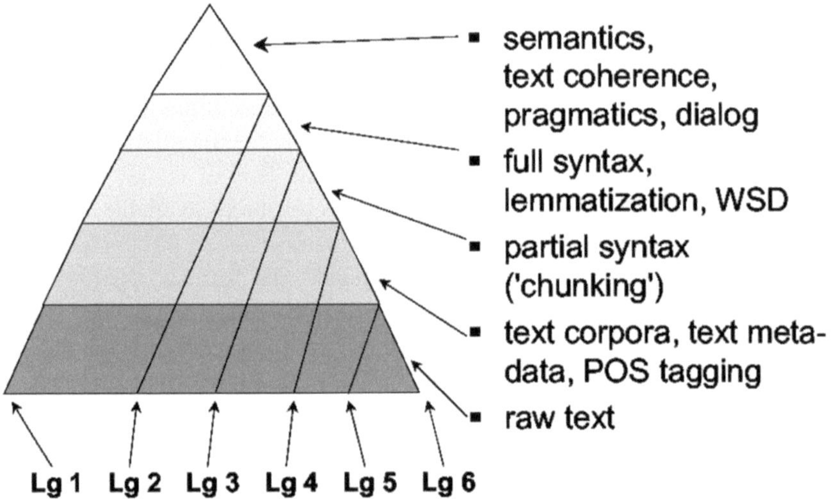

Figure 1. The language technology resource pyramid

2.3. The language technology "resource pyramid"

Most of the language technology resources listed in the appendix are typically not available for more than a few languages – very few if we reckon with all the approximately 6000 languages in the world – but still few even if we count only those languages having a written form using a standard orthography.

The resources cannot be created independently of each other. It turns out that there is an important "refinement" (or "linguistic ascent") relationship among the resources (or groups of resources). For instance, in the text corpus case, going from a "mere" collection of texts to a proper corpus entails a kind of selection process, where the main criterion is one of suitability for some particular purpose. The content of a text (corpus) in turn can be linguistically refined by providing it with successively more sophisticated annotations. When linguistic annotation is (at least in part) automated, this is normally 'ascent' in a more concrete sense, as well, since linguistic annotations on a lower level form the basis for those on a higher level, so that e.g. POS annotation of a text forms the basis for partial parsing of that text. Figure 1 illustrates this dependence among the resources discussed above in a more graphic manner, in the form of a "resource pyramid". In the main resource pyramid, there are smaller pyramids, delimited by the slanted lines. The full pyramid is available for languages of type Lg 1 (possibly only English), and many, many languages in the world are of the type Lg 6, having no language technology resources whatsoever. A fair number of languages are found in between, however, and of the four South Asian languages surveyed in more depth as part of the LDC LoDL survey (Bengali, Hindi, Panjabi, and Tamil), Bengali and Panjabi seem to be equipped with the foundations, i.e. they belong to type Lg 5 or Lg 4, whereas for Hindi and Tamil, there are more resources and more sophisticated resources, somewhere between types Lg 4 and Lg 3.

With automated annotation, there is a reciprocal relationship between annotated corpora and the tools used to annotate them. Typically, a pre-existing annotated corpus is used to train an automatic annotator (part-of-speech tagger, parser, etc.), which can then in turn be used to annotate other, previously unannotated corpora, or simply used in some language technology application. The initial annotated corpus does not appear out of thin air, however (cf. also Hardie et al., this volume). Most often, it is hand-annotated by (teams of) human linguists. This is a very time-consuming and labor-intensive effort. As an example, we can mention the Swedish SUC (Stockholm Umeå Corpus; Ejerhed and Källgren 1997), a morphosyntactically annotated and lemmatized one-million-word balanced corpus of modern published written Swedish. It took six years to compile SUC from scratch, and even then, there were still errors in the annotation in the first version, which have been corrected in the second version, which took another three years to complete (Britt Hartmann, p.c.).

Thus, both corpus compilation and above all corpus annotation seem to be very labor-intensive activities. The latter also requires a high level of linguistic training in the annotators, as well as a general agreement on what a linguistic description of the language should look like, i.e. an agreed-upon tradition as regards terminology, etc. It seems reasonable to assume that lesser-known language communities – especially those where language standardization is recent or in progress – will have few trained linguists and possibly no descriptive linguistic tradition to draw upon.

Even in the case of a well-described major language, however, the cost of annotation may be prohibitive – it is no coincidence that we find, for a number of languages, that there are at least some resources available belonging to the lower levels of the resource pyramid – there are many unannotated corpora and a fair number of POS-tagged corpora for many languages – but that even for English the number of treebanks can be counted on the fingers of one hand.[11] Hence, there is a fair amount of research in the language technology community addressing the issue of how to minimize the human effort in corpus annotation.[12]

3. Bootstrapping of language technology resources

A procedure where linguistic annotation of corpora is accomplished by starting out with a small amount of hand-annotated material (or none at all), and by (partly) automated means developing increasingly more sophisticated and correct automatic annotators from this basis, is known in the literature as "bootstrapping".[13] In the following, I will attempt to give an overview of the current state of the art in this research area, which has seen noticeably intensified activity in the last few years.

An essential and defining component of bootstrapping is *machine learning*, a subdiscipline of computer science (or of artificial intelligence) which concerns itself with self-learning computer systems. Rather than programming a computer with explicit rules for performing some task, machine learning algorithms are supposed to enable the computer to learn the regularities underlying the explicit rules, and, of course, to apply the knowledge thus acquired, in the same way as an explicitly programmed system would do it. Most – perhaps all – learning tasks in this area are logically reducible to the problem of *classification*, i.e. the computer learns to classify items, or instances, as belonging to particular types, or categories. For this reason, machine learning algorithms are often referred to as *classifiers* in the literature. For the purpose of classifying, or labelling, linguistic units in texts, there are at least seven kinds of linguistic units that have been subject to machine learning research:

- words (i.e., division into words;[14] level 0 in "the resource pyramid"; see Figure 1)
- parts of speech, or morphosyntactic categories in general (level 2)
- morphological regularities (i.e., learning morphology; level 2)
- syntax (level 3–4)
- (lexical) semantics, including word sense disambiguation and named entity (NE) recognition (level 4–5)
- text-level phenomena, such as dialogue acts and coreference (level 5–)
- translation equivalents (level 1)

Many machine learning methods are probabilistic, or statistical, in nature, so that the generalizations are made on the basis of the distributional properties (relative text frequency, frequency of context items, etc.) of large numbers of relevant instances. Translated into the linguistic domain, this generally means that large corpora are needed, where each item of interest is represented a number of times. For our purposes here, it is also important to know that machine learning methods are either *supervised* or *unsupervised*. In work on language technology, unsupervised generally means that the learning takes place on the basis of unlabelled data, although the label set itself may be known. The terms "labelled" and "unlabelled" here refer to the labels, or classificatory categories, to be learned by the machine learning method, e.g. part-of-speech tags, syntactic constituents or the senses of an ambiguous word, such as *bank*. The data (the corpus) can be labelled with labels in some other, "lower", linguistic domain, e.g. an unsupervised phrase structure syntax learner may work with data labelled for part of speech. Machine learning is a large and varied discipline, and also quite technical, so we will not go into details here (see Mitchell 1997 for a general textbook level introduction to the field, and Manning and Schütze 1999 for a more language technology-oriented treatment).

In this connection, it is important to note that, arguably, language technology – and consequently also the application of machine learning methods in language technology – has been shaped by the typological and other traits of the most explored language, namely English. These traits are, i.a.

- inflectional morphology with very few forms (two main and two marginal noun forms, four verb forms, uninflected adjectives, except for comparison forms in a few cases) [⇒ keeps type-token ratio[15] down]

- not much in the way of derivational morphology [⇒ keeps type-token ratio down]
- weak formal separation of parts of speech
- fairly rigid word order [⇒ works well with simple phrase structure formalisms]
- etymological spelling keeping homophonous items separate in writing [⇒ keeps (semantic) types separate]
- word separation markers in the orthography [⇒ keeps type-token ratio down]
- orthographic marking (capitalization) of proper nouns [⇒ keeps (semantic) types separate]
- compound parts written as separate words [⇒ keeps type-token ratio down]
- little non-concatenative morphology

However, English is in some respects an atypical language, and it would consequently be a mistake to believe that traits such as the ones listed and others will be characteristic of all or a large number of languages. I hasten to add that these traits are found in other languages, too, and not only in those genetically or geographically close to English. Thus, Chinese shares at least traits 1–5 with English, while 6–7 work differently (no word boundary markers and no special indication of proper nouns). My point – which I am not the first to make – is simply that there is an abundance of languages which work differently from English, and the question then rightly raises itself as to whether the same language technology methods which have worked so well for English will work equally well for languages drastically different from English in these and other respects.[16] Thus, a small experiment where Goldsmith's (2001) *Linguistica* program for unsupervised morphology learning was applied to a text in Greenlandic yielded rather poor results (Borin MS). The type-token ratio for the Greenlandic text was around 0.43 (3480 types / 8084 tokens), which can be compared to another language (chosen for convenience) – namely Finnish Romani – where 8048 word tokens correspond to 1147 word types, yielding a type-token ratio of approximately 0.14. English would presumably show an even lower figure.[17] As already mentioned, many machine learning methods rely directly or indirectly on statistics to do their work, and the basic unit that they work with is the orthographic word. In English, the orthographic word is very close in size to the basic lexical unit in linguistic descriptions of English, and it is arguably a great help to these automatic methods that English comes "pre-digested", as it were, i.e. pre-segmented into orthographic words of roughly the right size. The more instances of a word type the automatic methods can work with, the more certain their predictions about the behavior of that type will become, and conversely, if the type frequency falls below some threshold, they will be unable to say anything about it, basically.[18]

Thus, several traits of English conspire to make current probabilistic models work well even with quite small amounts of English text, by jointly "keeping down" the type-token ratio of English, as compared to many other (written) languages. On the other hand, English might lose some of its advantage if corpora were to come segmented into morphs instead of words, since derivational relationships in other languages tend to correspond to lexical relationships in English (e.g., an English noun corresponding to a Latinate adjective, as in *cat–feline*, etc.): "The proportion of unrecognizable morphemes [in the Finnish test data] is highest for the smallest corpus size (32.5%) and decreases to 8.8% for the largest corpus size" and "the proportion of unseen morphemes [in the English test data] that are impossible to recognize is higher for English (44.5% at 2000 words, 19.0% at 200 000 words)" (Creutz 2003 n.p.).

4. Three bootstrapping scenarios for lesser-known languages

In this section, we will have a closer look at three bootstrapping scenarios, both because they are fairly well-researched and because they seem promising for the problem of creating annotated language technology resources for lesser-known languages. At the same time, there are some theoretically interesting questions as to their general applicability, which I will address below.

The first scenario is one of *unsupervised learning of linguistic generalizations from corpora*. This is the scenario which would be most useful, were it to be realized even in part. Basically, we are talking about pure inductive (or possibly abductive) learning of linguistic regularities, of the kind envisioned by pre-Chomskyan American structuralists (e.g. in several of the papers reprinted in part 1 of Harris 1970; Garvin 1967; see also Borin 1991, ch. 6). In this area, the most intensive research has focused on the problem of learning inflectional morphology directly from an unannotated corpus. This is an interesting and important problem, since many of the languages of the world have more morphology than English – out of the 313 languages in the LDC LoDL survey, only 41 are listed as unequivocally having a "simple morphology" (54 have a non-simple morphology, and for the remainder apparently there were insufficient data). In a language having a "non-simple" morphology, morphological analysis will most certainly be useful or needed for carrying out other annotation tasks, such as lemmatization and syntactic analysis.

In the literature the problem of learning morphology automatically has been approached in a number of different ways. For a more thorough review of this literature, see Borin MS. Suffice it to say here that, while this field has seen some promising results, there still remain many unresolved research issues, e.g. the language dependence of the proposed methods. Generally, a "Standard European" kind of inflectional morphology is posited (e.g. explicitly by Goldsmith 2001), and it is not mentioned how more "exotic" morphologies are to be dealt with. Also, the general issue of evaluation tends not to be addressed at all, so that accuracy figures are at best impressionistic. Sharma et al. (2002) describe an experiment on automatic learning of the morphology of a South Asian language, Assamese.

Another frequently envisioned bootstrapping scenario is that of *cross-language annotation transfer*. Given the common situation of a dominant language which has some language technology resources coexisting in one political entity with a lesser-known language which lacks some or all of these resources, but where there are fair amounts of (machine-readable) parallel texts in the two languages, the idea naturally introduces itself of trying to transfer dominant language annotations into the lesser-known language via an alignment of the parallel texts on some linguistic level (see e.g. Borin 2002b for a discussion of the general idea, although not applied to dominant–lesser-known language pairs). How well this will work out is dependent on a number of factors, e.g. the kind of annotation targeted and the closeness of the languages involved (see Trosterud 2002), but in some cases it could be used in order to get a first rough annotation which could then be refined using a mix of human correction and automatic methods, as discussed in the next section. This kind of approach could be applicable in a situation such as that described by Zeisler (this volume), where resources for Ladakhi and Classical Tibetan could be developed in concert, as it were, to boot in a way that would be useful for the creating of Classical Tibetan teaching applications for speakers of Ladakhi (and other modern Tibetan languages). A special case of this methodology would be to use another language indirectly, as it were, using an annotation tool trained on some language A for annotating a different language B. Maynard, Tablan and Cunningham (2003) do exactly this when they apply an English named entity recognizer to Cebuano (an Austronesian language of the Philippines).

Although there has been considerably less research on this problem than on monolingual bootstrapping, researchers have endeavored to transfer at least the following kinds of annotation across languages in this fashion: lemmas (the extensive research on word alignment; see Borin 2002a); part of speech tags (Yarowski and Ngai 2001; Yarowski, Ngai and Wicentowski 2001; Borin 2002b); base NPs (Yarowski and Ngai 2001; Yarowski, Ngai and Wicentowski 2001); morphological analyses (Yarowski, Ngai and Wicentowski 2001); and morphemes (Johnson and Martin 2003).

The third and most realistic scenario is that of *computer-assisted human annotation* (or human-assisted computer annotation), representing a more sober assessment of the present capabilities of induction of linguistic regularities by machine learning than the first scenario, stating that its proper role is as an assisting technology for human annotators, rather than as a fully automated process.[19]

Especially promising is the simultaneous use of more than one source of (linguistic) information in concert, thereby achieving a result that is more than the sum of the parts. In this vein, there has been work on combining small amounts of linguistic knowledge (elicited from native speakers or taken from reference works written for humans) with various kinds of machine learning (e.g. Oflazer and Nirenburg 1999; Oflazer, McShane and Nirenburg 2001; Cucerzan and Yarowsky 2002; Neuvel and Fulop 2002), on how to best present language data to machine learning algorithms and how to select the most useful data items for human annotation (Engelson and Dagan 1996; Abney 2002; Steedman et al. 2003), and on the most cost-effective combination of human and machine annotation (Ngai and Yarowsky 2000). In practice, this third scenario seems to be the method of choice for rapidly creating language technology resources for a new language; recently this was attempted for Hindi as part of the DARPA/TIDES "surprise language exercise" (see Strassel et al. 2003 for a detailed account of this exercise and the work with Hindi).

5. The promise of language technology

A research program which follows quite naturally from the above would look roughly like this: Begin (systematic) testing of the methods proposed for rapid language technology resource collection and annotation with some "non-English" language as the target language. In the South Asian area there are any number of good candidate languages, and the choice here obviously will depend on which particular linguistic traits are considered most incompatible with the methods in question. If morphological complexity is considered important, as is arguably the case with the various morphology induction algorithms proposed in the literature, then probably some Dravidian language will make up the most appropriate testing ground, but the most important from a methodological point of view would be to work with a number of structurally different languages. It would be ideal if such experiments could be coordinated with general documentation efforts such as the documentation projects involving South Asian languages mentioned by Saxena in her introduction to this volume.

The purpose of such exercises would be first and foremost purely scientific: We would like to get a better understanding of the generality or language-specificness of these methods. At the same time, we might conceivably get the embryo of some language technology resources for a number of lesser-known South Asian languages, as well as general methods for turning language documentation into linguistic description in the most economical way.

Appendix: Language technology resources and tools

This listing of language technology resources and tools represents a synthesis of Sarasola (2000) and the LDC LoDL survey, discussed in section 2. LDC LoDL survey criteria are given in square brackets.

0. Prerequisites
- [language written]
- [standard digital encoding]
- [words separated in writing]
- [simple orthography]
- [sentence punctuation]
- [simple morphology]
- [(existence of) dictionary]
- [(existence of) newspaper]
- [(existence of) Bible (translation)]

1. Foundations
- Corpus: collections of raw text (untagged) [100 000 words of news text]
- Text corpus proper (untagged)
- [100 000 words of parallel text]
- Lexicon: Raw lists of forms, lemmas and affixes
- Machine-readable dictionaries (monolingual, bilingual, thesaurus, other) [10 000 word translation dictionary]
- Morphology: Description and formalization of morphological phenomena
- Speech databases (collections of digitized speech)
- Formal description and dictionaries of units for speech synthesis

2. Basic resources and tools
- Statistical tools for corpus treatment (bigram and trigram frequencies, word-count, collocations ...)
- Part-of-speech (POS) tagger
- POS-tagged corpora
- Lexical database containing information about parts of speech and morphology
- Morphological analyzer/generator [morphological analyzer]
- Speech recognition systems recognizing isolated words

3. Medium-complexity resources and tools
- Environment for (available) tool integration – using a standard for representation of linguistic knowledge: XML/SGML, etc.
- Spelling checker and corrector
- Structured lexical databases including multiword lexical units
- Surface syntax analyzer (“chunker”) recognizing simple (nonrecursive) constituents and phrases (NP, PP, verb)
- Web crawler managing language X

4. Advanced resources and tools
- Syntactically annotated corpora ("treebanks")
- Lemmatizer
- Grammar and style checkers
- Integration of dictionaries in text editors
- Lexical-semantic knowledge base. Concept taxonomy, e.g. *WordNet*
- Word sense disambiguation
- Speech processing at sentence level

5. Multilinguality and general applications
- Semantically annotated corpora
- Information retrieval and extraction
- Machine translation systems; translation of NPs and simple sentences
- Dialog systems
- Multilingual lexical-semantic knowledge base
- Language learning systems using human language technology

Notes

1. Actually, it was Max Weinreich (not Uriel) who coined the aphorism that Ostler refers to in the quote.
2. Mainly western European languages, but also Japanese and Chinese.
3. A largely synonymous term, with a somewhat longer history of use, is *language engineering*.
4. On the other hand, it *may* be a piece of language technology, but you have to know or to be able to infer something about its internal workings in order to be able to state this for a fact. In fact, it is my firm belief that language technology will make, e.g., multimedia programs for language learning or language renewal (such as the *Taitaduhaan* CD-ROM for Western Mono described by Kroskrity and Reynolds 2001) better than the ones produced at present, which as a rule do not use any kind of language technology.
5. Another way of saying this is that the knowledge of language encoded in language technology applications is *generative* in the same sense that this word is used in, e.g., "generative grammar", i.e. they model an infinite set of linguistic objects by finite means, although in language technology part of the knowledge may be probabilistic rather than categorical in nature. Even in those cases you can still often "reconstruct" some kind of generative system, with probabilities added to, say, the symbols and productions in a formal grammar. Note that the requirement that language technology mirror linguistically statable regularities in a faithful manner is best understood as pertaining to linguistic *behavior* and the observable *products* of this behavior – i.e., "performance", in the broad sense of the word – rather than to some system of linguistic rules supposedly underlying and accounting for this behavior in humans. In other words, most language technology work is theoretically uncommitted as to the nature and shape of (human) underlying linguistic knowledge, while still requiring that regular linguistic behavior (and its products) be captured by law-like generalizations and mechanisms in language technology applications. Note also that the common demand placed upon language technology applications, that they be *robust*, tallies perfectly with this construal of the ontological status of the linguistic knowledge embodied in such applications. But see Trosterud (this volume) for a slightly different view on this issue.

6. This is similar to – but not exactly the same as – the distinction made in general linguistics between a grammar formalism (sometimes somewhat high-handedly referred to as a "theory of grammar" or even "theory of language"), and a grammar of a particular language expressed using that formalism.
7. Some large languages are missing from this survey (which is no longer available on the LDC website, but which will be published eventually, according to Strassel et al. 2003), notably English, but also German, French, and some others; presumably these languages are considered high-density languages by default.
8. In the same way that literacy is not a necessary requirement for using e.g. digital telephony, you could make a case for a purely speech-based kind of language technology.
9. This figure is a ballpark estimate based on personal communications from Eckhard Bick (for the Danish treebank) and Jan Hajič (for the Prague dependency treebank), and on published figures pertaining to the work on the ICE-GB parsed corpus of British English (Nelson, Wallis, and Aarts 2002).
10. A telling remark in this context is the following, quoted by Phillipson (2003): "The most serious problem for the European Union is that it has so many languages, this [sic] preventing real integration and development of the Union. *The ambassador of the USA to Denmark, Mr Elton, 1997* [endnote I.1:] This remark was made at an informal lunch at the University of Roskilde, Denmark." (Phillipson 2003: 1, 208)
11. And now, the web can be used as a source of (a kind of) corpora, for assembling them on the fly, as it were (Ghani, Jones and Mladenić 2001a, 2001b; De Schryver 2002; Nilsson and Borin 2002; Maynard, Tablan and Cunningham 2003; Nilsson 2003; Oard et al. 2003). Of course, the languages must be written languages, and there must be a sufficient number of web publications in them. Of the two, the first is the more restrictive requirement, since only a modest fraction of the world's languages are written. Even written languages are quite unevenly represented on the web, however; in another connection, I endeavored to survey the availability on the web of material in two official minority languages in the Nordic area, Sámi and Finnish Romani. It turned out that while Sámi (mainly North Sámi) had a small web presence, Finnish Romani was, for all practical purposes, not represented at all on the web (Nilsson and Borin 2002: 416–417).
12. Note that this actually means minimizing human effort in the development of all sorts of language technology applications, since many such applications can be seen as special cases of corpus annotation.
13. The metaphor behind the term comes from the the English expression *to pull oneself up by one's own bootstraps*: "The term *bootstrapping* here refers to a problem setting in which one is given a small set of labeled data and a large set of unlabeled data, and the task is to induce a classifier. The plenitude of unlabeled natural language data, and the paucity of labeled data, have made bootstrapping a topic of interest in computational linguistics." (Abney 2002: 360) Here, we will assume the term to also include the setting where no labelled data are available, i.e., pure inductive (or abductive) learning.
14. A division into words is necessary at least in two separate cases: (1) in dealing with orthographies which do not regularly mark word boundaries, the best-known example being Chinese; (2) in segmenting continuously transcribed speech, e.g., certain kinds of automatic transcription.

15. The *type-token* ratio of a text is calculated by dividing the number of *different* words (and sometimes punctuation signs) – the *types* – by the total number of words (and sometimes punctuation signs) – the *tokens*. This is a measure (although indirect) of the vocabulary diversity of the text, and its inverse gives the sample mean, i.e., the average number of occurrences of a text word. Importantly, the type-token ratio is not constant, but decreases nonlinearly with text length. See Baayen 2001.
16. It could be argued that many of the researchers working in the field of machine learning of linguistic regularities are either English speakers or computer scientists, or both. In both cases there may be an lack of awareness – a benevolent interpretation – or disregard – seeing matters a bit more cynically – on the part of these researchers of such matters as language diversity, language typology, etc. On the other hand, it is often difficult to get linguists to take in, let alone voice an opinion on heavily mathematical work on statistical machine learning. Sparck Jones puts it very aptly when she says: "It has also to be recognized that the arrogance so characteristic of those connected with IT – the self-defined rulers of the modern world – is not merely irritating in itself, it is thoroughly offensive when joined to ignorance not only of language, but of relevant linguists' work" (1996: 13), and: "On the practical side, it is impossible not to conclude that many linguists are techno- and logico-phobes." (1996: 13–14).
17. With respect to the linguistic characteristics most relevant for our purposes here, Finnish Romani – a modern Indo-Aryan language – is more like Russian or German than English, but still much closer to English than to Greenlandic (see e.g. Vuorela and Borin 1998; Borin 2000). Cf. the figures given by Creutz (2003) for his 200 000 word English and Finnish corpora, with 17 000 and 58 000 word types, respectively, yielding type-token ratios of 0.085 for English, but 0.29 for Finnish.
18. Goldsmith's program is indirectly affected by this, too, since it relies on recurring "stems" and recurring "suffixes" in order for it to do its job properly – in the same way that a POS tagger relies on recurring word forms and word form sequences – and the morphological characteristics of a language such as Greenlandic ensure that most "stems" (in this sense) will be unique and, for the same reason, that most word forms will be unique.
19. This is reminiscent of the development in the field of machine translation (MT). In the beginning, there were high hopes for fully automatic high-quality MT, which were never realized. Only with the "ideological" reorientation of the field towards machine-aided human translation as the focal application area has MT started enjoying some kind of commercial success.

References

Abney, Steven
2002 Bootstrapping. *Proceedings of the 40th Annual Meeting of the ACL*, 360–367. Philadelphia: ACL.

Baayen, R. Harald
2001 *Word Frequency Distributions*. Dordrecht: Kluwer.

Borin, Lars
1991 The automatic induction of morphological regularities. Reports from Uppsala University Linguistics (RUUL) #22. Dept. of Linguistics, Uppsala University.
2000 A corpus of written Finnish Romani texts. *LREC 2000. Second International Conference on Language Resources and Evaluation. Workshop Proceedings. Developing Language Resources for Minority Languages: Reusability and Strategic Priorities*, 75–82. Athens: ELRA.
2002a ... and never the twain shall meet?" In *Parallel Corpora, Parallel Worlds*, Lars Borin (ed.), 1–43. Amsterdam: Rodopi.
2002b Alignment and tagging. In *Parallel Corpora, Parallel Worlds*, Lars Borin (ed.), 207–218. Amsterdam: Rodopi.
ms One in the bush: Low-density language technology. Göteborg University.

Borin, Lars, and Klas Prütz
2004 New wine in old skins? A corpus investigation of L1 syntactic transfer in learner language. In Guy Aston, Silvia Bernardini, and Dominic Stewart (eds.), *Corpora and Language Learners*, 69–89. Amsterdam: John Benjamins.

Creutz, Mathias
2003 Unsupervised segmentation of words using prior distributions of morph length and frequency. *Proceedings of the 41st Annual Meeting of the Association for Computational Linguistics*. Sapporo: ACL.

Cucerzan, Silviu, and David Yarowsky
2002 Bootstrapping a multilingual part-of-speech tagger in one day. *Proceedings of CoNLL-2002*. Taipei: ACL.

De Schryver, Gilles-Maurice
2002 Web for/as corpus: A perspective for the African languages. *Nordic Journal of African Studies* 11 (2): 266–282.

Ejerhed, Eva, and Gunnel Källgren
1997 Stockholm Umeå Corpus (SUC) version 1.0. Research report. Department of Linguistics, Umeå University.

Engelson, Sean P., and Ido Dagan
1996 Sample selection in natural language learning. In *Connectionist, Statistical and Symbolic Approaches to Learning for Natural Language Processing*, Stefan Wermter, Ellen Riloff, and Gabriele Scheler (eds.), 230–245. Berlin: Springer.

Garvin, Paul
1967 The automation of discovery procedure in linguistics. *Language* 43 (1): 172–178.

Ghani, Rayid, Rosie Jones, and Dunja Mladenić
2001a. Building minority language corpora by learning to generate web search queries. Technical Report CMU-CALD-01-100. Carnegie Mellon University Center for Automated Learning and Discovery.
2001b Mining the web to create minority language corpora. *Proceedings of the Tenth International Conference on Information and Knowledge Management (CIKM 2001)*.

Goldsmith, John
2001 Unsupervised learning of the morphology of a natural language. *Computational Linguistics* 27 (2): 153–198.

Harris, Zellig S.
1970 *Papers in Structural and Transformational Linguistics.* Dordrecht: Reidel.
Johnson, Howard, and Joel Martin
2003 Unsupervised learning of morphology for English and Inuktitut. *Companion Volume of the Proceedings of HLT-NAACL 2003 – Short Papers.* Edmonton: ACL.
Kroskrity, Paul V., and Jennifer F. Reynolds
2001 On using multimedia in language renewal. Observations from making the CD-ROM *Taitaduhaan.* In *The Green Book of Language Revitalization in Practice*, Leanne Hinton, and Ken Hale (eds.), 316–329. San Diego: Academic Press.
Manning, Christopher D., and Hinrich Schütze
1999 *Foundations of Statistical Natural Language Processing.* Cambridge, MA: MIT Press.
Maynard, Diana, Valentin Tablan, and Hamish Cunningham
2003 NE recognition without training data on a language you don't speak. *Proceedings of the ACL 2003 Workshop on Multilingual and Mixed-language Named Entity Recognition.* Sapporo: ACL.
McEnery, Tony, and Andrew Wilson (eds.)
2001 *Corpus Linguistics.* 2nd ed. Edinburgh: Edinburgh University Press.
Mitchell, Tom M.
1997 *Machine Learning.* New York: McGraw-Hill.
Nelson, Gerard, Sean Wallis, and Bas Aarts
2002 *Exploring Natural Language: Working with the British Component of the International Corpus of English.* Amsterdam: John Benjamins.
Neuvel, Sylvain, and Sean A. Fulop
2002 Unsupervised learning of morphology without morphemes. *Morphological and Phonological Learning: Proceedings of the 6th Workshop of the ACL Special Interest Group in Computational Phonology (SIGPHON)*, 31–40. Philadelphia: ACL.
Ngai, Grace, and David Yarowsky
2000 Rule writing or annotation: Cost-efficient resource usage for base noun phrase chunking. *Proceedings of the 38th Annual Meeting of the Association for Computational Linguistics.* ACL.
Nilsson, Kristina
2003 A meta search approach to locating and classifying reading material for learners of Nordic languages. Master's Thesis in Computational Linguistics, Dept. of Linguistics, Uppsala University.
Nilsson, Kristina, and Lars Borin
2002 Living off the land: The web as a source of practice texts for learners of less prevalent languages. *LREC 2002. Third International Conference on Language Resources and Evaluation. Proceedings*, 411–418. Las Palmas: ELRA.
Oard, Douglas W., David Doermann, Bonnie Dorr, Daqing He, Philip Resnik, Amy Weinberg, William Byrne, Sanjeev Khudanpur, David Yarowsky, Anton Leuski, Philipp Koehn, and Kevin Knight
2003 Desperately seeking Cebuano. *Companion Volume of the Proceedings of HLT-NAACL 2003 – Short Papers.* Edmonton: ACL.

Oflazer, Kemal, Marjorie McShane, and Sergei Nirenburg
2001 Bootstrapping morphological analyzers by combining human elicitation and machine learning. *Computational Linguistics* 27 (1): 59–85.

Oflazer, Kemal, and Sergei Nirenburg
1999 Practical bootstrapping of morphological analyzers. *Proceedings of Language Learning Workshop 1999*. ACL.

Ostler, Nicholas
n.d. Review: Workshop on language resources for European minority languages, Granada, Spain; 27 May 1998 (morning). <http://www.cstr.ed.ac.uk/~briony/SALTMIL/review.html>. (Accessed on 10 June 2003)

Phillipson, Robert
2003 *English-only Europe? Challenging Language Policy*. Andover: Routledge.

Sarasola, Kepa
2000 Strategic priorities for the development of language technology in minority languages. *LREC 2000. Second International Conference on Language Resources and Evaluation. Workshop Proceedings. Developing Language Resources for Minority Languages: Reusability and Strategic Priorities*, 106–109. Athens: ELRA.

Sharma, Utpal, Jugal Kalita, and Rajib Das
2002 "Unsupervised learning of morphology for building lexicon for a highly inflectional language". *Morphological and Phonological Learning: Proceedings of the 6th Workshop of the ACL Special Interest Group in Computational Phonology (SIGPHON)*, 1–10. Philadelphia: ACL.

Sparck Jones, Karen
1996 How much has information technology contributed to linguistics?. Paper presented at the *British Academy Symposium on Information Technology and Scholarly Disciplines*, 18–19 October 1996. The page references in the text are to the electronic version available via <http://xxx.lanl.gov/cmp-lg/9702011/>.

Steedman, Mark, Rebecca Hwa, Stephen Clark, Miles Osborne, Anoop Sarkar, Julia Hockenmaier, Paul Ruhlen, Steven Baker, and Jeremiah Crim
2003 Example selection for bootstrapping statistical parsers. *Proceedings of HLT-NAACL 2003*, 236–243. Edmonton: ACL.

Strassel, Stephanie, Mike Maxwell, and Christopher Cieri
2003 Linguistic resource creation for research and technology development: A recent experiment. *Association for Computing Machinery transactions on Asian language information processing (TALIP)*, vol. 2, Issue 2, 101–117.

Trosterud, Trond
2002 Parallel corpora as tools for investigating and developing minority languages. In *Parallel Corpora, Parallel Worlds*, Lars Borin (ed.), 111–122. Amsterdam: Rodopi.

Vuorela, Katri, and Lars Borin
1998 Finnish Romani. In *Minority Languages in Scandinavia, Britain and Ireland*, Ailbhe Ó Corráin, and Seamus Mac Mathúna (eds.), 51–76. Uppsala: A&W.

Yarowsky, David, and Grace Ngai
2001 Inducing multilingual POS taggers and NP bracketers via robust projection across aligned corpora. *Second Meeting of the North American Chapter of the Association for Computational Linguistics*. ACL.

Yarowsky, David, Grace Ngai, and Richard Wicentowski
2001 Inducing multilingual text analysis tools via robust projection across aligned corpora. *Proceedings of the First International Conference on Human Language Technology Research*. ACL.

Worrying about ethics and wondering about "informed consent": Fieldwork from an Americanist perspective

Colette Grinevald

1. Introduction

This paper is meant as a contribution to the ongoing discussions of the ethics component of documentary work on endangered languages today. Such discussions have become pressing in the context of a changing world characterized by on the one hand, an increased ethnic consciousness and the politicization of indigenous languages, and on the other, an increased sense of responsibility on the part of linguistic professionals working with communities of endangered languages. At this point, the issues of technological developments for that documentation are finding their way into meetings and conferences, with some discussion of the ethical and legal aspects linked to those technologies, such as issues of intellectual property and "access" for the materials produced and to be archived. But the fact is that much remains to be done to identify all the ethical issues involved, from the beginning to the end of such field projects and in terms of all the actors concerned, and then to articulate them, in a realistic and concrete manner, for academics and funding agencies not familiar with the nature of fieldwork in general, or with this kind of fieldwork in particular.

The task of how to tackle a discussion of the ethical issues as they present themselves today can feel admittedly not only challenging, but rather daunting in its complexity, and certainly somewhat overwhelming to talk and write about. This is particularly true for the majority of us field linguists who have been trained and have acquired experience in describing the grammatical structure of such endangered and un(der)-described languages (which by the way, is already a dauntingly complex and sometimes overwhelming challenge in themselves!), but have not been particularly prepared by our profession to handle and articulate to ourselves or the outside world the ethical issues embedded in the socio-political dimensions of our field projects. Meanwhile, those ethical concerns are becoming omnipresent in the enterprise of fieldwork, and have become one of the major concerns of the new sources of funding for documentation projects.

The aim of this paper is therefore threefold. First, it uses the opportunity of writing for a volume on the documentation of languages of South Asia to reflect on some of the specifics of the Americanist and Australianist perspectives on the issues of endangered languages and ethics of fieldwork. Second, it considers how ethical issues have been articulated in the last decades from those perspectives, starting by suggesting various ways of organizing the multiplicity of ethical issues that need consideration, and focusing in the end on the particular issue of "informed consent", one of the requirements of all of the new agencies who fund fieldwork on endangered languages. The third section is a case study of such a process of obtaining informed consent for a grant proposal for an endangered language of Central America – the Rama language of Nicaragua – that offers a critical assessment of what such a process entails.

To the extent that there may be a lack of discussions on ethics in fieldwork and language documentation in South Asia, this paper will hopefully provide a framework in which discussions on ethics of fieldwork and language documentation of languages of South Asia could be conducted. Although issues of language endangerment have not been high on the agenda for linguists working in the South Asian region or for South Asian language policy makers, for reasons which will be considered below, language endangerment and language shift is just as real there as in America, as evidenced by several of the other contributions to this volume, and consequently such discussions should be considered necessary at this point.

2. An Americanist perspective on endangered languages

This section is an attempt at accounting for why the growing concern about the fate of endangered languages was first raised, and most adamantly, by field linguists from two specific continents, Australia and America, while it has been perceived neither with the same intensity nor in the same terms by colleagues from Africa, and, more specifically for our purpose here, by fellow linguists working in South Asia.

2.1. Viewed from America, all minority languages are "endangered languages"

It is interesting to note how the choice of phrasing used to refer to the languages we are concerned with is regionally marked: while one talks generally of "minority tribal languages" for South Asia, and more commonly of "lesser-known languages" for Europe, the most common label is that of "endangered languages" for America.[1] This difference in terminology points to the various approaches to the issue, such as the European model dealing with the construction of Europe and the recognition of its "regional languages", and the Asian (and African) model dealing with widespread multilingualism in the context of postcolonial times. The American model, in contrast, wrestles with the past and present massive proportions of endangerment and death of languages, and the increasingly widespread demands for revitalization efforts on behalf of those languages by their linguistic communities (a model shared with Australia). Below, then, are some of the basic reasons as to why the "minority languages" of America have been cast into a profile of "endangered languages", and of why the sense of urgency and commitment is shared by so many Americanists, the present author included.

2.2. Massive loss of Native American people and languages

The story of Native America is a story of a massive loss of people and languages. Regarding the loss of people, the numbers are staggering. It is considered, for instance, that the loss of population was on the order of 90% in the early decades after the arrival of the Europeans. For instance, in Mexico alone, the population went down from 25 million in 1519 to 2 million by 1580. Massive death occurred mostly by contact disease, but also by deportations, slavery, massacres, and warfare. The phenomenon of the death of indigenous people by contact disease, massacres, or suicides following capture and sedentarisation is actually still ongoing, with cases regularly reported for the Amazon region.

While a massive loss of languages came therefore, early on, along with a massive loss of people, another equally massive and accelerated loss of languages is being documented today, although this time generally occurring as a result of language shift.[2] The figures circulating for the American continent are telling, particularly for its northern part. For the United States alone, the estimate of the number of languages at the time of colonization is 300 languages, of which only 175 are known today, but with only 20 still spoken by some children. And in terms of sheer vitality, i.e. of chances for the language to survive until the end of this century, the figures come down to only 5 of these 175 for the United States, and 6 for the 60 still spoken today in Canada. The case of the state of California is emblematic of this situation of loss, since it has by now reached the point of close to a total extinction of its languages,[3] after having been one of the most heavily populated indigenous areas of the continent at the time of colonization (much denser at that time than any part of Europe, for instance), with one of the most striking cases of genetic diversity of languages.

2.3. Language loss in America as loss of genetic diversity

The sense of massive loss can also be considered from the angle of the loss of genetic linguistic diversity, which inherently also means loss of typological diversity. A linguistic argument for the urgency of describing and documenting the Amerindian languages has therefore been the impact of this double loss (genetic and typological) for the studies of language origin and evolution, migration and population patterns, and cognitive diversity of the human capacity for language. In this perspective of diversity, the most telling figures become, then, not those of the absolute number of languages, but rather those of the number of language families, and language stocks (families of families).[4] And as shown in Table 1, the most striking figures for the entire world are in North and South America, when compared with figures of Europe, South and East Asia, Africa, and Australia:

Table 1. Global distribution of languages and stocks (figures from Nettle and Romaine 2000: 37)

	languages	stocks	avg. no. of lgs/stock
Europe	209	6	34.8
South and S.E. Asia	1400	10	140
Africa	1995	20	99.7
Australia	234	15	15.6
North America	230	**50**	4.6
Meso America	300	14	21.4
South America	419	**93**	4.5

The striking figures of North America correspond to the extraordinary linguistic variety of the languages along the Pacific Coast and those of South America to the wealth of languages of the Amazon region detailed in Queixalos and Renault-Lescure (2000).[5] The high figure of stocks and families in the Amazon region actually comes from a long list of isolates: in Bolivia, for instance, 19 of the 34 identified languages of the Amazonian region are isolates; while the rest of the languages belong to five different families

Taking into account these figures of number of stocks of America in general and South America in particular helps place this continent in perspective alongside the rest of the world, in particular if one compares these figures to those of Asia and Africa, where the situation seems reversed, with a very high number of languages but a relatively low number of stocks.[6]

2.4. The implications of language death by language shift in America

Essential to the understanding of the sense of doom and urgency felt by Americanists, in the face of the figures mentioned above, is that it is a situation of the confrontation of single indigenous minority languages, many times genetic isolates in fact for South America, with one dominant and all-powerful colonial language (be it English, French, Spanish, Portuguese or even Dutch). In such a context, language shift means giving up "the" indigenous language in favor of the colonial language. It means indeed moving away from an indigenous monolingualism, and moving through an imbalanced bilingualism with a not fully mastered dominant language of a very different linguistic structure and conveying a very foreign colonial culture, to arrive at a monolingualism in the colonial language but without full mastery of that colonial language. The experience of many Americanist fieldworkers has been to witness the consequences of such a widespread phenomenon of quickly accelerating language shift, in which people are lured away from their language and culture with promises of an integration that in fact never materializes, creating masses of doubly marginalized populations, never really integrated, yet also having lost their language and culture and the identity and dignity that went with them.

Two remarks will be added about this phenomenon of language shift in the American context. First, that in earlier times in the USA (Canada and Australia), such shift has been traumatically induced by public policies specifically meant to eradicate the languages. This was done, for instance, through kidnapping children against their parents' will, then keeping them in boarding schools with the expressed goal of making the children lose their language. The memory of such treatments still lingers on in native communities. Second, language shift is happening even with the largest indigenous languages of America, those that could seem still vital because of their (relatively) large number of speakers, but which are actually in danger of tipping over into widespread language shift. In this case, language shift is worrisome because it would seem to point to the possibility of a very generalized loss of the native languages of America in a foreseeable future, through an insidious but nevertheless very real process of shift. Such is the case of Kechua, the largest Native American language in the Andean region of South America, spoken today by 8 to 12 million speakers, and also for Navajo, the largest language in the USA, with over 120 000 speakers, but under pressure in spite of its bilingual education programs.[7]

2.5. Ethnic consciousness and politicization

Various dynamics that played themselves out on the American continent contributed to raising consciousness about the centrality of language to ethnic identity, and to politicizing the issue of ethnic languages. They consisted of successive initiatives of Whites during the last decades that more or less openly threatened the survival of indigenous languages and against which indigenous communities organized themselves. In the eighties in the USA it was the threat to the Native languages of the English Only movement that led to the organization of Native communities to press for what became the Native American Languages Act of 1990. This bill "guarantees the right of Native Americans to use and support their languages, and hails the indigenous languages as a unique part of American heritage that the government has the duty to assist native communities in preserving" (Hinton 2004: 20).[8] For Latin America, the turning point was around 1992, and the official celebrations of the supposed "discovery" of America sponsored by Spain and Latin American governments. At that time, the mobilization of the Native populations protesting this celebration and affirming their identity as well as their linguistic and cultural rights reached all of America. In the following years, one after the other, most countries of Latin America recognized their multiethnic and multilingual nature, and modified their Constitution to that effect.[9]

The politicization of Native American languages and the mobilization of many indigenous communities was matched by that of linguists working with such communities on native languages. Linguists became advocates for the Native American Languages Act in the USA,[10] and supported the indigenous movement surrounding the 500th anniversary of the supposed "discovery of America". Craig (1992a) retraces some of the dynamics of this movement that swept through the American continent and changed forever the relation of most academic linguists to the indigenous communities. The rhetoric about "endangered languages" in America was, for instance, first developed by Americanist linguists at the annual LSA meetings of December 1991, with a special panel on the subject of endangered languages, that led at the same meetings to the creation of the LSA Committee on Endangered Languages, and the Declaration of the LSA on behalf of endangered languages. The papers from the panel were published in the following issue of *Language* (Hale et al. 1992).[11] At about the same time, linguists from all over the world had been working to raise the issue with the UNESCO, and the 1992 XVth International Congress of Linguists in Quebec sponsored a panel on endangered languages.[12]

The point to be made is that the mobilization and politicization of a great number of academic Americanists echoed that of the indigenous communities with whom they were working. This sensitization took various forms, from participating in language institutes for speakers or community members, to developing new forms of negotiated fieldwork, to raising the issue of language endangerment in academia and beyond, to lobbying for legislation to protect those languages. All of this occurred while continuously working on describing some of these endangered languages.[13]

2.6. Similarities with Australia

All that has been said above about language loss, the sense of urgency, the politicization of languages, and the involvement of field linguists applies equally to Australia. On that continent, one finds the same decimation of population and loss of languages with shift of unique ethnic language to local English, official policies of language extermination by mistreatment of the population including sequestering of children into boarding schools, and a growing commitment by field linguists to acknowledge the concerns and needs of the Aboriginal populations. As in a number of Latin American countries, Australian fieldworkers (linguists and anthropologists) are being asked to help delimit traditional territory, which is frequently based on traditional narratives and dreams that are often part of the language material collected. They are also being asked to work with the communities on their bilingual education programs.

In fact, the mobilization of Australian linguists occurred somewhat before that of the Americanists, and their radicalization is even more marked. Australian linguists are often considered pioneers in the way they developed codes of ethics together with the Aboriginal people; as a professional group of field linguists they can even be considered as more sophisticated in their reasoning and consideration of the ethical issues raised by working on endangered languages, as will be mentioned later.

2.7. Contrast with South Asia

The various factors of the situation of Native American languages considered above in this first section, from the massive loss of people and languages, to genetic linguistic diversity, the direct impact of massive language shift on indigenous language loss, to the politicization of indigenous languages were assembled in order to account for why such a mobilization concerning endangered languages came from Americanist linguists. Such an exercise in outlining the specificities of the American continent was actually prompted by the occasional denigrating or simply puzzled remarks from field linguists of other regions of the world confronted with the attitude of many Americanist linguists on the issue of endangered languages and with their relations to communities of speakers. The clear stance of advocacy of the Americanists that is sometimes taken to be excessive or academically inappropriate when seen from other parts of the world was also said here to be largely shared by Australianist colleagues for similar reasons, with different[14] but parallel sociopolitical developments.[15]

In contrast, specific factors of other regions of the world could be put forward in order to explain how the issue of language endangerment may have been perceived there with less intensity, and viewed with less or different concerns by linguists who are specialists of those regions. In the case of South Asia, which is of interest here (and to some extent in the case of Africa, too) the politicization of languages and the ethnic consciousness of the phenomenon of language endangerment would seem to be less intense because of the following factors, among others:

(a) an active and multilayered multilingualism, in multicultural societies where language loss is less visible and tangible to the extent that it occurs in the midst of the maintenance of some other configuration of multilingualism,
(b) a different colonial history that has meant lower numbers of languages lost, and less active policies to eradicate the languages,
(c) larger numbers of languages, but in dialect chains of genetically related languages, with a common under-identification of "mother tongues" of less than 10 000 speakers,
(d) an identification of language more in terms of their role as expression of social identities than for their strictly linguistic differences
(e) less politicization of language issues *per se*, on a background of more developed politicization of religious issues and affiliation to a region, religion being stronger than language issues as a basis for identity[16]
(f) language shift normally to other indigenous languages rather than to former colonial languages.

These factors and the fact that very little attention has been paid to lesser-known languages in South Asia has had the effect that discussions on issues of fieldwork and ethics of fieldwork have been lacking so far in the discussions of the linguistic situation of that region of the world . One of the aims of this paper is therefore to present a framework under which these crucial questions could be discussed in a South Asian perspective.

3. On ethics

Ethics of fieldwork is at once an old and a relatively new subject. Today, in the context of new documentation projects of endangered languages, the issues to be included in a "code of (good/ethical) conduct" seem to have diversified and multiplied. They need to include issues linked to the use of new technologies, as well as a reconsideration of traditional data elicitation methods for linguistic analysis, and, crucially, need to take into account the complexities of the sociopolitical context of endangered languages.

3.1. Concerning methodological, technological and ethical issues

Ethical issues are embedded in a host of other -"ical" issues, such as methodological and technological ones, and permeate the whole enterprise of fieldwork, from its conception to its finalization, and involve more of the totality of its aspects than easily sensed or accepted. They are embedded in complex chains of issues, ranging from what kind of linguistics for which kind of work; to what kind of field methods to use to collect data, for what purpose, with what kind of speakers; to what technologies should be used to capture the data, analyze it, store it, archive it even, and then disseminate it, and to whom. Some are old questions, concerning the relation to the speakers and to the communities, which are always raised about fieldwork (see Craig 1992a, 1992b, 1993a; Grinevald 1997, 1998, 2000, 2003, to appear), but others are newer ones introduced or highlighted by new technologies and the archiving on the net with the potential of the latter for unlimited access, including such issues as intellectual property, the right to privacy, and the right of access.

In order to tackle the somewhat diffuse and overwhelming issue of "ethics", linguists can start by learning from the code of ethics of close professional fields, such as anthropology, sociology or psychology, for instance, disciplines which have a stronger tradition than the discipline of linguistics has had of attending to the ethical aspect of their work, including fieldwork. Among linguists, however, one must take note of the *Guidelines for Ethical Research in Indigenous Studies* of the Australian Institute of Aboriginal and Torres Strait Islander Studies, to be considered here later. There are in fact multiple ways of construing the ethical issues one is faced with while doing fieldwork, as will be quickly sketched out below.

3.2. Ethics: A time-line across two worlds

One way of looking at ethical issues is to consider them along the time-line of the different phases of a project, while acknowledging that much of the process is cyclical and elements of different phases may overlap. The question is what there is to think about and to handle in the three major phases of a fieldwork project: its before, during (entering, being in, leaving the field), and after. This was the approach presented in Craig (1992a)[17] and reproduced almost identically below. It is worth noting that this list was proposed before the present discussions of the specifics of the documentation of endangered languages, and seemed therefore much more radical at the time than it appears today. It was also crucially meant to give a first impression of the variety of issues of an ethical nature involved when a project is already cast into a negotiated fieldwork framework, one of working WITH speakers (rather than ON a language, or FOR a community).[18]

BEFORE: preparing from academia

– Choice of project and field site. This includes checking who is working on the ground, including identifying native linguists, national (as opposed to foreign) linguists already involved with the language and the linguistic community
– Nature of the initial contacts. The issue is how to introduce oneself, how to present one's project to the parties concerned, and how to seek permissions and support from institutions and individuals. This aspect of the preparation may vary greatly in complexity from one country to another

DURING: while in the field

– Choice of consultants. One of the issues related to the hiring of consultants to be considered is that of luring consultants away from national institutions and projects, because of better financial support from foreign grant money, for instance
– Informed consent. This is a very key issue in social sciences about which linguists have been mum for a long time. It would require seriously taking into consideration the perception of paperwork by indigenous people in addition to the legalistic concerns of the First World institutions. And it needs to include the sense that consent is to be rechecked at every stage of the project
– Disclosure of the purpose of study. This issue was particularly key with respect to missionary work, a highly sensitive issue in politically sensitive areas.[19] Today the issue includes negotiations as to the usefulness of the study for the community itself

- Disclosure of sources of funding. This was a particularly sensitive issue in politically sensitive areas (the CIA spy syndrome widespread in Latin America, for instance, in the second half of the twentieth century)
- Relations to indigenous communities. The issue deals with the necessity of recognizing consultants as members of a linguistic community, even if they live isolated out in the field or in towns away from their community of origin. It involves issues about the level of involvement and advocacy of the researcher and of the nature of some mode of reciprocity (what money can't buy), particularly important to establish when the communities are mobilized
- Relations to individual consultants. The issue is one of the potential empowerment of the individuals, contemplating their formal education and training so they become agents of the process of documentation and description of their own languages
- Relation to local scholars. It is of utter importance to identify the local scholars who have often worked in total isolation, without support or recognition, to consider how to eventually engage them in collaborative research. This can be a slow process, as the arrival of foreign linguists can be seen at first as a threat or an invasion, generally with good reason
- Relations to regional and local indigenous institutions. Depending on the country, dealing with the "gate keepers" can be cumbersome and time-consuming, but unavoidable. The representatives to deal with are often indigenous people themselves but of varying language groups, and sometimes are non-indigenous people
- Relation to governments, for visas and general work permits. In terms of time-line, some permits are usually solicited before leaving, while others must often be applied for once in the field for access to certain regions of the country

AFTER: Back in academia

- Returning copies of data and analysis to the community. The issues involved are those of data ownership, which has become of great concern today, and of appropriate modes of returning materials and presenting analysis to the community, which is still not as well attended to today as it probably should[20]
- Publishing. An expectation if not an obligation of academics, particularly when supported by foundations. The issues that linguists must consider are those of the anonymity vs recognition of the identity of the consultants vs co-authorship; of rechecking the material to be published with consultants and local authorities, and, as demanded in some cases, considering the choice of illustrative linguistic examples to be used[21]

- Producing materials of use to the community, writing for the community. This point deals with the so-called "applied linguistics" aspect of the work more and more demanded by communities and raises the issues, on one hand, of the manpower to do the work where and when there is often nobody else but the academic linguist to do it, and, on the other hand, the lack of academic recognition for this type of work
- Following up, staying in touch, keeping the field open for future researchers. The important idea is that such projects develop over years, decades even, with multiple field stays, and that they must be seen as evolving processes

This listing was just meant to give an idea of the variety of issues that involve the concept of ethics. In many ways it could be updated today as the discussion has progressed, but essentially touches on most of the unavoidable themes. What is important here is the concept that the period(s) of fieldwork proper are sandwiched between two phases of academic residence, the "before" of preparation and planning from a distance, and the "after" of returning to and reintegrating the academic environment. Implicit in this repeated back and forth movement between the field and academia is the fact that fieldworkers generally feel the pressures of often conflicting demands, those from the field and those from academia, two types of worlds that know little of each other.[22]

3.3. Another way of looking at ethics: Multiple allegiances

Another way to go about capturing the set of ethical questions to keep in mind when doing fieldwork on endangered languages is to approach them through the prism of the multiple allegiances that link the fieldworker to different entities and agents. This appears to be the approach of foundations such as the Volkswagen-DOBES and the Hans Rausing Endangered Languages Documentation Program based at SOAS in London, as pointed out in Austin (2003).

The overall picture of the network of responsibilities to different entities is sketched out on the DOBES web site. It offers a First World and TOP-DOWN view of the relations involved and the constraints and commitments that fieldworkers sign on to, with respect to their different partners: funding agents, technical archiving agents, academic agents. Funding agents have explicit requirements of outputs delivered in a timely fashion without risk of legal pursuit, archiving agents have specific technical requirements of data processing, and, while not appearing clearly in DOBES MPI schema, although a reality of the life of academic fieldworkers, the academic requirements of scientific quality of the output and of demands for certain types of production if one is to secure or pursue an academic career must also be taken into consideration.

The agents identified in the DOBES schema on the front page of its *Code of Conduct* include:

1. The funding agency (such as the Volkswagenstiftung in Germany, the Hans Rausing Endangered Languages Programs in England, the National Science Foundation in the United States)
2. The host academic institution for the project and the fieldworker
3. The archivers of the materials of the project (such as DOBES at the MPI Nijmegen, HRELAP in London as SOAS, and the Archives of Indigenous Languages of Latin America (AILLA), at the University of Texas at Austin)
4. The national and regional institutions of the country of fieldwork, including the indigenous organizations
5. The local community organization
6. The individual linguistic consultants
7. The users of the documentation

It is probably an understatement to say that the different agents involved create a maze of commitments of often conflicting nature, and that one of the major challenges of fieldwork is to juggle all these constraints, requirements, and commitments. What would also need to come forward is a stronger voicing on the part of experienced fieldworkers about the nature of the work of description and documentation of an endangered language and the kind of relations that field linguists establish with the speaker communities and individual consultants. Newman and Ratliff (2001) is a good place to begin to read about such relations on the part of some of the best-known field linguists from around the world.

3.4. A focused look at the relation of a linguistic fieldworker to a linguistic community: The Australian model

What follows is extracted from the *Guidelines for Ethical Research in Indigenous Studies* published by the Australian Institute of Aboriginal and Torres Strait Islander Studies. It replicates the "Principles of Ethical Research" spelled out (2002: 5–11) and is organized around issues that to a certain extent arise at the different phases of the time-line presented earlier, of before, during and after the actual fieldwork, but emphasizes the continuous process of establishing and maintaining a collaborative working relationship with the community.

A. Consultation, negotiation and mutual understanding
1. Consultation, negotiation and free and informed consent are the foundations for research with or about Indigenous peoples
2. The responsibility for consultation and negotiation is ongoing
3. Consultation and negotiation should achieve mutual understanding about the proposed research

B. Respect, recognition and involvement
4. Indigenous knowledge systems and processes must be respected
5. There must be recognition of the diversity and uniqueness of peoples as well as of individuals
6. The intellectual and cultural property rights of Indigenous peoples must be respected and preserved
7. Indigenous researchers, individuals and communities should be involved in research as collaborators

C. Benefits, outcomes and agreement
8. The use of, and access to, research results should be agreed upon
9. The researched community should benefit from, and not be disadvantaged by, the research project
10. The negotiation of outcomes should include results specific to the needs of the researched community
11. Negotiation should result in a formal agreement for the conduct of a research project, based on good faith and free and informed consent

If taken literally, there is no doubt that these principles of ethical research would substantially change the mode of operation of the majority of the linguistic fieldworkers around the world. As mentioned earlier, this Australian model is considered as the most socio-politically engaged, and outlines what is being discussed today more and more as "good practices" for all types of development work with indigenous populations (see Grinevald to appear).

3.5. Conclusion on ethics of linguistic fieldwork

No lists and categorizations will really capture what is at the heart of the question of ethics in fieldwork because much of it is a question of attitude, of posture, of a certain quality of the process that cannot be reduced to rules to be applied in a rigid manner. Needless to say, the way to approach the discussion of ethics is necessarily cast into a certain political and ideological framework, the one considered here being clearly one that demands academia to pay attention to the demands and needs of the linguistic communities of endangered languages, and to not be uniquely self-serving, as has been traditionally the case. Perhaps the most striking aspect of the Australian document is how it repeats several times, in the set of principles reproduced above but also in the guidelines for implementation of those principles not presented, how it is imperative to maintain flexibility, to allow for continuous reassessments and modifications of goals and ways of working. In this context, the issue of obtaining "consent", is, for instance, something to work at establishing, maintaining and nurturing, something that needs to take shape over time, by relations of trust and through shared experiences, as will be considered in the next section.

4. On getting free and informed consent

This aspect of ethics has come to the forefront of concerns of field linguists today, as the new foundations who finance the documentation of endangered languages have made it a requirement of all applications. What follows are some notes and musings on what free and informed consent might mean, seen from the field, with a case study to illustrate some of the complexities of obtaining such documents.

4.1. About consent and a case study

Informed consent is a well-known concept in the hard sciences, which rely on experimental procedures with human subjects who are asked to sign standard forms. But the discussion is relatively new for linguists, as the field of endangered language documentation is still developing.

4.1.1. Some thoughts on the notion of informed consent

Here again, the question will be approached from different perspectives, to try to cover the major issues involved. One could start by returning to the Australian guidelines already mentioned, and considering the following propositions, among other suggestions:

- Identify appropriate individuals and communities who should be consulted about your research project. There is almost always someone to speak for a particular place or area

- Identify community, regional or other indigenous umbrella organizations
- Communicate with relevant individuals and organizations by appropriate means. Face-to-face meetings are always desirable. The budgetary and funding implications of such visits should be considered
- From the outset, objectives should be clear, while maintaining flexibility and a willingness to modify your goals and ways of working

To follow such recommendations is no doubt much to ask of linguists fresh out of graduate linguistics courses who want to embark on such projects. They require getting to know the community well enough to judge who the "relevant" interlocutors are, as well as the "appropriate" means of communication, and a positive attitude towards collaborative research that always means a slower and more complex process, while academia and financing institutions are looking for demonstrated and marketable results and products. The concerns could alternatively be labelled as:

- Consent about what and for what
  The concerns range from permission to be in that particular place, to collect data, under what circumstances, and then to treat the data later on, and to make it public through standard publication or virtual internet distribution. They remind us of consent forms for experimental sciences, and property rights issues

- Consent from whom
  Clearly one must deal with a variety of constituencies from whom to get consent. There are individuals identified as leaders or representatives and collectivities organized in identifiable institutions or not; and there are speakers who provide the data and community members who may or may not be speakers of the language being documented but who consider themselves as much the owners of the language as the speakers themselves

- Getting consent when
  The foundations demand consent prior to actual fieldwork, as part of grant applications, but consent is a matter of process, of negotiations, of re-evaluation and updating. The initial consent is already the result of some interaction and fieldwork, and usually covers the process of planning. But as the project takes place and develops, it is likely that dynamics also develop that require periodic reviews and re-assessments

- Consent obtained how and in what form
  One needs to consider the validity of a written document, a source of some reassurance for the funding agencies operating in a legalistic First World,

> but a written document that is actually a foreign object in the oral tradition language communities that are specifically the target of those projects. The lack of literacy tradition means that the written medium is not easily mastered by members of the endangered language communities, who are usually, precisely, the most marginalized sectors of the society. And beyond the actual act of writing and signing, one needs to rethink the significance of a written document across cultures. Recent consideration of this cross-cultural challenge has led to the possibility of non-written forms of consent, such as video recordings of the discussions and agreements reached between the principals concerned

One should not expect that there will ever be a set guideline to informed consent. This is because fieldwork projects are not laboratory experiments. They are always embedded in specific socio-political circumstances that vary at infinitum, in the context of which informed consent is a question of establishing a working relationship cross-culturally and then maintaining it throughout the project. It is one of the most challenging aspects of all the multiple responsibilities that fall on the fieldworkers, who are academics usually raised and trained far away from the realities of the field and little prepared in the fields of sociology, political science, or anthropology, where such issues are addressed more regularly.

4.1.2. A story of letters of consent

What follows is a case study in the complexity of the concept, and in the practical issue of getting a letter of consent for a grant application that illustrates, among other things, the evolution in field approaches and financial institutional demands over a 20 year span. It offers the perspective on the subject of a long time fieldworker as committed from the start to the concept of collaborative research and informed consent as she feels now puzzled and challenged by the realities of their implementation.

The story is that of the Rama Language Project (RLP) of Nicaragua, a project of description, revitalization, and documentation of a very endangered language, the ethnic language of the smallest and most marginalized ethnic group within the country. It contrasts the conditions of the first phase of the project (1985–93) to those of its present second phase (2004–6) and its evolution in the need to obtain letters of consent from community and institutions. It focuses in particular on the process of obtaining successive letters of consent from the community for two successive applications to the Hans Rausing Endangered Language Program, in order to give a sense of all of the energies and emotions, positive and negative, that accompanied these processes.

4.2. RLP Phase 1: the Autonomy Project for the Atlantic Coast of Nicaragua

This first Rama Language Project took shape in the context of three dichotomies that needed constant readjustment and adaptation on the part of the academic linguist involved. One was the dichotomy of working in the midst of a revolutionary process with academic financial backing and being faced with different sets of demands; the second was that of a two track project of linguistic description and language revitalization at the same time, and the third the dichotomy of those who wanted the project, the visionary old woman who was one of the last speakers, and the community leaders (non-speakers) who had officially requested it.

4.2.1. The political context of a two track project

The Rama Language Project was one of the linguistic projects of "linguists for Nicaragua" and was cast from its origin into the political context of the revolutionary Sandinista government then in power. It was conceived and requested during the discussions to grant autonomy to the eastern half of the country – then known as the Atlantic Coast and now as the Caribbean Coast – where local indigenous and creole populations had maintained five ethnic languages with different degrees of vitality: Creole English of the Miskitu Coast Creole variety, Miskitu, Sumu, Rama, and Garifuna (see Craig 1992a, 1992b). Autonomy was granted in 1987, and new laws granted linguistic rights to all the ethnic groups of the region.

The project was a response to actual demands presented to the Sandinista authorities on the part of the Rama community through representatives, to revitalize their ethnic language, Rama. This was therefore an interesting case of politicization of a language from outside, in this case the outside force being the political discourse of multi-ethnicity and plurilingualism used in the whole region at the time. The Rama leaders spoke of the shame of standing up in multiethnic assemblies and of addressing the assembly in the regional form of English Creole, the language the community had shifted to, instead of in Rama, their ethnic language they had all but abandoned. The task given to the linguist who offered her services at that time to work in the region was therefore to respond to the demands of the Rama leaders for help in revitalizing their Rama language. Fieldwork was, however, financed by the National Science Foundation starting in 1986 for the production of a linguistic grammar of the language.

As a university academic, the linguist was expected to produce a linguistic analysis of the language. This was the time, twenty years ago, when linguistic circles were mostly preoccupied with internal theoretical wars, and when there was no discussion of the issue of endangered languages, nor of the concept of (or, even less the need or responsibility for) language documentation and language revitalization. But the reality in the field was that of a request for a community project of language revitalization, request from both representatives of the Rama community and from Sandinista authorities without a clue as to how to go about it, yet wanting results.

4.2.2. Speakers and non-speakers

Early on it became obvious that there was a hiatus between the speakers of Rama able and willing to work with the linguist and the "community" that had requested the project but held a very ambivalent to negative attitude toward those speakers.

The linguistic work on the Rama language had been officially requested by the Rama community living on the island of Rama Cay, relatively close to the main town of the coast, and the seat of the regional government. But that community had practically totally shifted to English Creole, and the request had been specifically presented by non-speakers. Meanwhile, the actual linguistic work had to rely entirely on Rama speakers from the more distant mainland further south, where the language and culture had been somewhat protected by the isolation of the tropical forest. The principal consultants of the project, beyond being two women, were not considered part of the community of Rama Cay, but as being from a jungle community considered then downcast. Rama Cay people talked of the members of that mainland community as "Tiger people" who spoke the "Tiger language". Hence the difficulties early in the project when materials produced for the community with these speakers were publicly rejected by some of the key people of Rama Cay.

There is no doubt that the Rama Language Project could not have happened without the leadership of Miss Nora Rigby (1923–2001) who persisted in the face of the multiple pressures she was subjected to in not only making the linguistic study of the language possible but in master-minding language revitalization activities on the island. On several occasions, when the linguist reached the point of wondering if it was all worth it and whether to continue in the face of much hostility from members of the community towards her,

Miss Nora Rigby clearly stated her decision to continue, to fulfil her vision of seeing her language written down and documented so it would not disappear forever. Craig (1992b) and Grinevald and Kauffmann (2003) were written as a homage to the power of that woman rescuer of her language.

4.2.3. Thinking through the notion of "consent" in this context.

This first phase of the Rama project, requested by the community, evolved therefore in the midst of major difficulties and multiple allegiances to be managed. One was working in a revolutionary time with its demands (and exhilaration); in the midst of a raging (anti-Sandinista) "Contra war" translating into common war time constraints. The other was handling the new divisions in the Rama community due to wartime, creating yet another layer over the already profound division of the community between islanders – non speakers and mainland – speakers. And finally, on top of everything, came the devastating hurricane Joan that destroyed 80% of Bluefields and left the island of Rama Cay momentarily under water, meaning among other things the loss of all the materials of the project.

In this context, the issue of consent mostly took the shape of getting a permit to enter the zone of sometimes active military conflict, of associating with a new regional research institute (Center for Research and Documentation of the Atlantic Coast known as CIDCA), and of seeing how to respond to community demands as transmitted by Sandinistas authorities. As for the issue of consent from the main speaker, Miss Nora, an old illiterate woman, it was simply affirmed in the grant application. It was clear from very early on in the project that the situation was one of mutual benefit between the main linguist and the main linguistic consultant for the language, with very evenly balanced demands for services. The linguist wanted time and attention for linguistic description, and the speaker the same for support in her initiatives on the island for some Rama language revitalization, culminating in her teaching Rama to kindergarten children daily for almost ten years.

This first phase of the RLP was certainly a learning experience and good training ground for reflections on projects of description and documentation of endangered languages. The challenges included the balancing act of doing straightforward linguistics, with few speakers including one fluent but semi-speaker as it were, in the midst of a war, and of balancing the demands of a community ambivalent about where the language was coming back from, with those of the main speaker and leader of the project who had her own agenda. It included the added layer of challenge from the Sandinistas asking how the Rama language was going to be revived and refusing to engage in a discussion of whether it was at all possible to revive it. Back in academia, the pressures came from financing foundations, including the negative evaluation for the renewal of the project, two years into it, by theoretical linguists that deemed the work produced (a census of the last speakers, a collection of texts transcribed, glossed and analyzed, an initial dictionary and an initial grammar sketch), as amounting to not having involved any "linguistics" yet, i.e., "theoretical" linguistics.[23]

4.3. RLP phase 2: HRELP Dictionary and archiving project 2004–6

As it were, the first phase of the project ended up with a draft of an extensive grammar of the language available to interested linguists in pdf format (but yet to be reviewed, completed, and published); data for a dictionary that was almost entirely entered into an early version of a new and promising software for dictionary making, 4th Dimension, and a collection of texts in the IT format, the ancestor of the now very widely used Shoebox program and its successor, Toolbox. And then the life of the two main linguists of the project, the present author and main assistant Bonny Tibbitts, took several turns and major changes so that the project material ended up boxed and left to sit for ten years. This is another reality of many such projects, boxes of data in linguists' offices or garages that did not materialize into final publications, for any number of reasons, from a sense of perfectionism or of overwhelming responsibility of being so few and isolated, working on such an impressive quantity of new data, with the feeling that the analysis will improve with time and additional data. There is the guilt, there is the sense of unfinished business, there is dealing with pressures of life that do not leave time or energy to even think about it.

But fieldwork projects of this sort are life-long adventures. Dictionaries are often stories of decades of the life of a linguist, and circumstances can change. In the case of the RLP, two different changes of circumstances made the reactivation of the project possible. One was new impetus from the community asking for more attention to the Rama language, and the other new developments in linguistics about documenting and archiving endangered languages, with an opportunity to obtain financial support to complete the project.

4.3.1. The set-up of this second phase

The second phase of involvement of the linguist with the Rama Language Project came again from a request from the community, and brought out the need for a letter of consent to be included in a grant application.

4.3.2. Once again, a request from the Rama community

Renewing contact with the community in view of returning to work on the language was made possible by a collaborative project of the University of Tromsø and the new regional university of the Caribbean Coast of Nicaragua, URACCAN. The first phase of the program focused on development programs for the Ramas, but as it was itself dedicated to the university training of coast people, this constraint eliminated from the start the possibility of much participation on the part of the Rama people, few of whom manage to complete high school level education. And when asked for their priorities for development projects, the Rama community had placed their linguistic and cultural rights second to the need of securing titles and protection for the traditional Rama territory. By then, the leadership of the first phase of the Rama language project had given way to a new generation of Rama Cay leaders, so that the defense of the Rama language was now articulated by some of the younger adults that over the years had observed the events of the Rama project in the school of Rama Cay. They had attended Sunday meetings and seen Miss Nora teach kindergarten children words, songs, and phrases of Rama for years. There were touching moments for the linguist when young leaders gave her back the basic lines of her speeches on Rama of ten to fifteen years before, about Rama being a real language one can be proud of, worth writing, worth studying, worth speaking and keeping.[24]

And that time the project was framed, not by the establishment of the new Autonomy statute of the region and implementation of new linguistic rights of the Sandinista agenda, but by the beginning of legal fights to protect the traditional Rama territory, another intense political moment of demands for the application of rights granted by the Autonomy Statute. The ongoing fight is to protect the Ramas against the invasion of their traditional territories, from the inland by poor white peasants, from the coast by land speculators. This land speculation is exacerbated by the threatening plans of building a "dry canal" (Canal Seco) that would cut right through the middle of the Rama territory.[25] This is an all-too-common case of the endangered language of an endangered people in endangered territories, and an example of renewed interest in a moribund language being tied to land issues vital to the survival of the communities.

4.3.3. An application to the new HRELP program from SOAS London[26]

The original application for a grant to produce a large dictionary and to archive the materials from the first phase included plans for a large training component for the Ramas to participate as much as possible in the production of the dictionary and to learn to use the database of the dictionary and of the archived materials. The project was conceived as including non-speakers with literate skills, particularly the Rama new secondary schools graduates. So in the fall of 2002, the issue arose of obtaining a letter of consent for a grant application.

4.3.4. The story of the letter of consent

The complexity of this episode of "informed consent" will be explored from the perspectives of its three different protagonists: the field linguist, the Rama community, and the financing foundation. It will consider the field linguist's first actual experience of asking "the community" for a "letter of informed consent" with observations about the mobilization of the language community to comply and produce the first letter of consent as a written document.

It will offer additional notes about the reaction of the same linguist after having the proposal turned down, and having to resubmit and obtain a second letter for a project in which support for community involvement had been cut out. The story of these two successive letters of consent illustrates some of the complexities of simultaneously handling the relations between linguist and community on the one hand, and linguist and foundation on the other, and how difficult and delicate the issue of getting a letter of "consent" can be, seen and lived from the field.

4.3.5. The letter itself

The idea of setting up a new project concerning the Rama language was presented at an official meeting on the island Rama Cay with members of the Rama community. The actual letter of consent was produced during a workshop held at the local university URACCAN under the auspices of the URACCAN-Tromsø project that financed the encounter between the linguist and community representatives.

The production of the actual document took half a day of discussion, elaboration and production, after initial reiteration by the linguist of the need for such a document and exposition of the work plan. It was a rather involved process, with hours of discussions typical of the indigenous style of collective decision making processes. The production of the actual letter of consent was an elaborate process that was concluded by the formal signing of the document by the community leaders, elected officials, and literate Ramas present.

Those were Ramas that had been in contact with the outside world and had been receiving training in western-style leadership in recent years. They were the ones to handle the official discourse on the need to save the Rama language for ethnic identity purposes. In an interesting way, no Rama speakers were present at the meeting that took place on university grounds.

The letter reads as follows (reproduced as is):

> Bluefields 15 of november 002
> Collet craig
> Rama documentation proyect
>
> To the community of Rama Cay is an opportunity to look back to our history where the language is in extintion. We then the leaders of the community of Rama Cay are willing to sopport the proyect; because is part of our identity, as an indigenous group. The teacher the leaders the student will surly participate and support this proyet and make it bright in our community. The access of the young people in participating in training will have to do with the development of our community. So then once more we omit ourself to willingly work, participate and help in the proyect.
> So then to end we only want to tell yo we hope that this wil be a reality.
>
> [Signed by seven people: *the president for the comunal, the secretary, two teachers, the pastor, and two community leaders.*]

4.3.6. Project rejection and reapplication: A second letter of consent

Things became more complicated when in fact the project was turned down in its initial form. The linguist was advised to reapply, but to concentrate on the production of the dictionary without including the training component for the community. Although a second proposal was accepted on principle, the foundation then requested further drastic cuts in the budget, to a final amount of a quarter of the original one, through elimination of most of the field components and interaction with the community. And it requested in addition a new letter of consent from the community dated from that year.

It is difficult to describe the kind of frustration this request for a second letter of consent created for the linguist. How to tell the community that the project had been turned down, that only a part of it had finally been granted, with no component of community work or fieldwork, but that another letter of consent from them was needed? And how to negotiate this from a distance and in a short time frame? In the end a short form letter was prepared by the linguist that reproduced some of the basic language of the original letter produced by the Ramas themselves. It read as follows:

> We the leaders of the Rama community know and understand about the Rama Dictionary Project presented to you by Colette Grinevald (we know her as Miss Colette). We consider that it is a project of great importance to the Rama community and we willingly support it. We believe it will help us in our efforts to save the Rama language and culture which is part of our ethnic identity. We hope that with your help this project can become a reality.

Through the miracle of email and the good will of local contacts that went to the island of Rama Cay to obtain signatures, ten Ramas signed this letter, not all the same ones of the first letter. There was this time a combination of teachers, official representatives, and community members. The signatures included this time some Rama speakers from the mainland community, as well as others unknown to the linguist. This letter of consent was filed away and satisfied the needs of the foundation, and support was granted for a dictionary production and archiving project. But it certainly raised a number of issues for the linguist about the process of requesting and producing letters of consent, and the significance of this written document at both ends.

4.4. Musings on obtaining letters of consent

This story was told in order to illustrate the kind of possibly extremely complex and difficult situations field linguists can find themselves in, caught between the demands, the needs, and the world views of the linguistic communities and those of the financing institutions. It was meant to give some concrete example of the kind of stressful field situations academics are little prepared to face. In this case, witnessing the extremely demanding and elaborate process of the production of the first letter of consent allowed the linguist to reflect on how much we first world academics demand of indigenous communities to conform to our needs for written documents and for representativeness. And considering the hopes raised earlier about involvement of the community through some training, the process of requesting a second letter for a project that did not officially include any community involvement and basically financed the production of a linguistic dictionary and the setting up of a web-based archive on French and US university sites produced both anger and anguish in the linguist. Without putting in question the necessity of including getting "informed consent" in the list of ethical issues to be handled in doing such kind of fieldwork, this experience certainly brought to the fore for the linguist the questions of consent from whom, consent to do what, consent obtained through what process, consent in what form, and ultimately consent for whom.

5. Conclusion

Fieldworkers working on endangered languages with endangered language communities face a complexity of pressures which academia and financing foundations may have very little sense of as yet. This paper was meant to bring out some of the issues which fieldworkers must deal with today in terms of meeting new standards of ethics, in the context of the recent explosion of new technologies and the possibilities of new forms of documentation and archiving of endangered languages. Among them is the issue of obtaining "consent" for the fieldwork to be carried out, an issue that is being discussed more and more openly and that has become a crucial component of grant applications. Raising the issue and being convinced of its validity does not mean that we linguists feel ready and prepared to attend to it appropriately and efficiently. Much remains to be thought through about the complex relation between legalistic and First World views and the realities of the field, particularly if one considers that working on very endangered languages means dealing with illiterate and marginalized populations and individuals. While there is not and never will be any pat solution to offer, no recipe for success, no guide on how to do it, it is hoped that some of the essential ingredients of the process of obtaining consent have been identified. It is certainly at the core a question of attitude and approach, one of establishing trust and being willing to negotiate and remain flexible, trust that is best built through relations over a period of time. In the end, consent is to be thought of more as a process that permeates the dynamics set in motion than as a question of obtaining some signed document to be filed away.

As stated at the beginning of this paper, this discussion of ethics in the fields was shaped by the American experience of the fieldworker writing it. The issues of ethics have indeed been brought up more pressingly by linguists working with communities of Australia and America where the phenomenon of language endangerment has been felt for decades now to be very pressing and of utmost urgency. As argued, the disappearance in the future decades of the vast majority of the remaining native languages of those two continents is considered a catastrophe, in terms of the immense loss of the diversity of human knowledge and genius that their disappearance will represent. This is not as much the case of the situation in South Asia, where multilingualism is more widespread, and where shifts of languages happen within shared cultural areas and less diverse language families. It remains that great numbers of languages are indeed endangered in that region of the world, too, and that linguists facing now the same task of documenting those languages will find themselves having to consider in turn not only the technological but also the ethical implications of this kind of work.

Notes

1. In fact, the term used by Hinton and Hale (2001) in their *Green Book of Language Revitalization in Practice* is that of "local" languages, not "minority" nor "endangered languages", in keeping with the terminology used by the speakers of such languages involved in revitalization and maintenance efforts.
2. Not that there were not cases of language loss due to colonization of indigenous people by other indigenous people prior to the arrival of the white people, such as the case of the spread of Kechua in the Andean regions or the Lengua Geral in Brazil. See Craig (1989).
3. See, on one hand, the story of ISHI, the last speaker of Yana, survivor of repeated massacres of his people in the early XXth century in northern California, as told by Kroeber's wife (Kroeber 1961), see also the striking figures for the languages of California given for 1992 in Hinton (1994). At the same time it is worth noting that California is one of the states with some of the most active revitalization projects today (Hinton 2001, 2004).
4. In that light, the massive loss of the great genetic diversity of the languages of California was partially compensated by a systematic study of these languages by linguists, including many doctoral students, of the University of Berkeley, starting in the middle of the twentieth century .
5. The documentation of the Pacific Coast languages is now limited by the extinction or near extinction of its speaker populations, while that of the Amazonian languages that remain is still very partial, in spite of commendable recent efforts, particularly the recent training of a cohort of Brazilian non-missionary linguists. See Grinevald (1998) for an overview of the study of the native languages of South America.
6. How to interpret the high figure of isolate languages that is at the origin of the high figure of stocks in South America is a matter of ongoing debate, as due to either a process of isolated diversification, or massive language loss of sister languages, or both. See the discussion in Adelaar (2000) for instance, and mention of the phenomenon in Grinevald (1998).
7. In both cases, the issue of how to talk about this language loss can be delicate in that the sensibilities of the communities concerned are such that this is not the view of the language they wish to see emphasized.
8. Now counteracted by the recent Bush administration; the No Child Left Behind Act of 2002 disadvantages all children taught in the bilingual school programs when put through the English only testing process (See Hinton 2004).
9. See Craig 1992a and Grinevald 1997 for an overview of this situation. The latest country to officially recognize its indigenous languages is Mexico (see the *Ley General de Derechos Lingüísticos de las Lenguas Indígenas de México: Diario Oficial de la Nación*. March 2003)
10. In particular Ken Hale and Akira Yamamoto, whose tireless efforts on behalf of the indigenous communities of the USA must be saluted, this work remaining so often invisible in academic circles. The least we can do as we write the story of these intense times is to keep the memory of Ken Hale alive, as we promised ourselves to do.
11. The movement is kept alive as the LSA announces the creation of a Ken Hale chair for fieldwork courses at LSA Summer Institutes, and the National Science

Foundation launched a special funding program in 2005 for the documentation of endangered languages (DEL).

12. With linguists from the UNESCO efforts and from the LSA winter panel. See Uhlenbeck (1992) and later Grenoble and Whaley (1998) for the earlier assessments of the situation of endangered languages by linguists.
13. See, for a Latin American example, England (1992, 1998), Cojti (1990) and Grinevald (2002) for a sense of the "Mayan movement" and the changing relations of linguists to native speakers of Mayan languages of Guatemala. For the role of linguists in revitalization programs, see Hinton and Hale (2001), for instance, and for a sobering and humbling consideration of the relation of academic linguists to native speaker communities, see Gerdts (1998).
14. Anju Saxena (personal communication) and Annamalai (2003).
15. My personal experience, for instance, is to have, on several occasions and as I perceived it, somewhat condescendingly, been treated as a "bleeding heart" or "social worker" Americanist, and so, notably, by fellow Africanists. See also, for instance, Ladefoged's criticism of the LSA panel and the publication of its proceedings (Ladefoged 1992, an Africanist in fact), and another kind of response (Dorian 1993).
16. Thanks to Anju Saxena for a discussion of this list of differences and the information that SIL was actually kicked out of India, in contrast with its own omnipresence, until recently, in most Latin American countries.
17. Based at the time on various documents from different disciplines, such as, the American Anthropological Association (*Handbook on Ethical Issues in Anthropology* [1987], and *Principles of Professional Responsibilities* [1990]), the American Sociological Association (*Revised ASA Code of Ethics* [1980], and the American Psychological Association (*Ethical Principles of Psycholinguistics* [1990]).
18. Different fieldwork frameworks are discussed in Cameron et al. (1992) and presented in Craig (1993), Grinevald (1998, 2003). The simplest model, when there is little ethnic consciousness and no politicization of the language, is that of work ON the language, with the linguist entering the community and finding a speaker or speakers with whom to work and largely ignoring the community, except to live in it, but without developing work relations with the community. This was the so-called "ethical" framework, typical of much fieldwork until the latter part of the twentieth century (such as many PhD dissertation work by graduate students in the seventies, as Craig 1977). The most complex model, with multiplication of responsibilities and allegiances, and increased likelihood of need for negotiation and handling of conflict of interests, is a model said to be WITH speakers, at best BY speakers, and is the standard framework today wherever the ethnic groups and the languages are politicized. It is certainly the dominant model for fieldwork in Colombia, Bolivia and Peru today, for instance. The most advanced model, BY the speakers, is best represented by the Guatemalan model, in particular the trained native Mayan linguists of OKMA, submitting their own documentation projects to the foundations today. (See England 1992 and Grinevald 2002 about the state of affairs in Guatemala today.)
19. The ghosts of Project Camelot and the Vietnam era, and the issue of the Peace Corps infiltration in certain countries marked the period that saw the developments of these codes of conduct. The concern remains today, such as during the

Sandinista times when doing fieldwork in the Contra War area, or in most border areas of the Amazonian region for instance.

20. An updated version of this list would actually situate this issue of return of materials and preparation of materials appropriate for community use toward the end of the DURING in the field phase, with as much community involvement as possible and evidence of the work done being left in the field in the first place. This is the case even if little evidence of the work is still available on the next visit.
21. An issue raised for instance by native Mayan linguists in the late 1980s, who objected (with good reasons in the opinion of the present author) to the indeed extremely high frequency of the transitive verbs "to kill" and "to hit" in the grammatical examples of the multiple PhD theses of North American academics being produced at that time.
22. Grinevald (2000) was an attempt at explaining to non-academics the kinds of pressures felt by fieldworkers upon return from the field, and the nature of the academic dynamics which are unavoidable if one is to stay in academia. Gerdts (1998) already mentioned, is a well articulated presentation of what communities want of linguists.
23. Financial support was again provided later by the National Science Foundation, the Wenner-Gren Foundation and the National Endowment for the Humanities. Paul Chapin, then director of the linguistics program of the National Science Foundation, is to be commended for considering applications for fieldwork in Nicaragua at the time the United States government was orchestrating the demise of the Sandinista government and financing the Contra War.
24. This in the context of seemingly no recognition of the source of those speeches now completely appropriated by the community, nor any concrete evidence left on the island of any of the multiple series of publications and materials produced. This is the common and very humbling experience of the systematic disappearance of all materials produced for over a decade, disappearance observed regularly at each new field trip and dealt with for the years of the project by constant reprinting and redistribution of materials. First World linguists need to learn to deal with the non-materiality of oral tradition cultures, meaning no familiarity with written materials, as well as the weather conditions of tropical humidity that easily destroys anything one might want to keep. Maybe even more disturbing is the realization of the total absence of consultation of the materials still kept in the library of the research institute CIDCA by any of the URACCAN students and faculty (non-Rama) directly or remotely involved with research projects with the Ramas on Rama Cay.
25. Such a "dry canal" would be an alternative to the practically obsolete Panama Canal, and would function as a railroad line cutting across southern Nicaragua to transport containers from big tankers between two deep sea ports, the one on the Caribbean coast being built near and on Rama settlements.
26. This is the Hans Rausing Endangered Language Documentation Program, one of three major sources of funding for documentation of endangered languages today, the others being the Volkswagen-DOBES program and the NSF-NEH-Smithsonian Documentation of Endangered Languages program (DEL).

References

Adelaar, Willem F. H.
2001 Descriptive linguistics and the standardization of newly described languages. In *Lectures on Endangered Languages From Kyoto Conference 2000*, Osamu Sakiyama (ed.), 69–80. Kyoto: ELPR Publication Series Coo2.

Annamalai, E.
2003 The opportunity and challenge of language documentation in India. In *Language Documentation and Language Description*, vol. 1, Peter R. Austin (ed.), 159–167. London: Hans Rausing Endangered Languages Project.

Austin, Peter R. (ed.)
2003 *Language Documentation and Description*, vol 1. London: The Hans Rausing Endangered Languages Project.

Cameron, Deborah, Elizabeth Frazer, Penelope Harvey, B. H. Rampton, and Kay Richardson
1992 *Researching Language: Issues of Power and Method*. London: Routledge.

Craig, Colette
1977 *The Structure of Jacaltec*. Texas, Austin: University of Texas Press.
1989 Review of N. Rigby & R. Schneider: *Dictionary of the Rama Language: Rama-English-Rama-Creole-Spanish/English-Rama, Speaking with the Tiger*, vol. 2, Berlin: Dietrich Reimer Verlag. *The International Journal of American Linguistics* 56 (2): 293–304.
1992a A constitutional response to language endangerment: The case of Nicaragua. *Language* 68 (1): 11–16.
1992b Miss Nora, rescuer of the Rama language: A story of power and empowerment. In *Locating Power, Proceedings of the Second Berkeley Women and Language Conference*, vol. 1, Kira Hall, Mary Bucholtz, and Birch Moonwomon (eds.), 80–89. Berkeley: University of California, Berkeley.
1993a Fieldwork on endangered languages: A forward look at ethical issues, *Proceedings of the XVth International Congress of Linguists*, vol. I, 33–42. Sainte-Foy, Quebec: Les Presses de l' Université de Laval.
1993b Comment [on: Ethics, advocacy and empowerment: Issues of method in researching language]. *Language and Communication* 13 (2): 97–100.

Dorian, Nancy
1993 A response to Ladefoged's Other view of endangered languages. *Language* 69: 575–579.

England, Nora
1992 *Autonomía de los idiomas mayas: Historia e identitad*. Guatemala, Cholsamaj.
1998 Mayan efforts towards language preservation. In *Endangered Languages: Current Issues and Future Prospects*, Lenore A. Grenoble, and Lindsey J. Whaley (eds.), 99–116. Cambridge: Cambridge University Press.

Gerdts, D.
1998 Beyond expertise: The role of the linguist in language revitalization programs. In *Endangered Languages – What Role for the Specialists?*

Nicholas Ostler (ed.), 13–22. Bath: The Foundation for Endangered Languages.

Grenoble, Lenore A., and Lindsey J. Whaley (eds.)
1998 *Endangered Languages*. Cambridge: Cambridge University Press.

Grinevald, Colette
1997 Language contact and language degeneration. *Handbook of Sociolinguistics*, Florian Coulmas (ed.), 257–270. Oxford: Blackwell.
1998 Language endangerment in South America: A programmatic approach. In *Endangered Languages*, Lenore A. Grenoble, and Lindsey J. Whaley (eds.), 124–160. Cambridge: Cambridge University Press.
2000 Los lingüistas frente a las lenguas indígenas. In *As línguas amazônicas hoje*, F. Queixalos, and O. Renault-Lescure (eds.), 35–53. Saõ Paolo: IRD MPEG Instituto Socioambiental.
2002 Linguistique et langues mayas du Guatemala. *Faits de Langues* 20: 17–27. [Meso-Amérique, Caraibes, Amazonie volume 1] Paris: Ophrys.
2003 Speakers and documentation of endangered languages. In *Language Documentation and Description*, vol. 1, Peter K. Austin (ed.), 52–73. London: The Hans Rausing Endangered Documentation Project.
2004 Les langues amérindiennes risquent de disparaître, lenguas amerindias en peligro de desparecer, Amerindian languages in danger of extinction, llengües amerindies en peril de desaparèixer. Dans *Voix, Voces, Voices* 176–177. Forum de Barcelone: Lunwerg editores.
to appear Encounters at the brink: Linguistic fieldwork among speakers of endangered languages. In *Vanishing Languages of the Pacific Rim*, Osahito Miyaoka, Osamu Sakiyama, and Michael E. Krauss (eds.), Chapter 3. Oxford: Oxford University Press.

Kroeber, T.
1961 *ISHI in Two Worlds: A Biography of the Last Wild Indian in North America*. Berkeley: University of California Press.

Hale, Ken, Colette Craig, Nora England, Laverne Masayesva Jeanne, Michael Krauss, Lucille Watahomigie, and Akira Yamamoto
1992 Endangered languages. *Language* 68 (1): 1–42.

Hinton, Leanne
1994 *Flutes of Fire: Essays on California Indian Languages*. Berkeley: Heyday Books.
2004 The death and rebirth of Native American Languages. In *Endangered Languages and Linguistic Rights: On the Margin of Nations. Proceedings of the Eighth FEL Conference, Barcelona (Catalonia), Spain 1–3 October 2004*, J. Argenter, and R. McKenna Brown (eds.), 19–25. Bath: Foundation for Endangered Languages

Hinton, Leanne, and Ken Hale (eds.)
2001 *The Green Book of Language Revitalization in Practice*. New York: Academic Press.

Ladefoged, Peter
1992 Another view of endangered languages. *Language* 68: 809–811.

Mithun, Marianne
1999 *The Languages of Native North America*. Cambridge: Cambridge University Press.

Nettle, Daniel, and Suzanne Romaine
2000 *Vanishing Voices: The Extinction of the Worlds Languages*. New York: Oxford University Press.
Newman, Paul, and Martha Ratliff (eds.)
2001 *Linguistic Fieldwork*. Cambridge: Cambridge University Press.
Queixalos, F., and O. Renault-Lescure (eds.)
2000 *As línguas amazônicas hoje*. Saõ Paolo: IRD MPEG Instituto Socioambiental.
Robins, R. H., and E. M. Uhlenbeck (eds.)
1991 *Endangered Languages*. Oxford: Berg.

Subject index

Language index

Bei Fragen zur Produktsicherheit wenden Sie sich bitte an:
If you have any questions regarding product safety,
please contact:

Walter de Gruyter GmbH
Genthiner Straße 13
10785 Berlin
productsafety@degruyterbrill.com